AF786002

ND | Springer Series in **Nonlinear Dynamics**

ND | Springer Series in **Nonlinear Dynamics**

Series Editors: F. Calogero, B. Fuchssteiner, G. Rowlands, M. Wadati, and V. E. Zakharov

Solitons – Introduction and Applications
Editor: M. Lakshmanan

What Is Integrability?
Editor: V. E. Zakharov

Rossby Vortices and Spiral Structures
By M. V. Nezlin and E. N. Snezhkin

Algebro-Geometrical Approach to Nonlinear Evolution Equations
By E. D. Belokolos, A. I. Bobenko, V. Z. Enolsky, A. R. Its and V. B. Matveev

Darboux Transformations and Solitons
By V. B. Matveev and M. A. Salle

Optical Solitons
By F. Abdullaev, S. Darmanyan and P. Khabibullaev

Wave Turbulence Under Parametric Excitation
Applications to Magnetics
By V. S. L'vov

Kolmogorov Spectra of Turbulence I Wave Turbulence
By V. E. Zakharov, V. S. L'vov and G. Falkovich

Nonlinear Processes in Physics
Editors: A. S. Fokas, D. J. Kaup, A. C. Newell and V. E. Zakharov

A. S. Fokas D. J. Kaup
A. C. Newell V. E. Zakharov (Eds.)

Nonlinear Processes in Physics

Proceedings of the III Potsdam – V Kiev Workshop
at Clarkson University, Potsdam, NY, USA
August 1–11, 1991

With 41 Figures

Springer-Verlag
Berlin Heidelberg New York
London Paris Tokyo
Hong Kong Barcelona
Budapest

Professor A. S. Fokas
Professor D. J. Kaup
Clarkson University, Potsdam, NY 13699-5815, USA

Professor A. C. Newell
University of Arizona, Tucson, AZ 85721, USA

Professor V. E. Zakharov
Landau Institute for Theoretical Physics, ul. Kosygina 2,
117334 Moscow, Russia and
University of Arizona, Tucson, AZ 85721, USA

ISBN-13:978-3-642-77771-4 e-ISBN-13:978-3-642-77769-1
DOI: 10.1007/978-3-642-77769-1

Library of Congress Cataloging-in-Publication Data. Nonlinear processes in physics / A. S. Fokas . . . [et al.]. p. cm. – (Springer series in nonlinear dynamics) Includes bibliographical references and index. ISBN-13:978-3-642-77771-4 1. Nonlinear theories–
Congresses. 2. Soliton theory–Congresses. 3. Mathematical physics–Congresses. I. Fokas, A. S., 1952– . II. Series. QC20.7.N6N662 1993 530.1'4–dc20 92-32811

This work is subject to copyright. All rights are reserved, whether the whole or part of the material is concerned, specifically the rights of translation, reprinting, reuse of illustrations, recitation, broadcasting, reproduction on microfilm or in any other way, and storage in data banks. Duplication of this publication or parts thereof is permitted only under the provisions of the German Copyright Law of September 9, 1965, in its current version, and permission for use must always be obtained from Springer-Verlag. Violations are liable for prosecution under the German Copyright Law.

© Springer-Verlag Berlin Heidelberg 1993
Softcover reprint of the hardcover 1st edition 1993

The use of general descriptive names, registered names, trademarks etc. in this publication does not imply, even in the absence of a specific statement, that such names are exempt from the relevant protective laws and regulations and therefore free for general use.

Typesetting: Camera ready copy from the authors/editors

57/3140 - 5 4 3 2 1 0 – Printed on acid-free paper

Preface

In 1979, a historical meeting took place at the Institute for Theoretical Physics in Kiev, USSR, where 48 American Scientists, specialists in nonlinear and turbulent processes, met for two weeks with their soviet counterparts. This meeting provided the unique opportunity for USA and USSR participants to directly interact personally and scientifically with each other. This interaction was of great importance not only for the individuals involved but also for the science of nonlinear phenomena in general.

At the end of the meeting, it was agreed that this exchange should continue, and it was decided to have the next meeting in the USA in 1981. Unfortunately, due to the political situation at that time, the second meeting in the USA never materialized. However, in 1983, the Soviet scientists organized in Kiev a second Workshop. This second meeting was again quite successful. Similar meetings, with growing success were organized at Kiev in 1987, and 1989. It should be noted that 405 participants from 22 countries participated at the fourth Kiev workshop on Nonlinear and Turbulent Processes. The Chairman of this workshop was V. Zakharov, who has also been a co-chairman of all the previous workshops.

Even earlier, in 1972, there had been a Potsdam workshop (the first Potsdam meeting) on nonlinear waves which was organized by Alan Newell. This workshop had served as a valuable precursor for much of the work on nonlinear waves in the USA. It was then repeated in 1978 (the second Potsdam workshop) which was a time in the midst of the recent explosive growth in nonlinear waves.

Except for the above four Kiev Workshops, there had been no scientific meetings where a large number of Soviet and USA scientists working in nonlinear and turbulent processes had met. At the closing of the Fourth Kiev Workshop, it was decided to have the next meeting in the USA. Because of the general political climate, it appeared that such an endeavor could now be feasible.

The III Potsdam-V Kiev Workshop on Nonlinear Processes in Physics was held at Clarkson University, Potsdam, NY from August 1-11, 1991. It was sponsored by the Clarkson School of Science and was funded by the National Science Foundation, the Department of Energy, the US Air Force of Scientific Research, the US Office of Naval Research, the Sloan Foundation and the School of Science of Clarkson University. The organizing committee was A.S. Fokas, D.J. Kaup, A.C. Newell and V.E. Zakharov.

This was the first major scientific workshop in the USA where the Soviet scientists formed a major contingent of the participants. There were 31 USSR participants out of a total of 106 participants. The emphasis of the workshop was on the interaction between mathematical techniques and problems of physical interest. It was particularly successful in mixing plasma physicists, fluid physicists and soliton theorists. The exchanges between these groups, particularly with the participation of the Soviet delegation, were quite stimulating.

The lectures ranged from algebraic features of integrable systems and vortex dynamics to applications in plasma physics, ionospheric physics, nonlinear optics, oceanic studies and solid state. It is not possible to give full justice to all the excellent lectures here, but we shall mention some results. Lax presented a survey of the work on the zero dispersion limit for several types of dispersive systems and Levermore, Tian, and Venakides presented new important developments and applications of the Lax–Levermore theory and the associated Whitham's equation. Explicit solutions of this equation, using algebraic-geometric techniques, were presented by Krichever and Dubrovin. Krichever also reviewed the recent appearance of Whitham's equation in the minimal models of 2D quantum gravity. Another connection between soliton theory and 2D quantum gravity was presented by Its who discussed the role of discrete Painlevé equations and gave a rigorous description of their continuous limit. Deift described a rigorous methodology for studying the long time behavior of the Riemann–Hilbert problems arising in the inverse spectral theory. Santini discussed the possibility of solving purely algebraic equations by the algebraic-geometric techniques developed in soliton theory. New results in the inverse spectral theory of evolution equations in two spatial dimensions were described by Boiti (dromions for DSI), Zhou (KPI), Sung (DSII), and Pogrebkov (KPI). McKean described the spectral theory associated with bi-Hamiltonian structures in classical mechanics, and Dorfman reviewed Hamiltonian and symplectic structures for evolution equations in one and two spatial variables. McLaughlin described algebraic-geometric aspects of the perturbation theory of certain soliton equations. Applications of the dressing method to nonlocal nonlinear evolution equations and to nonlinear evolution equations in multidimensions were discussed by Degasperis and Sabatier respectively. Takhtajan introduced a reversible soliton cellular automaton. Korepin derived and analyzed the integrable PDE's satisfied by quantum correlation functions. Beals described the action angle formulation of the Gel'fand-Dikii hierarchies. Shulman talked about new results regarding degenerate dispersion law. Alber analyzed certain complicated but integrable Hamiltonian systems. Bogoyanlenskij analyzed a certain $2+1$ version of the KdV and showed that it exhibits the phenomena of breaking of solitons. Conte and Fordy explained how to implement the Painlevé test in the presence of negative resonances.

Grunbaum discussed a novel inverse problem arising in low energy medical imaging (diffuse tomography) and Monk described analytical and numerical aspects of a certain inverse scattering problem. Hasegawa described some of the technical points involved with using solitons as pulses in the proposed transatlantic optical cable and Rupasov described new rigorous results in the quantum theory of stimulated Raman scattering. Bona showed how a rough wave model could give a respectable explanation of the appearance of underwater sand ridges seen on sloping beaches.

Important new results in vortex dynamics were presented by Zabusky, Majda, Ichikawa, Horton and Petviashvili. Zabusky showed excellent detailed visual descriptions of the collisions and reconnection of vortex filaments while Majda and Ichikawa each independently described the stretching and kinking of vortex filaments in irrotational fluids. Both Horton and Petviashvili discussed the rotational case and in particular, showed that the strength of the vortex determines whether the vortex behaves as a point vortex or a KdV soliton in collisions. Langmuir turbulence and collapsing cavitons in the ionosphere were discussed by Don DuBois while Rao and Kaup described how the mode conversion into electron Bernstein waves could cause the observed quenching of the downshifted peaks seen in the Tromsø and Arecibo ionospheric modification experiments. Morales discussed the nonlinear refraction of an rf wave in the ionosphere while Hada and Hamilton both presented studies of nonlinear Alfvén waves in space plasmas. One of the major presentations in collapse physics was made by Malkin who presented new analytical results for the self-focusing problem of the two-dimensional nonlinear Schrödinger equation. Zakharov and Rubenchik each discussed aspects of weak turbulence.

Most of these lectures are summarized in these proceedings.

Potsdam, NY
Moscow
Summer 1992

A.S. Fokas
D.J. Kaup
A.C. Newell
V.E. Zakharov

The III Potsdam – V Kiev Workshop

1. Cindy Smith
2. Vladimir Malkin
3. Irene Dorfman
4. Xiaoying Dong
5. Dmitrii Yu. Manin
6. Alexander R. Its
7. Xin Zhou
8. Susanne Brenner
9. V. A. Marchenko
10. Norman J. Zabusky
11. David A. Coker
12. Boris Dubrovin
13. Li-Yeng Sung
14. Michael Isichenko
15. Ildar Gabitov
16. Grover Swartzlander
17. Robert Conte
18. David J. Muraki
19. Keith Blow
20. Marthe Sabatier
21. Heinz Steudel
22. Yukio Kaneda
23. Robert Hamilton
24. Louise Bullough
25. David McLaughlin
26. Pierre Sabatier
27. Tohru Hada
28. Robin Bullough
29. Andrew Majda
30. Fei Ran Tian
31. Gendi Pang
32. Evgenii Kuznetsov
33. Stephanos Venakides
34. Michael Spector
35. Ladislav Trlifaj
36. Alexander Dykhne
37. Oleg Bogoyavlenskij
38. Efim Kudashev
39. Vladimir Korepin
40. N. S. Erokhin
41. A. M. Rubenchik
42. Semen Moiseev
43. Solomon Alber
44. E. Z. Gusakov
45. Wendell Horton
46. Percy Deift
47. David Gurarie
48. Alejandro Aceves
49. David Newman
50. K. A. Gorshkov
51. Alexander Mikhailov
52. Elena Villalon
53. Stefan Wabnitz
54. Anne DeBouard
55. Negasha Rao
56. Richard Beals
57. Jerry Bona
58. D. F. DuBois
59. Vadim Shvets
60. Fred Abergel
61. Martin Kruskal
62. E. Schulman
63. A. P. Fordy
64. Einar Mjolhus
65. Yanguang Li
66. Piotr Grinevich
67. A. Yu Orlov
68. A. S. Fokas
69. Andrei Pogrebkov
70. Darryl Holm
71. G. Fraiman
72. Paolo Maria Santini
73. Marco Boiti
74. Charles Doering
75. Jyh-Hao Lee
76. Mie Kodama
77. Valery Rupasov
78. Yujikodama
79. Lay Kodama
80. P. Kaliappan
81. Boris Kupershmidt
82. Pedro Ripa
83. Roberto Camassa
84. Andrea Ripa
85. Antonia Degasperis
86. John Weiss
87. George Morales
88. Vladimir Yan'kov
89. L. M. Pismen
90. V. Petviashvili
91. Yoshi Ichikawa
92. Carson Chow
93. David Kaup

Contents

Part I Nonlinear Equations

Part II Inverse Scattering Transforms

Part V Hydrodynamics and Turbulence

Part I

Nonlinear Equations

Multisoliton-like Solutions of Wave Propagation in Periodic Nonlinear Structures

A.B. Aceves[1] and *S. Wabnitz*[2]

[1]Department of Mathematics and Statistics, University of New Mexico, Albuquerque, NM 87131, USA
[2]Fondazione Ugo Bordoni, Via B. Castiglione 59, I-00142 Rome, Italy

Multisoliton-like solutions of the coupled mode equations describing waves propagating in nonlinear gratings are obtained by imposing a constraint on the multisoliton solutions of the integrable Massive Thirring Model.

Wave propagation in periodic structures is an important and well-developed field of classical physics and has several applications ranging from solid state physics to optics. A simplified approach to this problem which has been of wide applicability in optics is given by deriving coupled mode equations for the slowly varying envelopes of the field with opposite phase velocities.

In this paper we construct multisoliton-like solutions of these coupled equations by assuming a particular form of the solution which includes a scaling and a phase transformation on the solutions of an integrable model in classical field theory known as the Thirring model. We find that this ansatz indeed gives solutions if one adds a constraint on the solutions of the integrable model; the constraint being that the intensity of one mode has to be proportional to that of the second mode, which in turn translates into particular choices of the parameters in the multi-soliton solutions.

The relevant coupled equations for the complex amplitudes of the field, $E(z,t) = e_1 e^{i\omega T - i\beta Z} + e_2 e^{i\omega T + i\beta Z}$ read in dimensionless units

$$\partial_t e_1 + \partial_z e_1 - ie_2 - i(|e_2|^2 + \sigma|e_1|^2)e_1 = 0 \tag{1.1a}$$

$$\partial_t e_2 - \partial_z e_2 - ie_1 - i(|e_1|^2 + \sigma|e_2|^2)e_2 = 0 \tag{1.1b}$$

In the absence of the self-phase modulation term ($\sigma = 0$), the above equations are integrable by means of the inverse scattering transform (IST) [1,2] from which soliton solutions can be obtained. An alternative construction of multisoliton solutions by use of Bäcklund transformation (BT) was given in [3] and we will use this approach to etablish a possible connection between the solutions of the model studied here and those of the masssive Thirring model.

We first notice that three invariances of the integrable model,

(a) *phase invariance* $\quad e_i \to e_i e^{i\phi_0}$

(b) *translational invariance in t and z* $\quad (z \to z + z_0, \quad t \to t + t_0)$

are preserved in the general case $\sigma \neq 0$, thus one expects at least three conserved quantities. The first conserved quantity of (1.1) is, $\int(|e_1|^2 + |e_2|^2)dz$, which comes from the conservation equation

$$\partial_t(|e_1|^2 + |e_2|^2) = \partial_z(|e_2|^2 - |e_1|^2) \tag{1.2}$$

It gives a relation between the total energy and its flux. This conservation law, which is independent of σ, will be an important tool in the construction of soliton-like solutions [4,5]. Other two conserved quantities are, the Hamiltonian and the total momentum flux. Whether there exist more conserved quantities, is not known, but one only needs equation (1.2) for this work.

The first step in our construction is to assume a solution of the form $e_j = \alpha\psi_j e^{i\theta}$, where ψ_j is a solution of the Thirring model and θ depends on z and t. Substutition of this ansatz in equations (1.1) gives two equations for θ,

$$\partial_\tau\theta = \sigma\alpha^2|\psi_1|^2 + (\alpha^2-1)|\psi_2|^2 \tag{1.3a}$$

$$\partial_\eta\theta = \sigma\alpha^2|\psi_2|^2 + (\alpha^2-1)|\psi_1|^2 \tag{1.3b}$$

where $\tau = \frac{1}{2}(t+z)$ and $\eta = \frac{1}{2}(t-z)$ are the cone coordinates.

Therefore, for the ansatz to work, a consistency condition on (1.3) must be imposed. Let us obtain this condition by first integrating (1.3a) and then differentiating the result with respect to η, $\partial_\eta\theta = \sigma\alpha^2\int^\tau \partial_\eta|\psi_1|^2 d\tau' + (\alpha^2 - 1)\int^\tau \partial_\eta|\psi_2|^2 d\tau'$

The right hand side can now be simplified to

$$\sigma\alpha^2\int^\tau \partial_\eta|\psi_1|^2 d\tau' - (\alpha^2-1)|\psi_1|^2 \tag{1.4}$$

if we use equation (1.2), which in the cone variables reads $\partial_\tau|\psi_1|^2 = -\partial_\eta|\psi_2|^2$. Then from equations (1.3b) and (1.4) the consistency condition now reads

$$\int^\tau \partial_\eta|\psi_1|^2 d\tau' = 2\frac{(\alpha^2-1)}{\sigma\alpha^2}|\psi_1|^2 + |\psi_2|^2 \tag{1.5}$$

and the condition for which it is satisfied is [5], $|\psi_1|^2 = K|\psi_2|^2$, $\alpha = \sqrt{\frac{2}{2+\sigma(K+\frac{1}{K})}}$.

To determine the conditions under which the consistency condition is satisfied for the multisoliton solutions of the integrable model, we follow the construction procedure of multi-solitons by use of multi-Bäcklund transformations given in [5]. One has that the n-soliton solution of the integrable model is, $\psi_1 = -\nu e^{-i\beta}$, $\psi_2 = \rho e^{-i\kappa}$ where $e^{i\kappa} = \sum_{i=1}^n[\frac{\lambda_i}{\bar\lambda_i}]$, $e^{i\beta} = (1 - 2\sum_{i,j=1}^n \frac{f_i\gamma_{ij}\bar f_j}{\lambda_j})$. Finally,

$$\nu = 4i\sum_{i,j=1}^n f_i\gamma_{ij}\bar g_j\ ,\ \rho = 4i\sum_{i,j=1}^n \frac{f_i\gamma_{ij}\bar g_j}{\lambda_j^2} \tag{1.6}$$

In the above expressions $f_i = a_i e^{i(\bar\lambda^2\tau + \bar\lambda^{-2}\eta)}$, $g_i = b_1 e^{-i(\bar\lambda^2\tau+\bar\lambda^{-2}\eta)}$ where a_i, b_i are arbitrary complex constants as well as λ_i. Finally the matrix $\gamma = \mathbf{\Gamma}^{-1}$, where $\Gamma_{ij} = \frac{2\lambda_i\bar f_i f_j - 2\bar\lambda_j\bar g_i g_j}{\lambda_i^2-\bar\lambda_j^2}$. The parameters λ_i are the eigenvalues of the associated scattering problem and their amplitudes give the velocity of each soliton component, so if all $\lambda's$ are of different amplitudes, the initial state will decompose into n separated one soliton solutions. It is clear that equations (1.1) also have an asymptotic solution that has the form of n separated soliton-like solutions , therefore we now consider the case of the bound n-soliton solution where $|\lambda_i| = |\lambda|, i = 1, 2, ...n$.

Although the actual computation of the solution is rather cumbersome, we can easily see the conditions under which $|\psi_1|^2 = K|\psi_2|^2$. From (1.6), one can see that if we impose the $n-1$ constraints, $g_i = 0, i = 2, 3, ...n$ then $|\psi_1|^2 = |\nu|^2 = 16|g_1|^2|\sum_{i=1}^{n} \gamma_{i1} f_i|^2$, $|\psi_2|^2 = |\rho|^2 = \frac{16}{|\lambda|^4}|g_1|^2|\sum_{i=1}^{n} \gamma_{i1} f_i|^2$. In other words $|\psi_1|^2 = \frac{1}{|\lambda|^4}|\psi_2|^2$, so in the solution form for $e_i = \alpha\psi_i e^{i\theta}$, α takes the value $\sqrt{\frac{2}{2+\sigma(|\lambda|^4+\frac{1}{|\lambda|^4})}}$.

These constraints on the coefficients g_i, indeed give nontrivial solutions as it can be found in [5], where we carry out the calculations for $n = 2$. In the particular case where $|\lambda_1| = |\lambda_2| = 1$, the intensities are proportional to $F(z)[1 - cos(ct)]$, where F and c are given in [5]. In other words, breather-like solutions are obtained for this nonintegrable system.

In conclusion, we have obtained multisoliton-like solutions of a coupled mode set of equations that in nonlinear optics, describe the envelopes of two counterpropagating modes in a nonlinear waveguide with a periodic refractive index. These solutions were obtained by applying a constraint of proportionality of intensities to the solutions of the Thirring model which is integrable in the IST sense.

Work of the first author has been supported by an Air Force grant (No. AFOSR-91-0009). This work was carried out in the framework of the agreement between the Fondazione Ugo Bordoni and the Istituto Superiore Poste e Telecomunicazioni.

References

1. E.A. Kuznetsov and A.V. Mikhailov, Teor. Mat. Fiz. (1977), **30** 303.
2. D.J. Kaup and A.C. Newell, Lett. Nuovo Cimento (1977), **20** 325.
3. D. David, J. Harnad and S. Shnider, Lett. in Math. Phys. (1984) **8** 27.
4. A.B. Aceves and S. Wabnitz, Phys. Lett. A. (1989), **141** 37.
5. A.B. Aceves and S. Wabnitz, Preprint (1991).

Complex Deformation of Integrable Hamiltonians over Generalized Jacobi Varieties

S.J. Alber

Department of Mathematics, University of Nevada, Reno, NV 89557, USA

A class of algebraic and nonalgebraic completely integrable Hamiltonian systems defined on the symplectic manifolds is constructed. In particular, the meromorphic systems over the generalized Jacobians of the Riemann surfaces are considered in both scalar and matrix cases. Non-canonical action-angle variables are used to linearize the corresponding Hamiltonian flows and to establish the commutativity of the obtained finite (or infinite) collections of the integrable Hamiltonian systems.

1 Introduction

The present paper is devoted to constructing and studying a wide class of algebraic and nonalgebraic completely integrable Hamiltonian systems which are defined on symplectic manifolds. In most of the problems under consideration the symplectic structure is taken to be of the standard type (except for Relativistic Toda lattices [35] and associated continuous systems [10]).

Classical algebraic-geometric methods based on complex analysis and symplectic geometry are used as traditional and most powerful tools for the studies of the integrable dynamical systems in algebraic case. This approach enables one to split an integrable system into a collection of finite (or infinite) number of commuting integrable Hamiltonian systems, which are defined on cotangent bundles of Jacobi (or Prym) varieties. It is shown that in appropriate action-angle variables the solutions of Hamiltonian systems are expressed as linear flows with respect to parameters $(x_0, x_1, ...)$ along the trajectories. As a result, the problem is reduced to a classical Jacobi problem of inversion and the solutions of initial integrable system are expressed as symmetric functions of the solutions of the Jacobi system.

The solution of the initial integrable system may also be found by constructing [28, 36] the single-valued Baker-Akhiezer wave functions on smooth Riemann surfaces. Since the classical Jacobian of smooth Riemann surface is isomorphic to an n-dimensional torus, both methods provide the solutions in terms of Θ -functions.

On the other hand, integrable problems were found to have the so-called separatrix solutions. In this case the Riemann surfaces are degenerate and corresponding Jacobians are noncompact. (Soliton and rational solutions of the integrable systems are shown [7, 2, 1, 31, 18, 29] to be examples of this kind).

In spite of the fact that the noncompact Jacobi varieties can be constructed as a limit from Jacobi varieties over smooth Riemann surfaces, the analysis of the generalized Jacobi varieties gives rise to difficult algebraic, analytic and geometric problems. Therefore in the case of a singular Riemann surface a much more complicated theory of generalized Jacobi varieties (studied by Rosenlicht [34],Serre [37]; Mumford and van Moerbeke [30],Fay [21],Altman and Kleiman [15]) has to be used instead of classical theory of Θ -functions. Recently some algebraic problems connected with elliptic and rational solutions of integrable systems defined over Riemann surfaces with singular points have been investigated by Treibich and Verdier [38, 39]. The general case of algebraic completely integrable systems was studied in [24, 25]. Note that in order to find canonical action-angle variables for integrable Hamiltonian systems over Riemann surfaces it was suggested in [32] that meromorphic differentials of second and third kind can be used as well.

Some known integrable systems are coupled with generalized Jacobi varieties. An important case of the hierarchy of Nonlinear Schrödinger equations (NSE) was analysed in [33] using the theory of generalized Jacobi varieties. A very important linkage was found between integrable systems and the representation theory of Kac-Moody algebras. In particular, the Nonlinear Schrödinger equation was studied [23] in the framework of Kac-Moody theory. This equation was also investigated as an example of a Hamiltonian system determined by matrix loop algebras [3, 4, 6]. The general theory of hierarchies of integrable systems and corresponding series of Hirota equations were constructed by Kac and Wakimoto [27] using the methods of representation theory. As a special case a hierarchy of Nonlinear Schrödinger equations together with τ-solutions were found.

The present paper is organized as follows. The main result is described in Section 2, where a general class of completely integrable Hamiltonian systems is constructed. We make use of the construction of non-canonical action-angle variables that has been used for Hamiltonian systems over hyperelliptic curves. It is shown that the corresponding Hamiltonian flows are linearized. In particular, one obtains both scalar and matrix integrable Hamiltonian systems defined over algebraic Riemann surfaces. On the other hand, the Flaschka-McLaughlin Hamiltonian system [22] of periodic Toda lattice can be considered as an important example of nonalgebraic integrable system. In Section 3 we study new classes of meromorphic completely integrable Hamiltonian systems determined over *smooth* Riemann surfaces, when the number $(n+k)$ of variables in Hamiltonians is greater than genus n of the corresponding Riemann curve Γ. It is shown that in this case the Hamiltonian system is also equivalent to Jacobi system of inversion. This time the system of inversion is found to have meromorphic differentials. The poles of meromorphic differentials can be viewed as marked points (*singularities*) of the Riemann curve Γ. In this sense, one can consider a smooth Riemann surface Γ together with the Hamiltonian system as a *singular* curve. This means that the Jacobi variety of the meromorphic Hamiltonian system over smooth Riemann surface Γ has to be studied as a generalized Jacobi variety. One can show that this Riemann surface Γ may be obtained as a limit of smooth Riemann surfaces of higher genus with holomorphic Hamiltonian systems.

In the same manner the case where the number of variables in a Hamiltonian system is less then the genus of Riemann surface is studied.

The first hierarchy in obtained set of meromorphic systems is the hierarchy of Nonlinear Schrödinger equations [14]. A connection between NSE and two-potential KdV equations is established in Section 4 using direct operator between the spaces of solutions. Thus, the solution of the meromorphic system NSE is reduced to Jacobi systems with holomorphic differentials.

2 Nonalgebraic Integrable Hamiltonian systems

This Section is devoted to establishing a general class of completely integrable Hamiltonian systems. Throughout this section we consider complex Hamiltonians determined by general (not necessary algebraic) functions. Consequently, the corresponding Hamiltonian systems will be viewed as the systems defined on complex symplectic manifolds. As a special result we obtain algebraic Hamiltonian systems defined on Jacobians of algebraic curves. In some cases, these will be generalized Jacobians.

Let $R_j(z_j, w_j), (j = 1, ..., n)$, be n complex functions of variables (z_j, w_j), defined in C_j^2, and let

$$\Gamma_j: R_j(z_j, w_j) = 0 \tag{2.1}$$

be the Riemann surface of the implicit multi-valued function $w_j = w_j(z_j)$.
In order to describe the phase space of the problem we take the product

$$C^{2n} = C_1^2 \times ... \times C_n^2$$

with the coordinates $(z_1, ..., z_n; w_1, ..., w_n)$
The manifold C^{2n} is considered to be a complex symplectic manifold. We assume that symplectic structure in C^{2n} is determined by standard complex 2-form

$$\omega^2 = \sum_{j=1}^{n} dz_j \wedge dw_j.$$

Next, to find the most general class of the Hamiltonian functions defined over given $\Gamma^n = \Gamma_1 \times ... \times \Gamma_n$ we choose n^2 auxiliary complex functions

$$s_{jk}(z, w), \quad 1 \leq j, k \leq n \tag{2.2}$$

of two complex variables z, w.
Let D be a determinant of the matrix with entries $d_{jk}(z_j) = s_{jk}(z_j, w_j(z_j))$. Then cofactor of any element d_{jk} will be denoted by D_{jk}. Let also $\bar{m} = (m_1, ..., m_n) \in C^n$ be an n-dimensional vector.
In what follows we consider n-dimensional linear space of multi-valued Hamiltonians

$$H = \sum_{j=1}^{n} R_j(z_j, w_j) M_j(z_1, ..., z_n; \bar{m}) \tag{2.5}$$

defined on the complex symplectic manifold C^{2n}. Here $M_j(z_1, ..., z_n; \bar{m})$ is a linear function

$$M_j(z_1, ..., z_n; \bar{m}) = \sum_{r=1}^{n} m_r \frac{D_{jr}}{D} \tag{2.6}$$

of complex parameters m_r.

The main goal of this paper is to prove that for any choice of n^2 functions $s_{jk}(z_j, w_j(z_j))$ all Hamiltonian systems (2.5) are completely integrable and in appropriate action-angle variables corresponding Hamiltonian flows can be linearized. To achieve this we construct complex deformation of Hamiltonian (2.5) with parameters m_r kept fixed. First of all, we consider embedding of the Hamiltonian system (2.5) in the n-dimensional space of Hamiltonian systems

$$H = \sum_{j=1}^{n} [R_j(z_j, w_j) + S_j(z_j, w_j)] M_j(z_1, ..., z_n; \bar{m}) \tag{2.7}$$

Here

$$S_j(z_j, w_j) = \sum_{k=1}^{n} c_k s_{jk}(z_j, w_j) \tag{2.8}$$

are functions with n complex coefficients $\bar{c} = (c_1, ..., c_n)$.

Definition 2.1 *The n-dimensional space of Hamiltonian systems (2.7) is called the n-dimensional deformation of the Hamiltonian system (2.5).*

Now we can formulate a general theorem

Theorem 2.2 *For every vector-function $M_j(z_1, ..., z_n; \bar{m})$ there exists a subdomain of C^{2n} such that the Hamiltonian system (2.7) is completely integrable.*

Proof. From coordinate and momentum parts of Hamiltonian system (2.7) we easily obtain a complete system of first integrals

$$R_j(z_j, w_j) + S_j(z_j, w_j) = 0 \tag{2.9}$$

Recall that the canonical action-angle variables cannot be determined in the general case. So, we will find non-canonical action-angle variables which reduce Hamiltonian system to the Jacobi system of inversion. In fact, action variables can be chosen equal to the parameters of the deformation (2.8) $H_r = c_r$. Now, in order to find conjugate angle variables we obtain from (2.9) $w_j = w_j(z_j; c_1, ..., c_n)$, as implicit functions of z_j defined on Riemann surfaces

$$\Gamma_j(\bar{c}) \colon R_j(z_j, w_j) + S_j(z_j, w_j) = 0 \tag{2.10}$$

Next, we introduce an action-function S

$$S(z_1, ..., z_n; H_1, ..., H_n) = \sum_{j=1}^{n} \int_{z_j^0}^{z_j} w_j(z_j) dz_j. \tag{2.11}$$

Then angle variables σ_r (conjugate to the action variables H_r) are determined by the partial derivatives

$$\sigma_r = -\frac{\partial S}{\partial H_r} = \sum_{j=1}^{n} \int_{z_j^0}^{z_j} \frac{s_{jr}}{\frac{\partial (R_j + S_j)}{\partial w_j}} dz_j \tag{2.12}$$

Using Hamiltonian equations one can linearize given Hamiltonian system. It follows from (2.12) that

$$\frac{d\sigma_r}{dx} = m_r \tag{2.13}$$

Integrating this expression with respect to parameter x one obtains the Jacobi system

$$\sigma_r = \sum_{j=1}^{n} \int_{z_j^0}^{z_j} \frac{s_{jr}}{\frac{\partial R_j + S_j}{\partial w_j}} dz_j = m_r(x - x_0) + d_r(x_0) \tag{2.14}$$

This system demonstrates that in the action-angle variables (H_r, σ_r) Hamiltonian flows are linear functions of parameter x

$$H_r = const., \; \sigma_r = m_r(x - x_0) + d_r(x_0) \tag{2.15}$$

To get the solutions of initial system (2.5) we put in (2.14) all action variables equal to zero $H_r = c_r = 0$. Q.E.D.

In the most general case we can establish a hierarchy of integrable systems as a combination of commutative Hamiltonian systems (2.5) or (2.7). As an example of nonalgebraic integrable system one can take the periodic Toda lattice with Flaschka-

Mclaughlin Hamiltonians [22]. Recall that the corresponding function R has the form

$$R = (-1)^{n+j+1} \cosh(\frac{W}{2}) + z^{n+1} + c_1 z^n + \dots + c_{n+1}. \tag{2.16}$$

The other example would be the systems based on the Hamiltonians of periodic relativistic Toda lattice and associated continuous systems [10].

Obtained theorem is valid for both algebraic and nonalgebraic functions R_j and has many important corollaries. The most important case is when all functions R_j are supposed to be the same

$$R_j(z, w) \equiv R(z, w)$$

In this case one can use symmetric product $SC^{2n} = S^n C^2$ and construct generalized Jacobi variety of the Riemann surface Γ_S of the function

$$R(z, w) + S(z, w) = 0 \tag{2.17}$$

As a result, we get that even in nonalgebraic case the Hamiltonian system is defined on the Jacobian.

Note that angle variables used [8, 9, 13] in all considered Hamiltonian systems over hyperelliptic curves are of the form (2.12). Similar expressions were found for the class of polynomial matrix systems [19, 20, 5] in the framework of the representation theory.This means that one can use the Hamiltonians (2.5) and (2.7) as explicit Hamiltonians for the polynomial matrix systems.

In conclusion we note that in analogy with algebraic case there exists an important problem of finding hidden symmetries and Kac-Moody representations for nonalgebraic completely integrable Hamiltonian systems.

3 Algebraic Integrable Hamiltonian systems over noncompact Jacobi varieties

As it was emphasized in Introduction classical integrable systems generally are corresponding to multi-Hamiltonian systems defined on compact Jacobi varieties. At the same time,there are examples of integrable systems defined on generalized or noncompact Jacobi varieties. Usually the noncompact Jacobi variety is constructed as an Abelian variety corresponding to a singular Riemann surface, obtained as the limit of smooth Riemann surfaces.

Here we describe a new class of Hamiltonian systems defined on noncompact Jacobi varieties.In this case the noncompact variety is constructed over smooth Riemann surface. Here the number $(n + k)$ of variables in the Hamiltonian system is greater than genus n of the corresponding Riemann curve Γ. The Riemann surface Γ can be represented as a limit of Riemann surfaces of higher genus with holomorphic Hamiltonian systems. In this sense a smooth Riemann surface Γ can be considered as a *singular* curve. The Jacobi variety corresponding to a meromorphic Hamiltonian system with $(n + k)$ variables is a generalized Jacobi variety.

We start with one-Hamiltonian systems or, in other words, with $(n+k)$ -dimensional meromorphic mathematical oscillators. Then the integrable system will be made up of h commutative one-Hamiltonian systems. The formal construction is the same as in the classical case [10, 11], but the structure of the generalized Jacobi variety is more complicated.

First we give several definitions and notations.
Let $C_N(z) = c_0 z^N + ... + c_N$ be a polynomial of degree N.

Definition 3.1 *A class of integrable systems over hyperelliptic curves is called odd if N=2n+1 and even otherwise.*

Assume that the polynomial $C_N(z)$ is factorized into the product of two polynomials

$$C_N(z) = R_L(z) S_{N-L}(z)$$

and

$$L \leq N - n - k + 1$$

Let

$$\Gamma: W^2 + \frac{S_{N-L}(z)}{R_L(z)} = 0 \tag{3.1}$$

be the Riemann surface of genus n and $k < n$.
To describe the phase space of the problem we consider the symmetric product $C^{2(n+k)} = S^{n+k}C^2$ with the coordinates $(z_1, ..., z_{n+k}, w_1, ..., w_{n+k})$.
Lastly we consider the systems with the meromorphic Hamiltonians

$$H_M = \sum_{j=1}^{n+k} \frac{R_L(z_j) w_j^2 + S_{N-L}(z_j)}{\prod_{r \neq j}(z_j - z_r)} M(z_j; \alpha, \beta) \tag{3.2}$$

defined on the complex symplectic manifold $C^{2(n+k)}$. Here $M(z_j; \alpha, \beta)$ is a rational function

$$M(z; \alpha, \beta) = \sum \bar{\alpha}_l G_l(z) + \sum \bar{\beta}_m \frac{G_m^*(z)}{z^{m+1}}. \tag{3.3}$$

Since the value of the polynomial $G_m^*(z)$ in point z_j is equal to

$$G_m^*(z_j) = -z_j^{m+1} G_{n-m-1}(z_j)$$

then all Hamiltonians (3.2) make up a $(n + k)$ - dimensional linear space with polynomials $G_l(z)$ chosen as a basis.
Now we formulate the main result of this section

Theorem 3.2 *For every rational function $M(z; \alpha, \beta)$ the Hamiltonian system with the Hamiltonian (3.2) is completely integrable and can be reduced to the Jacobi system of inversion*

$$\frac{1}{2} \sum \int_{z_j^0}^{z_j} \frac{z_j^{r-1} dz_j}{\sqrt{-C_N(z_j)}} = \left(\sum_{l=0}^{n+k-1} \bar{\alpha}_l \delta_{n+k-l}^r - \sum_{m=0}^{n+k-1} \bar{\beta}_m \delta_{m+1}^r \right)(x - x^0) + D_r(x^0) =$$

$$m_{\alpha,\beta,r}(x - x^0) + D_r(x^0) \tag{3.4}$$

Proof. This result follows from the general theorem considered in Section 2. Here we give a short proof for the meromorphic Hamiltonians to show the connection with generalized Jacobians.
First of all we will find a complete system of first integrals.
The coordinate part of the Hamiltonian system with the Hamiltonian (3.2) has the form

$$\frac{dz_j}{dx} = \frac{2w_j R_L(z_j)}{\prod_{s\neq j}(z_j - z_s)} M(z_j; \alpha, \beta) \tag{3.5}$$

Combining these equations with the second (momentum) part of Hamiltonian system we get the complete system of first integrals

$$R_L(z_j) w_j^2 + S_{N-L}(z_j) = 0 \tag{3.6}$$

Now we introduce action-angle variables and reduce the problem to the Jacobi problem of inversion.
Let S be the action-function

$$S(z_1, ..., z_{n+k}; H_1, ..., H_{n+k}) = \sum_{j=1}^{n+k} \int_{z_j^0}^{z_j} \sqrt{\frac{-S_{N-L}(z_j)}{R_L(z_j)}} dz_j \tag{3.7}$$

Here

$$S_{N-L}(z) = s_0 z^{N-L} + ... + s_{N-L}$$

We choose $H_r = s_{N-L-n-k+r}$ to be the action variables. Hence the conjugate angle variables need to be defined as the derivatives

$$\sigma_r = -\frac{\partial S}{\partial H_r} = \frac{1}{2} \sum_{j=1}^{n+k} \int_{z_j^0}^{z_j} \frac{z_j^{n+k-r}}{\sqrt{-C_N(z_j)}} dz_j \tag{3.8}$$

Using Hamiltonian equations we see that

$$\frac{d\sigma_r}{dx} = \frac{1}{2} \sum_{j=1}^{n+k} \frac{z_j^{n+k-r}}{\sqrt{-C_N(z_j)}} \frac{dz_j}{dx} =$$

$$\sum_{j=1}^{n+k} \frac{z_j^{n+k-r}}{\prod_{s\neq j}(z_j - z_s)} M(z_j; \alpha, \beta) = \sum_{l=0}^{n+k-1} \bar{\alpha}_l \delta^r_{n+k-l} - \sum_{m=0}^{n+k-1} \bar{\beta}_m \delta^r_{m+1}$$

Integrating this equations one gets the Jacobi system of inversion over the Riemann surface (3.1).Q.E.D.□.

Corollary 3.3 *The Abel's transformation defined by the action-angle map (3.4) maps the symmetric product $S^{n+k}\Gamma$ into the generalized Jacobi variety*

$$Jac_g(\Gamma) = C^{n+k}/Z$$

where Z is the lattice of the periods of holomorphic and meromorphic differentials including the periods of poles.

Using results of the theorem 3.2 we can get integrable systems as the commutative collections of meromorphic Hamiltonian systems. Let Γ be the Riemann surface (3.1) described above and

$$M_1(z; \alpha_1, \beta_1), ..., M_h(z; \alpha_h, \beta_h)$$

be any polynomials of the type (3.3) with distinct α_j, β_k

Definition 3.4 *A set of Hamiltonian systems with the Hamiltonians (3.2), $H_{M_r}(x_r)$, $r = 1, ..., h$ is called the integrable system of the type*

$$(\Gamma, x_r, M_1, ..., M_h, H_{M_1}, ..., H_{M_h}). \tag{3.9}$$

The integrable system (3.9) is equivalent to h Jacobi systems of inversion. All Hamiltonian flows, defined by H_{M_r}, commute.

Definition 3.5 *All systems (3.2) with the same $C_N(z)$ have common angle variables but different action variables. We call these systems the associated continuous integrable systems.*

The most important and well known example of meromorphic systems is the hierarchy of Nonlinear Schrödinger equations. In [14] the Lax representation with linear operator

$$L_2 = \partial_x + z\begin{pmatrix} 0 & -1 \\ 1 & 0 \end{pmatrix} - 1/2\begin{pmatrix} -(r+q) & i(r-q) \\ i(r-q) & (r+q) \end{pmatrix} \tag{3.10}$$

is used. It is shown that corresponding differential equations are defined over Riemann surface

$$\Gamma\colon W^2 + C_{2n+2}(z) = 0 \tag{3.11}$$

of genus n, and the corresponding commutative Hamiltonians are functions of $(n+1)$ variables

$$H_X = \sum_{j=1}^{n+1} \frac{w_j^2 + C_{2n+2}(z_j)}{\prod_{r\neq j}(z_j - z_r)}(G_1(z_j) + c) \ \textit{in stationary case} \tag{3.12}$$

$$H_T = \sum_{j=1}^{n+1} \frac{w_j^2 + C_{2n+2}(z_j)}{\prod_{r\neq j}(z_j - z_r)} G_2(z_j) \ \textit{in dynamical case} \tag{3.13}$$

Therefore the obtained Jacobi systems of inversion (3.4) contain meromorphic differentials and the corresponding Jacobian is the generalized Jacobi variety. Comparing both (3.12) and (3.13) with (3.2) one can observe that the hierarchy of NSE is just the first term ($k = 1$) in the found series of meromorphic Hamiltonian systems.

In the next section we establish a link between Nonlinear Schrödinger equations and two-potential KdV equations using operator between the spaces of solutions of these two problems. Therefore the NSE system will be reduced to Jacobi system of inversion with only holomorphic differentials.

4 Nonlinear Schrödinger equations

As it was pointed out in Section 3, the hierarchy of Nonlinear Schrödinger equations is the first term in the infinite set of meromorphic systems. In this section we will show that there exists an operator from the space of solutions of NSE on the space of solutions of two-potential KdV equations. However in the case of two-potential KdV both Jacobi integral systems contain only holomorphic differentials. So there arises a problem to find a direct transformation from the integral system with meromorphic differentials to the system with holomorphic differentials .

At the beginning we obtain the hierarchy of two-potential KdV systems[8, 13].

Theorem 4.1 *Let G_m be polynomials of degree $m = 0, 1, 2, \ldots$ with respect to a free complex parameter z. The equation of KdV type*

$$\frac{\partial U}{\partial t} = -\frac{1}{2}\frac{\partial^3 G_m}{\partial x^3} + 2\frac{\partial G_m}{\partial x}U + G_m\frac{\partial U}{\partial x} \tag{4.1}$$

with two-potential function

$$U = -z^2 + uz + v \tag{4.2}$$

generates the hierarchy with the first member in the form

$$\frac{\partial u}{\partial t} = \frac{3}{2} u \frac{\partial u}{\partial x} + \frac{\partial v}{\partial x} \tag{4.3a}$$

$$\frac{\partial v}{\partial t} = -\frac{1}{4} \frac{\partial^3 u}{\partial x^3} + v \frac{\partial u}{\partial x} + \frac{1}{2} u \frac{\partial v}{\partial x} \tag{4.3b}$$

This hierarchy is called two-potential KdV systems.

Proof. Substituting the polynomials G_m in the identity (4.1) and equating the corresponding coefficients we get complete hierarchy of two-potential systems.□
In order to split the two-potential system into two commutative integrable Hamiltonian systems we consider a set ($n = 1, 2, 3, \ldots$) of corresponding stationary KdV equations

$$\frac{\partial^2 G_n}{\partial x^2} G_n - \frac{1}{2} \left(\frac{\partial G_n}{\partial x}\right)^2 - 2G_n^2 U = 2C_N(z) \tag{4.4}$$

Here $C_N(z)$ is a polynomial of even degree $N = 2n + 2$.
A pair (4.1) and (4.4) is proved to be equivalent to the dynamical equation

$$\frac{\partial G_n}{\partial t} = G_n' G_m - G_n G_m'. \tag{4.5}$$

Using root variables z_j obtained from $G_n = \prod_{j=1}^{n} (z - z_j)$ we find desired Hamiltonian systems.

Theorem 4.2 *Each pair (4.4) and (4.5) is equivalent to two commutative completely intagrable Hamiltonian systems with Hamiltonians*

$$H = \sum_{j=1}^{n} \frac{w_j^2 + C_N(z_j)}{\prod_{s \neq j} (z_j - z_s)} \quad \textit{in stationary case} \tag{4.6}$$

$$H = \sum_{j=1}^{n} \frac{w_j^2 + C_N(z_j)}{\prod_{s \neq j} (z_j - z_s)} G_m(z_j) \quad \textit{in dynamical case} \tag{4.7}$$

The proof is similar to the proof of Theorem 3.2 . □
It follows from identity (4.1) and trace formulas that unknown potentials are found in the form

$$u = -2 \sum z_j - c_1 \tag{4.8a}$$

$$v = +2 \sum_{j<k} z_j z_k - 3\left(\sum z_j\right)^2 - 2c_1 \sum z_j - c_2 \tag{4.8b}$$

Using the classical relation between the solutions of the Jacobi systems of inversion and Θ-functions one expresses symmetric polynomials from (4.8) in terms of Θ-function. As a result, the function u has the form

$$u = i \frac{\partial}{\partial x} \ln \frac{\Theta(Ux + Vt + K - A(P_+))}{\Theta(Ux + Vt + K - A(P_-))} \tag{4.9}$$

Here $U, V, K, A(P_+)$ *and* $A(P_-)$ are *constants*, which depend on the following Riemann surface

$$W^2 + C_N(z) = 0$$

Analogous expression is found for the function v .
To establish a connection between Nonlinear Schrödinger equations and two-potential systems. we use the usual Lax-representation [1] for NSE with the linear operator in the form

$$L_1 = \partial_x - z\begin{pmatrix} -i & 0 \\ 0 & i \end{pmatrix} - \begin{pmatrix} 0 & q \\ r & 0 \end{pmatrix} \tag{4.10}$$

Eliminating second component ψ_2 from auxiliary linear system $L_1\Psi = 0$ we reduce the matrix auxiliary problem to scalar linear Schrödinger equation with potential (4.2). Hence the generating equation of NSE is found to be the same equation (4.1).

Lemma 4.3 *The link between Nonlinear Schrödinger equations and two-potential KdV systems is established in the form of the mapping P :*

$$u = i\frac{\hat{q}'}{\hat{q}} \tag{4.11a}$$

$$v = \hat{r}\hat{q} - \frac{1}{2}\left(\frac{\hat{q}'}{\hat{q}}\right)' + \frac{1}{4}\left(\frac{\hat{q}'}{\hat{q}}\right)^2 \tag{4.11b}$$

from the space of vector-functions $(\hat{q}, \hat{r})$ onto the space of vector-functions (u, v).
The operator P is invariant under the transformation

$$(\hat{q}, \hat{r}) \rightarrow (e^{iz}\hat{q}, e^{-iz}\hat{r}). \tag{4.12}$$

Here $z = z(t)$ is arbitrary complex function of the variable t .

This lemma can be easily verified by direct calculations.□
Note that in the case $z(t) = const$ this invariance gives rise [33] to C^*-extension of classical Jacobi variety $Jac(\Gamma)$ to generalized Jacobi variety of Nonlinear Schrödinger equation.

Theorem 4.4 *The transformation (4.11) yields that if the vector-function (u, v) satisfies (4.3) then the vector-function $(\hat{q}, \hat{r})$ is the solution of the generalized NSE*

$$i\frac{\partial \hat{q}}{\partial t} = -\frac{1}{2}\frac{\partial^2 \hat{q}}{\partial x^2} + \hat{r}\hat{q}^2 + \hat{q}\frac{dz}{dt} \tag{4.13a}$$

$$i\frac{\partial \hat{r}}{\partial t} = \frac{1}{2}\frac{\partial^2 \hat{r}}{\partial x^2} - \hat{q}\hat{r}^2 - \hat{r}\frac{dz}{dt} \tag{4.13b}$$

Proof is strightforward.□
Finally we establish a connection with NSE.

Theorem 4.5 *If a vector-function $(\hat{q}, \hat{r})$ is the solution of (4.13) then the functions $q = e^{iz}\hat{q}$ and $r = e^{-iz}\hat{r}$ satisfy the initial Nonlinear Schrödinger system of equations*

$$i\dot{q} = -\frac{1}{2}q'' + rq^2 \tag{4.14a}$$

$$i\dot{r} = \frac{1}{2}r'' - qr^2 \tag{4.14b}$$

Proof is strightforward.□

Thus the problem of solutions of Nonlinear Schrödinger equations is reduced to analogous problem for the two-potential KdV equations.
It follows from (4.11),(4.8) and (4.9) that the algebraic solutions of Nonlinear Schrödinger equations have the form

$$q = e_q \exp(ax + bt)\frac{\Theta(Ux + Vt + K - A(P_+))}{\Theta(Ux + Vt + K - A(P_-))} \tag{4.15a}$$

$$r = e_r \exp(-ax - bt)\frac{\Theta(Ux + Vt + K - A(P_-))}{\Theta(Ux + Vt + K - A(P_+))} \tag{4.15b}$$

that coinside with those found in [26, 33, 3, 4].
Note that comparing (3.13) and (4.10) one can observe that

$$L_2 = gL_1g^{-1} \tag{4.16}$$

where g is a constant matrix

$$g = 1/2\begin{pmatrix} -i & i \\ 1 & 1 \end{pmatrix} \tag{4.17}$$

The same relation can be used for pair operators

$$A_2 = gA_1g^{-1} \tag{4.18}$$

It means that the space of all solutions of Lax equations with L_1 coinsides with the space of all solutions of Lax equations with L_2.

5 Discrete Integrable Systems over Jacobi Varieties

For many years the discrete and continuous integrable systems have been investigated using algebraic-geometric methods .

At the same time there were a lot of attempts to establish a link between discrete and continuous integrable systems in the form of limit approximation. The result usually depends on the chosen limit process. In particular, the Boussinesq equation and Korteweg- de Vries (KdV) equation were shown to be two different limit continuous integrable systems for one of the most important discrete models, namely the Toda lattice .

On the other hand there is an important problem to find a regular procedure for the discretization of the integrable continuous systems.

Here one of the possible methods of such construction is described.

Every discrete problem considered as a Hamiltonian problem over Jacobi variety splits into a number of discrete transformations and a number of commutative Hamiltonian flows. As an example we consider three hierarchies of discrete systems, namely Toda lattices

$$\frac{dv_k}{dt} = v_k(w_k - w_{k+1}) \tag{5.1a}$$

$$\frac{dw_k}{dt} = (v_{k-1} - v_k), \tag{5.1b}$$

relativistic Toda lattices

$$\frac{dv_k}{dt} = v_k(v_{k-1} - v_{k+1} + w_k - w_{k+1}) \tag{5.2a}$$

$$\frac{dw_k}{dt} = w_k(v_{k-1} - v_k) \tag{5.2b}$$

and Volterra lattices

$$\frac{dv_k}{dt} = v_k(v_{k-1} - v_{k+1}), \tag{5.3}$$

using similar stationary and dynamical equations .

According to the general theory of Abelian differentials one obtains that the stationary X-equations are equivalent to the discrete Jacobi system of inversion

$$\frac{1}{2}\sum_{j=1}^{n} \int_{z_{j,k}}^{z_{j,k+1}} \frac{z_j^{n-m} dz_j}{\sqrt{-C_N(z_j)}} = U_m \tag{5.3}$$

here U_m does not dependent on k.

The dynamical system in root variables

$$\frac{d(z_{j,k})}{dt} = \frac{2G_k(z_{j,k})\sqrt{-C_N(z_{j,k})}}{\prod_{s\neq j}(z_{j,k} - z_{s,k})} \tag{5.4}$$

is a Hamiltonian system with Hamiltonian

$$H_T = \sum_{j=1}^{n} \frac{(w_{j,k}^z)^2 + C_N(z_{j,k})}{(z_{j,k})^{2q} \prod_{s\neq j}(z_{j,k} - z_{s,k})} G_k(z_{j,k}) \tag{5.5}$$

Notice that in the case of the relativistic Toda lattices ($q = \frac{1}{2}$) the Poisson structure is determined by a skew-symmetric matrix

$$P_{i,n+j} = \delta_j^i z_{j,k}, for 1 \leq i \leq n \tag{5.6a}$$

$$P_{n+i,j} = -\delta_j^i z_{j,k}, for 1 \leq j \leq n \tag{5.6b}$$

Here we describe the general method of constructing the discrete systems corresponding to the continuous ones.

Let $(\Gamma, x_r, M_r, H_{M_r})$ be an integrable system (9).

Definition 5.1 *We replace the first q continuous systems of inversion (4) by the following discrete systems of inversion*

$$\frac{1}{2}\sum \int_{z_{j,l}}^{z_{j,l+1}} \frac{z_j^{k-1} dz_j}{\sqrt{-C_N(z_j)}} = m_{\alpha_r,\beta_r,k}\Delta x_r, r = 1, ..., q. \tag{5.7}$$

The obtained discrete system is called **the discrete system associated with the continuous system** *and vice versa.*

The discrete transformations of the cotangent bundle T^*T^n defined by the discrete systems are canonical transformations. They preserve the Hamiltonian structure of the remaining continuous part. So we can say that the discrete flows commute with the continuous Hamiltonian flows.

In accordance with our approach the following general definition gives a link between discrete and continuous integrable systems.

Definition 5.2 *The discrete and continuous integrable systems are associated with each other:*

1.) If they have the same spectrum (Riemann surface Γ) as well as the same Hamiltonians for the continuous part of each system;

2.) If the discrete Jacobi systems of inversion are in fact a lattice-discretization of the continuous Jacobi systems.

As examples we consider hierarchies of Toda lattices, relativistic Toda lattices and Volterra lattices. The simplest case is the case of Toda lattices.

Comparing spectrum, Hamiltonians and Jacobi systems of inversion for the Toda lattices and the two-potential systems (4.3) we obtain the following conclusions

Theorem 5.3 *The hierarchy of the two-potential systems is the systems associated with the hierarchy of Toda lattices.*

Theorem 5.4 *The hierarchy of Toda lattices is associated with the hierarchy of Nonlinear Schrödinger equations.*

The continuous systems associated with the hierarchies of relativistic Toda lattices and Volterra lattices are also found.

Acknowledgements. The author would like to thank I.M.Sigal for helpfull discussions.

References

[1] M. Ablowitz and H. Segur,*Solitons and Inverse Scattering Transform,* SIAM,Philadelphia,(1981)

[2] M.J. Ablowitz and J. Satsuma, *Solitons and rational solutions of nonlinear evolution equations,* J. Math. Phys.,16(1978),2180

[3] M. Adams, J. Harnad and E. Previato , *Isospectral Hamiltonian flows in finite and infinite dimensions 1.Generalized Moser systems and moment maps into loop algebras,*Comm.Math.Phys.,117(1988),451-500

[4] M. Adams, J. Harnad and J. Hurtubise, *Isospectral Hamiltonian flows in finite and infinite dimensions 2.Integration of flows,* Preprint,CRM,1635 (1989)

[5] M. Adams, J. Harnad and J. Hurtubise, *Coadjoint orbits, spectral curves and Darboux coordinates,* Preprint, CRM, Montreal, 1648(1989)

[6] M. Adams, J. Harnad and J. Hurtubise, *Liouville generating functions for isospectral flow in loop algebras,* Preprint,CRM,1652(1990)

[7] H. Airault, H.P. McKean and J. Moser, *Rational and elliptic solutions of the KdV equations and related many-body problems,* Comm. Pure Appl. Math.,30(1977),95

[8] S. Alber, *Investigation of equations of Korteweg-de Vries type by the method of recurrence relations,* ICP,USSR Academy of Sciences(1976),J. London Math. Soc. 2(1979),467-480

[9] S. Alber, *On stationary problems for equations of Korteweg-de Vries type,* Comm. Pure and Appl. Math.,34(1981),259-272

[10] S. Alber, *Associated integrable systems,* J.Math.Phys., 4(1991)

[11] S. Alber, *Hamiltonian systems on the Jacobi varieties,* Proceedings of MSRI "Geometry of Hamiltonian systems",Berkeley (1991)

[12] S. Alber, *Dynamical systems over noncompact Jacobi varieties,* Proceedings of the International symposium "Nonlinear dynamics in Mathematics and Science" , Amherst, (1991)

[13] S. Alber and M. Alber, *Hamiltonian formalism for finite-zone solutions of integrable equations,* C.R.Acad.Sci. Paris,301,Ser. 1, 16(1985), 777-780

[14] S. Alber and M. Alber, *Hamiltonian formalism for Nonlinear Schrödinger equations and sine-Gordon equations,* J.London Math. Soc.,(2) 36(1987),176-192

[15] A. Altman and S. Kleiman, *Compactifying the Picard Scheme,* Advances in Mathematics, 35(1980),50-112

[16] E. Arbarello, *Periods of Abelian integrals,theta functions and differential equations of KdV type,* Proceedings of the International Congress of Mathematicians,Berkeley,USA,(1986)

[17] V. Arnold, *Mathematical methods of classical mechanics,* Springer-Verlag,New York,(1978)

[18] F. Calogero and A. Degasperis,*Spectral Transform and Solitons,* Amsterdam,North-Holland,(1984)

[19] P. Deift, L.C. Li and C. Tomei, *Matrix factorization and integrable systems,* Comm. Pure Appl. Math.,42(1989),443-521

[20] P. Deift and L.C. Li, *Generalized affine Lie algebras and the solution of a class of flows associated with the QR eigenvalue algorithm,* Comm. Pure Appl. Math.,42(1989),963-991

[21] J. Fay, *Theta functions on Riemann surfaces,* Springer Lecture Notes,(1973)

[22] H. Flaschka and D. Mclaughlin, *Canonically conjugate variables for the Korteweg-de Vries equation and the Toda lattice with periodic boundary conditions,* Prog. Theor. Phys.,55(1976),438

[23] H. Flaschka, A.C. Newell and T. Ratiu, *Kac-Moody Lie algebras and soliton equations,* Physica 9D,(1983),300-323

[24] J.-P. Francǫise, *The Arnold formula for algebraically completely integrable systems,* Bulletin of the Amer. Math. Soc.,17(1987),301-303

[25] J.-P. Francǫise, *Calculs explicites d'action-angles,* Séminaire de Mathématiques Supérieures de Montréal,102(1986),101-120.

[26] A. Its, *Inversion of Hyperelliptic integrals and integration of nonlinear differential equations,* Vest. Leningr. Univ., 7(1976),37

[27] V. Kac and M. Wakimoto, *Exceptional hierarchies of soliton equations,* Proceedings of Symposia in Pure Math. (1989)

[28] I. Krichever, *Integration of nonlinear equations by the method of algebraic geometry,* Funct.Anal.Appl. 11(1977),15-31

[29] I. Krichever, *Rational solutions of the K.-P. equation and integrable systems of particles on the line,* Funct.Anal.Appl.,12(1978), 76-78

[30] D. Mumford and P.van Moerbeke, *The spectrum of difference operators and algebraic curves,* Acta Math.,143(1979), 93-154

[31] S. Novikov,S. Manakov,L. Pitaevskii and V. Zakharov, *Theory of Solitons,* Plenum Publishing Co., New York,(1984)

[32] S. Novikov and A. Veselov, *Poisson brackets and complex tori,* Proc.Steklov Inst.Math.,3(1985),53-65

[33] E. Previato, *Hyperelliptic quasi-periodic and soliton solutions of the nonlinear Schrödinger equation,* Duke Math. Journal,52(1985),329-377

[34] M. Rosenlicht, *Generalized Jacobian varieties,* Ann.of Math., 59(1954),505-530

[35] S. Ruijsenaars and H. Schneider, *A new class of integrable systems and its relation to solitons,* Ann.Phys.(NY) 170(1986),370

[36] M. Sato, *Soliton equations as dynamical systems on infinite dimensional Grassmann manifolds,* Res.Inst.Math.Sci. Kokyuroku,439(1981),30

[37] J.-P. Serre, *Groupes algébriques et corps de classes,* Hermann,Paris,1959

[38] A. Treibich, *Courbes Elliptiques tracées sur des Jacobiennes Généralisées,* C.R.Acad. Sc.Paris,307(1988),Serie 1,531-534

[39] A. Treibich and J.L. Verdier, *Solitons elliptiques* , Birkhauser-Boston(1989)

Collective Coordinates by a Variational Approach: Problems for Sine Gordon and Φ^4 Models

J.G. Caputo[1] *and N. Flytzanis*[2]

[1]LESP, INSA and URA CNRS 230, B.P. 8,
F-76131 Mont-Saint-Aignan Cedex, France
[2]Department of Physics, University of Crete, Heraklion, Greece

A certain number of non-linear partial differential equations can be solved exactly by the method of inverse scattering [1]. When they are perturbed the evolution of the solution can be obtained by perturbation methods [2]. Parameters of the unperturbed problem become slowly varying because of the perturbation.
In many other cases however only a few analytical solutions are known and the equation only has a few conserved quantities so that the mathematical structure of the problem is much more limited. In those situations to obtain the evolution of collective coordinates a variational method can be used. This method introduced by Rice and Mele [3] in the study of lightly doped poly-acetylene has had some remarkable sucesses for the sine-Gordon model [4]. Legrand guided by an algebraic identity gave a collective coordinate ansatz which he used for the study of a perturbed breather [5]. The equation for the shape variable was undefined when the breather was flat. In this paper which is a condensed version of [6] it is shown why through a projection argument and new problem free ansatzes are presented.

Consider the perturbed sine-Gordon equation:

$$\phi_{tt} - \phi_{xx} + \sin\phi - F(\phi, \phi_t, t) = 0 \tag{1}$$

Assume that the solution of (1) can be written as:

$$\phi(x, y(t), k(t)) = 4arctan(\frac{\sinh(y(t))}{\cosh(k(t)x)}) \tag{2}$$

which can describe a breather or a kink antikink in the center of mass reference frame.
The evolution equations for y and k are obtained using a Lagrangian density for (1):

$$l = \frac{1}{2}{\phi_t}^2 - \frac{1}{2}{\phi_x}^2 - (1 - cos\phi) + P \tag{3}$$

where P corresponds to the perturbation terms assumed non dissipative. Because (1) can be seen as a condition for the variation of $\int\int ldxdt$ to be zero, the evolution of y and k are obtained by writing the usual Lagrange equations from the reduced Lagrangian:

$$L(y, \dot{y}, k, \dot{k}, t) = \int ldx \tag{4}$$

where (2) has been substituted for ϕ in the calculation. In terms of y and k and in the absence of perturbations:

$$L = \dot{y}^2 \frac{8}{k}(1 + \frac{2y}{\sinh 2y}) - \dot{y}\dot{k}16\frac{y}{k^2} + \dot{k}^2 \frac{2}{3k^3}[(\pi^2 + 4y^2)(1 - \frac{2y}{\sinh 2y}) + 8y^2]$$

$$-8k(1 - \frac{2y}{\sinh 2y}) - \frac{8}{k}(\tanh y)^2(1 + \frac{2y}{\sinh 2y}) \tag{5}$$

When $y = 0$ all terms vanish except for the coefficient of $\dot{y}^2$. In fact all standard ode integrators tried failed to get accross the $y = 0$ line and maintain reasonable accuracy. From the Hamiltonian it can be inferred that $\dot{k}$ explodes when $y = 0$ unless an algebraic condition is verified between $(y, \dot{y}, k, \dot{k})$. ¿From these observations it seems that the equation is ill-defined. It turns out that this fact can be established from (2).

It can be shown [5,6] that the Lagrange equations for the y and k collective coordinates:

$$\frac{\partial L}{\partial a_i} - \frac{d}{dt}(\frac{\partial L}{\partial \dot{a}_i}) = 0 \tag{6}$$

where $a_1 = y$ and $a_2 = k$ are equivalent to:

$$\int_{-\infty}^{+\infty} dx[\frac{\partial l}{\partial \phi} - \frac{\partial}{\partial t}(\frac{\partial l}{\partial \phi_t}) - \frac{\partial}{\partial x}(\frac{\partial l}{\partial \phi_x})]\frac{\partial \phi}{\partial a_i} = 0 \tag{7}$$

The quantity in the brackets is the right hand side of (1). The problem is that $\frac{\partial \phi}{\partial k} = 0$ when $y = 0$ and therefore at that instant the projection carries no information and the evolution of k is undefined. One could think that (2) was an unwise physical choice but any general ansatz like:

$$\phi(x, y(t), k(t)) = 4arctan(\frac{\sinh[f(y(t), k(t))]}{\cosh(g(y(t), k(t))x)}h(y(t), k(t))) \tag{8}$$

where f can be 0 gives rise to undefined evolution equations [6]. This problem also arises in different models like ϕ^4 and double sine-Gordon [6].

A way of fixing things is not to do a projection anymore. Introducing an a_i dependance in the solution ansatz leads to a Lagrangian:

$$L = L(a_i, \dot{a}_i, \ddot{a}_i) \tag{9}$$

The Lagrange equation for a_i is written as:

$$\frac{\partial L}{\partial a_i} - \frac{d}{dt}(\frac{\partial L}{\partial \dot{a}_i}) + \frac{d^2}{dt^2}(\frac{\partial L}{\partial \ddot{a}_i}) = 0 \tag{10}$$

It can be shown [6] that the Lagrange equation (9) is equivalent to:

$$\int_{-\infty}^{+\infty} dx[\Xi\frac{\partial \phi}{\partial a_i} - \frac{\partial}{\partial t}(\Xi\frac{\partial \phi}{\partial \dot{a}_i})] = 0 \tag{11.a}$$

where

$$\Xi = \frac{\partial l}{\partial \phi} - \frac{\partial}{\partial t}(\frac{\partial l}{\partial \phi_t}) - \frac{\partial}{\partial x}(\frac{\partial l}{\partial \phi_x}) \tag{11.b}$$

which is no longer a simple projection. Following this idea, a simple ansatz that can be introduced for the breather problem is:

$$\phi(x, y(t), k(t)) = 4arctan(\frac{\sinh[y(1+\dot{k})]}{\cosh kx}) \tag{12}$$

While this ansatz is rather arbitrary, it is very appropriate to make our point for the removal of the divergence in the equation for k. It is not also unreasonable since any shape oscillation can cause a small extension or contraction of the breather. When y is large the $\dot{k}$ term leads to a wobbling kink antikink pair. The reason for putting y as a factor is to avoid a forced time dependance of k. In the absence of perturbations $\dot{k} = 0$ so that the pure breather or kink antikink solutions still exist. A similar solution ansatz can be chosen for the ϕ^4 model [6].

The collective coordinate approach has allowed remarkable semi-quantitative agreement between the numerical solution of the full pde and the reduced dynamics [4]. However the method remains very empirical. For example for ϕ^4 resonnances have been observed on the pde simulations for kink antikink collisions by Campbell et al [7]. Using the ansatz suggested in [7] Fei and Vazquez [8] were able to find the resonnance values and widths on the collective coordinate odes. They truncated the reduced Lagrangian so as to eliminate the ill-definition of the equations and reduced the problem to a particle coupled to a harmonic oscillator. Choosing correctly the coupling they were able to get a good agreement with the pde. This is a very surprising fact considering the rather severe modifications of the Lagrangian. With our approach we hope to get an agreement with the pde without an adjustable parameter and better understand the relativistic aspects of the collision. The price to pay is a complication of the evolution equations.
As a general conclusion we think it is important before undertaking the complex calculations to get the evolution of the collective coordinates to check if the projection is not done on a zero or colinear modes.

REFERENCES

1. R. K. Dodd, J. C. Eilbeck, J. D. Gibbon and H. C. Morris , "Solitons and Nonlinear wave equations", Academic Press (1982)

2. A. C. Scott and D. Mclaughlin , "Perturbation analysis of fluxon dynamics", Phys. Rev. A **18**, nb. 4 (1978)
and references 17 and 18 therein.

3. M. J. Rice and E. J. Mele, "Phenomenelogical theory of soliton formation in lihgtly doped polyacetylene", Solid State Comm. **35**, p. 487 (1980)

4. G. Reinisch, J. C. Fernandez, N. Flytzanis, M. Taki and S. Pnevmatikos, "Phase lock of a weakly biased inhomogeneous long Josephson junction to an external microwave source", Phys. Rev. B **38**, nb. 16, p. 11284 (1988)

5. O. Legrand, "Kink-antikink dissociation and annihilation: a collective-coordinate description", Phys. Rev. A **36**, 5068 (1987)

6. J. G. Caputo and N. Flytzanis, "Kink-antikink collisions in sine-Gordon and ϕ^4 models: problems in the variational approach, submitted to Phys. Rev. A, December 1991

7. D. K. Campbell, J. F. Schonfeld and C. A. Wingate, "Resonance structure in kink-antikink interactions in ϕ^4 theory", Physica **9D**, p. 451 (1983)

8. Z. Fei and L. Vazquez, "Resonance phenomena in a dynamical system with two degrees of freedom ", submitted to Physica D, December 1990

Spatiotemporal Chaos in the Nonlinear Three Wave Interaction

C.C. Chow, A. Bers, and A.K. Ram

Plasma Fusion Center, Massachusetts Institute of Technology, Cambridge, MA 02139, USA

The nonlinear three wave interaction (3WI) has applications to plasma physics, nonlinear optics, and hydrodynamics [1]. We consider one form of these equations, which describes the saturation of a linearly unstable parent wave by nonlinearly coupling to two damped daughter waves [1, 3]. The system exhibits spatiotemporal chaos (STC) in the long time, large system limit. The term STC specifically refers to the chaotic dynamics of coherent structures or spatial patterns [2]. This is contrasted with fully developed turbulence where there is a cascade to small scales and with low dimensional chaos where spatial degrees of freedom are not involved. The conservative form of the 3WI is integrable by inverse scattering transforms (IST) and has soliton solutions. We use perturbation theory about those solutions to gain some understanding of the behaviour.

We study the dynamics in one spatial dimension x. The equations for the 3WI in a weakly dissipative medium are given by[1, 3]

$$\begin{aligned} \partial_t a_i &= -a_j a_k + \gamma_i a_i + D\partial_{xx} a_i \\ \partial_t a_j - \partial_x a_j &= a_i a_k^* - \gamma_j a_j \\ \partial_t a_k + \partial_x a_k &= a_i a_j^* - \gamma_k a_k, \end{aligned}$$

where the a's are complex wave envelopes, the γ's are growth or dissipation coefficients, and D is a diffusion coefficient. The diffusion term arises if we assume that the growth of the linear wave has a slow variation. It is then the lowest order reflection invariant term that provides a cutoff in wave number of the growth. We have transformed to the frame of the parent wave and normalized the daughter group velocities to one. The subscript i denotes the high frequency unstable parent wave. The other two waves are refered to as the daughters. The group velocities satisfy the condition $v_k > v_i > v_j$ (ie. the highest frequency parent wave has the middle group velociy). In the absence of growth, damping and diffusion ($\gamma_l = D = 0$) the IST solutions for this group velocity ordering include solitons [1, 4, 5].

We numerically simulated the system on the domain $x \in [0, L)$ with periodic boundary conditions. We measured the correlation function, $S_l(x,t) =< a_l(x-x', t-t')a_l(x',t') >$, where the averaging was over time. We began with random initial conditions and evolved until the transients died away before the system was analysed. We were interested in the large system, long time limit. We consider the case with parameters $D = .001$ and $\gamma_i = .1$, $\gamma_j = \gamma_k = 1$, and $L = 20$. The figure shows a sample of the spatiotemporal evolution profiles and the correlation functions of the parent and daughter envelopes. We only show

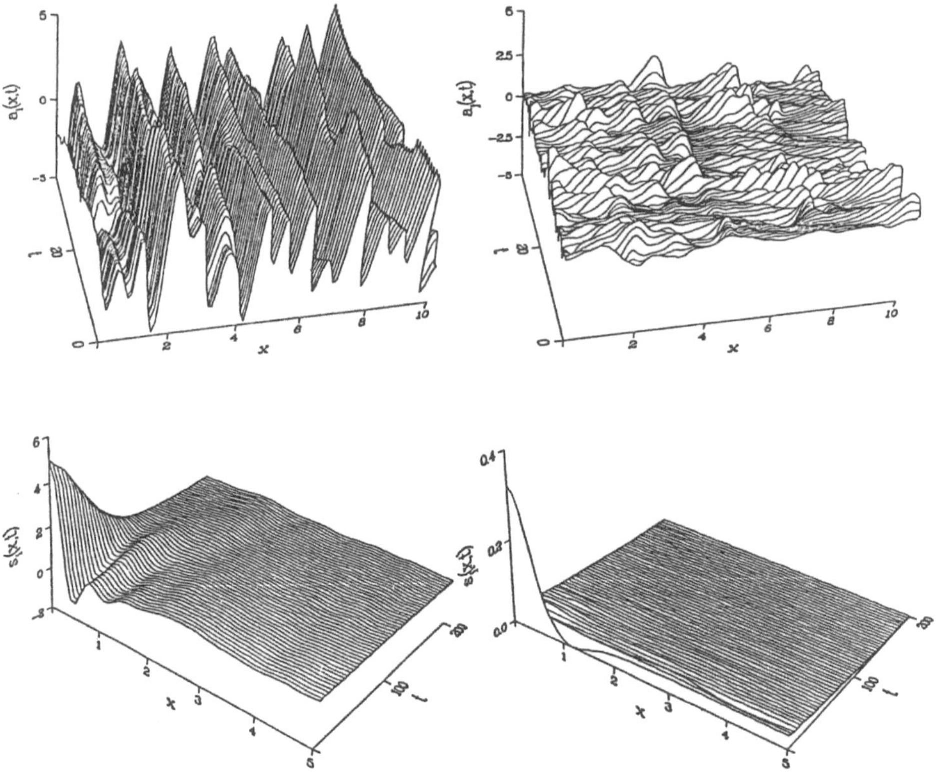

Figure 1

one daughter, the other will be similar but with the opposite parity. The profile of the parent wave is clearly irregular but some spatial and temporal scales can be observed. The profile of the daughter wave shows a sea of structures convecting to the left. The correlation functions for both the parent and the daughter clearly exhibit decorrelation in the long time and long length limit. The parent has a dominant spatial mode as observed in the spacetime profile. The daughter shows a definite spatial scale as well. The daughter correlation function shows two time scales. There is a fast decay with a long tail. The parent correlation function also exhibits two time scales. There is a slow decay, on the order of the daughter long time scale. On top of the long time scale is a ripple which would appear as a small peak at a much higher frequency in the power spectra. This corresponds to the short time scale which is manifested in the profile as a growth and depletion cycle.

The main features of the behaviour can be understood if we consider perturbations about the conservative 3WI. The IST solutions for the conservative case on the infinite domain [1, 5] show that solitons exist but they do not belong uniquely to a particular envelope. A soliton in the parent wave will be transfered to each daughter. The daughter solitons will be of the form $|a_j| = 2\eta \operatorname{sech} 2\eta$, where η is the eigenvalue of the bound state in the parent envelope. With growth and dissipation, the parent will deplete to quasi-solitons then grow back and repeat the process. This time scale t_c is the short time observed in the simulation. The parent wave with some initial η will deplete and be left with some effective eigenvalue η'. Then

$$t_c \sim \frac{1}{\gamma_j} \ln \frac{\eta}{\eta'}.$$

The damping coefficients for the two daughters are chosen equal. Ignoring the effects of diffusion on this short time scale, a multiple-time scale perturbation analysis about the IST soliton decay solution is used to estimate η'[6]. The threshold eigenvalue for decay, η is estimated by a perturbation expansion of the scattering data. This yields

$$\eta \sim \gamma_j + 2\xi_p \gamma_i,$$

where ξ_p is the parent correlation length and will be defined later. From η we are able to estimate the daughter correlation length. This is given by the quasi-soliton width $\xi_d \sim 1/\eta$. The short time scale of the daughter envelope is due to the damping time of the quasi-soliton and will be given by $t_d \sim 1/\gamma_j$.

The long time behaviour is governed by the diffusion. The trivial fixed point of the 3WI is given by

$$\partial_{xx} a_i + q_0^2 a_i = 0,$$

$$a_j = a_k = 0,$$

where $q_0 = \sqrt{\gamma_i / D}$. Modes with $q < q_0$ will damp and those with $q > q_0$ will grow. However, when a local area between two turning points contains a bound state with eigenvalue η it will deplete. The results are long wavelength distortions about the mode q_0. We thus obtain a correlation length for the parent, $\xi_p \sim 1/q_0$. If $D = 0$ there will not be any nonlinear saturation of the instability because the larger the q the larger the amplitude required to fulfill the area threshold. The long time scale τ_p for the parent is then given by the diffusion time across a length ξ_p and so $\tau_p \sim (2\pi)^2/\gamma_i$. This is the time scale where the local parent structures will shift postition, collide with other structures or diffuse away. As daughter quasi-solitons collide with the local structures they tend to seed further depletions, creating new quasi-solitons in the same spatial position. This gives a long time scale for the daughters related to that of the parent. The long time scale is the diffusion time across a quasi-soliton width $\tau_d \sim 1/(\eta^2 D)$. The constant feedback between the parent and daughters gives rise to the chaos. There is no particular route to STC because of the lack of straightforward bifurcations. Any value of growth will induce STC as long as the system is large enough. For weak growth and damping, perturbation theory is adequate in estimating the length and time scales.

This work was supported in part by NASA Grant No. NAGW-2048, LLNL subcontract B108475, and NSF Grant No. ECS-88-2475.

Also by DOE Grant No. DE-FG02-91ER-54109.

References

[1] D.J. Kaup, A. Reiman, and A. Bers, Rev. Mod. Phys. **51**, 915 (1979) and references therein.

[2] P.C. Hohenberg and B.I. Shraiman, Physica D **37**, 109 (1989).

[3] A. Bers in *Plasma Physics-Les Houches 1972*, edited by C. De Witt and J. Peyraud (Fordon and Breach, New York, 1975).

[4] V.E. Zakharov and S.V. Manakov, Zh. Eksp. Teor. Fiz. Pis'ma Red. **18**, 413 (1973) [Sov. Phys.–JETP Lett. **18**,243 (1973)].

[5] D.J. Kaup, Stud. Appl. Math. **55**, 9 (1976).

[6] C.C. Chow, A. Bers, A.K. Ram, to be published.

On the Instability of the Static Soliton-like "Bubbles"

A. de Bouard

Laboratoire d'Analyse Numérique, Université Paris Sud,
Bat. 425, F-Orsay Cedex, France

The following $\psi^3 - \psi^5$ nonlinear Schrödinger equation :

$$i\frac{\partial\psi}{\partial t} + \Delta\psi - \alpha_1\psi + \alpha_3|\psi|^2\psi - \alpha_5|\psi|^4\psi = 0 \tag{1}$$

possesses, in some domain of the parameters α_i, some localised solutions propagating with velocity v which can be interpreted as rarefaction bubbles in a Bose condensate (see [1]). These localised solutions satisfy the " boundary conditions" :

$$\psi(x,t) \to \sqrt{\rho_0}e^{i(\omega t+\mu(\frac{x}{|x|}))} \text{ when } |x| \to +\infty \tag{2}$$

where ρ_0 is a positive constant, μ depends on the velocity v and is equal to zero in the particular case of stationary solutions, i.e when $v = 0$. Such solutions of equation (1) where found explicitly in space dimension one, and numerically in higher dimension (see [1]).

Here, we consider the general nonlinear Schrödinger equation :

$$i\frac{\partial\psi}{\partial t} + \Delta\psi + F(|\psi|^2)\psi = 0 \text{ in } \mathbb{R}^n \tag{3}$$

and study the existence and instability of localised stationary solutions of (3) having the preceding behavior (with $\mu = 0$) as $|x|$ go to infinity. We may assume, by changing F if necessary, that $\omega = 0$; hence the solutions of (3) will be asked to satisfy :

$$\psi(x,t) \to \sqrt{\rho_0} \text{ when } |x| \to +\infty \tag{4}$$

The following theorem is proved by using the techniques of Berestycki and Lions [3] (or Berestycki, Gallouët and Kavian [4] in the case n $= 2$)

Theorem 1 *Assume F is continuously differentiable from $\mathbb{R}_+$ into $\mathbb{R}$ and that :*

(i) $F(\rho_0) = 0$

(ii) $F'(\rho_0) < 0$

(iii) *There is a nonnegative number $\rho_1 < \rho_0$ such that $U(\rho_1) < 0$ with $U(s) = -\int_{\rho_0}^{s} F(\tau)d\tau$*

(iv) *if* $n \geq 3$: $\limsup_{s \to +\infty} \frac{F(s)}{s^{\frac{(l-1)}{2}}} \leq 0$, $l = \frac{n+2}{n-2}$

if $n = 2$: $\forall \alpha > 0, \exists C_\alpha > 0, \forall s \geq 0, F(s) \leq C_\alpha e^{\alpha s}$.

Then, there is a real valued, twice continuously differentiable, radially symmetric and positive function ϕ *satisfying :*

(a) $\Delta\phi + F(\phi^2)\phi = 0$ *in* $\mathbb{R}^n$

(b) $\lim_{|x| \to +\infty} \phi(x) = \sqrt{\rho_0}$, $\frac{d\phi}{dr}$ *is positive for* $r \in (0, +\infty)$, $r = |x|$.

(c) $\exists\, C > 0, \exists\, \delta > 0, \forall\, |\alpha| \leq 2, |\partial^\alpha(\phi - \sqrt{\rho_0})| \leq C_\alpha e^{\delta|x|}, \quad x \in \mathbb{R}^n$.

Such a solution is called a "bubble" (see [2]).

Now, let ϕ be a bubble solution and let ϕ + u be a solution of (3). The linearized equation for $U = (Reu, Imu)^t$ then writes :

$$\frac{dU}{dt} = \mathcal{A}U, \quad \mathcal{A} = \begin{bmatrix} 0 & L_1 \\ -L_2 & 0 \end{bmatrix} \tag{5}$$

where :

$$\begin{cases} L_1 v = -\Delta v - F(\phi^2)v \\ L_2 v = -\Delta v - F(\phi^2)v - 2\phi^2 F'(\phi^2)v \end{cases}$$

Theorem 2 *Under the assumptions of theorem 1:*
.The continuous spectrum of $\mathcal{A}$ *lies on the imaginary axis.*
.All the discrete eigenvalues of $\mathcal{A}$ *with finite multiplicity are real an at least one of them is positive.*

The details of the proof of this theorem can be found in [5]. The idea is that if λ is an eigenvalue of $\mathcal{A}$, then the kernel of $L_2 + \lambda^2 L_1^{-1}$ is not trivial, where L_1^{-1} can be defined as an unbounded operator on $L^2(\mathbb{R}^n)xL^2(\mathbb{R}^n)$ with dense domain. Theorem 2 implies the following instability result, independantly of the nonlinear term F.

Theorem 3 *Assume in addition with (i)-(iv) of theorem 1 that F is sufficiently regular, and let* $s > \frac{n}{2}$, *then the following holds :*
There is a positive ϵ *such that for any positive* δ *there is a* u_0 *in* H^s *with* $\|u_0\|_{H^s} < \delta$ *such that if* $u(t) \in H^s(\mathbb{R}^n)$ *and* $v(t) = u(t) + \phi$ *is the solution of (3) with initial data* $u_0 + \phi$, *then there is a positive* t_0 *such that* $\|u(t_0)\|_{H^s} > \epsilon$

remark : If instead of stationary solutions of (3) satisfying (4), we consider solutions of the form $\phi_\omega(x,t) = e^{i\omega t}u_\omega(x)$, with $\lim_{|x| \to +\infty} |u_\omega(x)| = 0$, then it is known that the solutions can be stable or unstable, depending on the nonlinearity F. Hence, replacing the "boundary condition" $\lim_{|x| \to +\infty} \phi(x,t) = 0$ by condition (2) or (4) give rise to slightly different phenomena.

References

[1] I.Barashenkov, V.Makhankov
Phys. lett. A **128** (1988) 52-56

[2] I.Barashenkov, A.Gocheva, V.Makhankov, I.Puzynin
Phys. D **34** (1989) 240-254

[3] H.Berestycki, P.L.Lions
Arch.rach.mech.anal. **82** (1983) 313-345

[4] H.Berestycki, T.Gallouët, O.Kavian
C. R. acad. sc PARIS **297** (1983) Serie I 307-310

[5] A. de Bouard
PhD thesis (1991) , Universite Paris-Sud, Otsay France. and article in preparation

Symplectic and Hamiltonian Structures of Nonlinear Evolution Equations

I. Ya. Dorfman

Institute of Chemical Physics, Russian Academy of Sciences, ul. Kosygina 4, 117977 Moscow, Russia

Introduction. A Hamiltonian structure on a finite-dimensional manifold can be introduced either by endowing it with a (pre)symplectic structure, or by describing the Poisson bracket with the help of a tensor with two upper indices named the Poisson structure. Under the assumption of nondegeneracy, the Poisson structure is nothing else than the inverse of the symplectic structure. Also in the degenerate case the distinction between the two approaches is almost insignificant, because both presymplectic and Poisson structures split into symplectic structures on leaves of appropriately chosen foliations. Hamiltonian structures that arise in the theory of evolution equations demonstrate something new in this respect: trying to operate in local terms, one is induced to develop both approaches independently. Hamiltonian operators, being the infinite-dimensional counterparts of Poisson structures, were the first to become the subject of investigations[1-4]. A considerable period of time passed before the papers[5,6] initiated research in the theory of symplectic operators, being the counterparts of presymplectic structures. In what follows, we focus on the main achievements in this field.

1. Explicit description of symplectic operators. The operators under consideration are matrix differential ones

$$I_{ij} = \sum p_{ijk} (d/dx)^k , \qquad p_{ijk} = p_{ijk}(\dots, u_l^{(m)}, \dots) \tag{1}$$

Skew-symmetry means $I_{ij} = -I^*_{ji}$. Also the fact that the corresponding 2-form must be closed, can be explicitely expressed in terms of coefficients. Namely, it is equivalent to a system of polynomial equations

$$V_{ijks}(z) - z^s V_{ikjs}(1/z) = \overline{V}_{kjis}(1+z) , \tag{2}$$

with the coefficients of the polynomials involved being expressed via partial and higher variational derivatives of p_{ijk}. This allows to consider some special cases of interest. In particular, the set of equations in one-variable case can be derived explicitely for any prescribed order N. For instance, N=1 gives $I = (dp/dx) + 2p(d/dx)$, where $p = p(x, u, u_x, u_{xx}, u_{xxx})$ and $\partial p/ u_{xx} = (d/dx)\, \partial p/ u_{xxx}$. One of the consequences of (2) is also that for an N-th order operator being symplectic, the coefficient by $(d/dx)^k$ can only depend on the derivatives up to the order 3n+1-k (see[7]).

2. Symplectic operators whose coefficients depend on $u_1^{(m)}$ in a linear way. It occurs that for p_{ijk} in (1) linearly depending on $u_1^{(m)}$ we can write

$$\overline{V}_{ijks}(z) = V_{ijks}(-z)$$

and therefore the equation

$$V(z)-z^{s}V(1/z)=V(-1-z) \tag{3}$$

is basic for the description problem. A natural interpretation of (3) can be given in terms of group representations. Namely, solutions of (3) are in correspondence with irreducible representations of the group of triangle D_3 in the space of degree-s polynomials. This allows to describe in full symplectic operators of the class under consideration both in one-variable[7] and in many-variable cases.

3. Symplectic operators of differential-geometric type. Another particular shape of (1), namely

$$I_{ij}= a_{ij}(u)\ (d/dx) + \sum b_{ijk}(u)u_x^k\ ,\ u=(u_1,\ldots,u_n) \tag{4}$$

was investigated in[9]. The intrpretation given is differential-geometric in its nature: a_{ij} is a pseudo-Riemannian metric, b_{ijk} corresponds to a connection compatible with this metric and satisfying some additional requirements. Notably, the symplectic structure of Lagrangian systems with the action of the shape

$$S= \int [c_{ij}(u)u_x^i u_t^j + H(x,t,u,u_x,u_{xx},\ldots)]\ dt$$

is defined by operators of the form (4).

4. Multisymplectic systems. Approaches to (2+1)-dimensional theory. The Krichever-Novikov (KN) equation

$$w_t= w_{xxx} - 3/2\ \ w_x^{-1}w_{xx}^2$$

was the first nontrivial example of a bisymplectic evolution equation, i.e. an equation possessing two local symplectic structures

$$I= w_x^{-1}(d/dx)\circ w_x^{-1},\qquad J= (d/dx)\circ w_x^{-1}(d/dx)\circ w_x^{-1}(d/dx)$$

that are compatible and produce higher symmetries and conservation laws with the help of the Lenard scheme action.[5,6]

Important interrelations between the KdV, modified KdV and KN equations, their Hamiltonian and symplectic structures, and the corresponding Bäcklund transformations were revealed in[10]. A generalization to the case of many dependent variables can be found in[11].

Evolution systems possessing multisymplectic structure,i.e. a collection of compatible local symplectic structures, are considered in[12]. These are Schwarzian modifications of the ones with multiHamiltonian structures previously investigated. Schwarzian modification is the analog of the transformation u=S(w), S denoting the Schwarzian derivative, in the space of evolution systems.

Some new approaches to the theory of multidimensional Hamiltonian structures were recently discovered. In a joint paper with A.Fokas (yet unpublished) the author develops a noncommutative version of Hamiltonian formalism that allows to include the known biHamiltonian structures in (2+1) dimensions into the general Adler-Gelfand-Dikii scheme.

An attempt to understand what happens with the procedure of Schwarzian modification when we pass to the (2+1) dimensions, is undertaken in[13].The (2+1) version of the KN equation is considered, and the Bäcklund transformation that links it to the Kadomtsev-Petviashvili equation is presented. Still the problem cannot be named exhaustively investigated, as only the symplectic image of the structure (d/dx) has been found.

The concluding remark is the following. Typically, Hamiltonian structures originate from the spectral problem, while symplectic operators arise as a result of a guess, or of some special transformation from the Hamiltonian ones. It is of interest to know if there exists any kind of a canonical procedure of deriving syplectic pairs, like , for example, the Adler-Gelfand-Dikii scheme.

REFERENCES

1. Zakharov,V.E.& Faddeev, L,D. Funct.Anal Appl.1971, 5:4, 18
2. Gardner, C.S. J. Math. Phys.1971,12,1548
3. Magri,F. J.Math.Phys. 1978,19,1156
4. Gelfand,I.M.& Dorfman, I.Ya. Funct. Anal. Appl.1979,13:4,13
5. Dorfman,I.Ya. Phys.Lett.1987,A125,240
6. Dorfman,I.Ya. Dokl. Akad. Nauk SSSR, 1988,302,792
7. Dorfman,I.Ya. Phys. Lett. 1989, A140,378
8. Dorfman, I.Ya.& Mokhov, O.I. Local symplectic operators and structures related to them. Preprint 1990,subm. to J.Math.Phys.
9. Mokhov, O.I. Funct. Anal Appl. 1990,24:3,86
10. Wilson,G. Phys.Lett.1988, A132,445
11. Wilson,G. On the Adler-Gelfand-Dikii bracket. In: Proc. of CRM workshop (J. Harnad&J.Marsden eds.),1990, Montreal,77
12. Antonowicz,M.&Fordy,A.P. Multicomponent Schwarzian KdV hierarchies. Preprint 1990
13. Dorfman,I.Ya.&Nijhoff,F.W. On a 2+1 dimensional version of the Krichever-Novikov equation. Preprint 1990, subm. to Phys.Lett.A.

A Fuchs Extension to the Painlevé Test

R. Conte[1], *A.P. Fordy*[2], *and A. Pickering*[2]

[1]Service de physique de l'état condensé (DRECAM-SPEC), Centre d'études de Saclay, F-91191 Gif-sur-Yvette Cedex, France

[2]Department of Applied Mathematical Studies and Centre for Nonlinear Studies, University of Leeds, LS2 9JT, UK

We present a recent improvement [1] to the Painlevé test such that negative resonances can be treated. To this end we demand that the general solution of both the given nonlinear equation and its linearisation be single valued. This gives rise to compatibility conditions for every integer resonance, whether positive or negative. We present two illustrative examples: a generalised Chazy, and an anharmonic oscillator and indicate some future directions.

Introduction

In the times of Cauchy, analytic functions of a complex variable were defined either by their explicit expression or by some integral, series or other representation. Then Fuchs, Poincaré and (mainly) Painlevé studied the conditions under which the general solution of an ordinary differential equation (ODE) could define new functions (single valued by definition). This gives rise to the following mathematical problem, posed by Painlevé:

"Déterminer toutes les équations différentielles algébriques du premier ordre, puis du second ordre, puis du troisième ordre, etc., dont l'intégrale générale est uniforme."

An ODE is said to have the Painlevé property (PP) if its general solution is single valued, except perhaps at fixed critical points; a point at which branching occurs is called critical and is said to be fixed if its position does not depend upon initial conditions, being determined purely by the coefficients in the equation. A critical point is said to be moveable if its position depends upon initial conditions. An ODE should not have moveable critical points, since their positions cannot be predicted from inspection of the equation so that convenient cuts cannot be made in the complex plane. We emphasise that an isolated essential singularity is not a critical point, so is not an obstacle to the PP (contrary to a common belief). Bureau [2] uses *stability* as a synonym for the PP. Such equations are said to be stable. Stability is a global property not a local one.

Let R be a rational function of u(x) and its derivatives and analytic in x. The following classes of ODE have been studied in [2-11]:

$$R(u',u,x) = 0 , \tag{1.1}$$

$$u'' = R(u',u,x), \tag{1.2}$$

$$u''' = R(u'',u',u,x) . \tag{1.3}$$

The results of this classification are as follows:

First order (Fuchs, Poincaré and Painlevé).

Results concerning (1.1) are summerised in chapter 11 of [5]. Stable equations are Riccati or of Briot-Bouquet type, whose solutions can be written in terms of elliptic and other known algebraic or transcendental functions, so just have moveable poles.

Second order [3,6]. Each stable ODE of class (1.2) is either reducible to a first order stable equation or is equivalent to one of Painlevé's six new equations defining his transcendents. Again, the only moveable singularities at this order are poles. These results can be found in [2,5,7].

Third order [2,8-11]. Stable ODEs of class (1.3) have not yet been fully classified and no new transcendents have yet been found. An interesting feature encountered at third order is that some stable ODEs possess moveable singularities other than poles. The best known example is Chazy's equation [8]:

$$u'''-2uu''+3u'^2 = 0 , \tag{1.4}$$

which possesses a moveable natural boundary but is, nevertheless, stable.

General methods for tackling this problem were developed by Painlevé [4]. These and all later developments consist of two parts:

(i) build necessary conditions for stability

(ii) check sufficiency either by expressing the general solution in terms of 'known functions' (to date: solutions of linear equations, elliptic functions, the six Painlevé transcendents) or prove the transcendence (and single-valuedness) of the general solution.

General (but incomplete) methods [4,6,7] have been developed for tackling part (i), but part (ii) consists of a series of particular methods devised for individual equations.

However, in 1977 and 1978 Ablowitz, (Ramani) and Segur [12,13] showed that ODE reductions of well known 'soliton equations' (PDEs in 1+1 dimensions) have the Painlevé property. Indeed, in the simpler cases, these PDEs reduce to Painlevé's equations. They thus showed that Painlevé transcendent equations and others of Painlevé type could be solved by the inverse spectral transform (IST), inherited from the corresponding soliton equation. This gives a general method for part (ii) above for a large class of equations.

Ablowitz, Ramani and Segur [14] then developed an algorithm (more closely related to that of Kowalevskaya [15] than to the Painlevé school) for testing ODEs for the Painlevé property. They conjectured that all reductions of 'integrable' PDEs should have the PP so that by testing all reductions with the above mentioned algorithm one effectively tested the PDE. An unsatisfactory feature of this test for PDEs is the need to test all reductions (all being rather nebulous).

Weiss, Tabor and Carnevale [16] introduced a modification of the ARS test by considering expansions about 'singular manifolds' instead of singular points. This not only enabled them to tackle PDEs directly, but also enabled further methods to be developed which showed connections between the Painlevé method and other soliton techniques.

The Painlevé test has been an extremely successful method of testing both ODEs and PDEs for 'complete integrability' and has been the subject of a large number of papers (see the reviews [17-19], the conference proceedings [20] and references therein). Nevertheless, this method (as currently used) is incapable of decisively testing certain equations, such as those possessing several negative resonances. One such equation is that of Chazy which has only negative resonances. This equation has a movable natural boundary, but its general solution can be written down in terms of hypergeometric functions (see, for instance, [8]) and is single valued in its domain of definition. Even integrable equations, such as the members of the KdV hierarchy, have 'secondary' Painlevé branches which possess several negative resonances. There has been some discussion of negative resonances [19], but they were largely ignored as being harmless.

In this paper (following [1]) we maintain that negative resonances can contain important (sometimes decisive) information regarding the integrability of an equation. To this end we give a method of deriving compatibility conditions for negative resonances and show that it can be extremely hazardous to ignore them! Furthermore, we show that 'secondary' branches are by no means secondary; for many equations our method extracts the same information from all branches, even though the conventional Painlevé analysis could only derive full information from the principal branch.

Our approach is to simultaneously test a nonlinear equation and its linearisation, treated as a (rather weakly) coupled system. Applied to the linearised equation, the Painlevé analysis reduces to a Fuchsian analysis about a regular singularity. The roots of the indicial equation of the linearisation are just (up to a constant integer shift) the resonances of the nonlinear equation. We demand that the solutions to both the nonlinear and linear equations, be single-valued. We therefore require that all roots of the indicial equation (and thus all resonances of the nonlinear equation) be distinct integers, whether positive or negative. Since the roots of the indicial equation differ by integers, compatibility conditions arise.

2. The Method

For simplicity, we restrict attention to nonlinear evolution equations of the form:

$$u_t = K[u] \tag{2.1a}$$

where K[u] is a polynomial function of $u, u_x, \dots, u_{Nx}$. The case of ODEs is then easily incorporated by taking $u_t=0$. The linearisation of (2.1a) is:

$$w_t = K'[u]w \tag{2.1b}$$

where $K'[u]w = \frac{d}{d\varepsilon} K[u+\varepsilon w]\big|_{\varepsilon=0}$. Equation (2.1b) is the equation satisfied by generalised symmetries (commuting flows) of (2.1a), when they exist.

<u>Remark</u>

Our examples of section 3 are, in fact, all ODEs, but the method is applicable to all cases. In [1] PDEs arise in a discussion of integrable hierarchies.

We first carry out a standard Painlevé expansion of the nonlinear equation (2.1a), using the modification [21] of the WTC method. This 'invariant' approach simplifies many of the expressions, hiding the complications in the definition of S, C and χ (see the appendix). We seek a solution of the form:

$$u^{(p)} = \chi^{-\alpha} \sum_{l=0}^{\infty} u_l \chi^l . \tag{2.2}$$

A leading order analysis gives a number of possible choices of α, depending upon the nonlinearities. Each α corresponds to a possible choice of dominant terms $\hat{K}[u]$ of K[u] and this, in turn, leads to a number of possible starting terms u_0 as solutions of the <u>algebraic</u> equation:

$$\hat{K}(u_0\chi^{-\alpha})\Big|_{\chi'-1=u_0'=0} = P_\alpha(u_0)\chi^{-\beta} = 0 , \tag{2.3a}$$

where β is the <u>weight</u> of the dominant expression $\hat{K}$ and P_α a polynomial. The coefficients u_l are determined recursively by:

$$\left[\hat{K}'[u_0\chi^{-\alpha}]\chi^{l-\alpha}\Big|_{\chi'-1=u_0'=0}\right] u_l = \text{expressions involving } u_0, \dots, u_{l-1}, S, C . \tag{2.3b}$$

The coefficient of u_l on the left vanishes for certain values of l, called resonances, which must be integer for an 'integrable' equation:

$$\mathcal{R} = \{r_1, \dots, r_n\} , \quad r_1 \le r_2 \le \dots \le r_n . \tag{2.3c}$$

The number n depends upon the highest order derivative in the dominant expression $\hat{K}[u]$, not on the order of K[u]. The right hand side of (2.3b) must also vanish for these values of i, giving rise to compatibility conditions whenever i reaches a positive $r \in \mathcal{R}$. No conditions arises at negative resonances.

To pass the Painlevé part of our test an equation should have the following properties:

(PI) Each possible choice of α and all corresponding r_i must be integer, with r_i being distinct.

(PII) all branches should be such that at any positive resonance, the compatibility conditions are identically satisfied.

(PIII) there exists a branch with the number of resonances n=N, the order of the full operator K[u].

Remarks

(i) We have dropped the requirement of a principal branch with the consequence that it may not be possible to build the general solution as a finite pole Laurent expansion. However, we still require the existence of a branch with the 'correct' number of resonances.

(ii) Negative resonances of 'secondary' branches are allowed since they give no conditions which contradict integrability.

This weakened condition (PIII) allows some 'bad' equations through our net, but these are caught at step 2 below.

Our second step is to consider the linearised equation (2.1b), with $u=u^{(p)}$, the Painlevé expansion (2.2), for each of the branches. For each branch we can write (2.1b) as:

$$\hat{K}'[u^{(p)}]w = I'[u^{(p)}]w , \tag{2.4}$$

where $\hat{K}'$ and I' are respectively the linearisations of the dominant and inferior parts of $K-u_t$. $\hat{K}'[u^{(p)}]$ is scaled in such a way that $\chi=0$ is a regular singularity in the Fuchsian sense. We seek an expansion:

$$w = \chi^{\sigma}\sum_{i=0}^{\infty} w_i\chi^i , \quad w_0 \neq 0 , \tag{2.5a}$$

where $\sigma \in \{\sigma_1,\ldots,\sigma_n\}$ is a root of the indicial equation:

$$\hat{K}'[u_0\chi^{-\alpha}]\chi^{\sigma}\big|_{\chi'-1=u_0'=0} = 0 . \tag{2.5b}$$

Comparing this with (2.3), we see that:

$$\sigma_i = r_i-\alpha \text{ for } r_i \in \mathcal{R} , \quad i=1,\ldots,n . \tag{2.5c}$$

In order that the general solution of the linear equation (2.4) be single valued we require, in addition to PI-III,

FI that the indicial equation has n distinct, integer solutions $(\sigma_1,\dots,\sigma_n)$,

FII that the compatibility conditions arising at each σ_i, $1\le i\le n$ be identically satisfied.

Remarks

(i) FII gives compatibility conditions for r_i, regardless of whether they are positive or negative.

(ii) It should be noticed that since the position of the singularity is determined by the nonlinear equation, even this linear equation has moveable poles.

We are demanding much more than (2.4) being just Fuchsian when we ask for w to be single valued. This is a strong constraint which, together with the above weakened Painlevé property of (2.1a) enables us to distinguish integrable cases in a wide variety of equations hitherto untestable by the Painlevé method (see section 2 of [13]). A feature of equations which pass our test, but have no principal branch, is the presence of an essential singularity [1].

3. Examples

Here we present just two examples. More can be found in [1,22].

Example 1 : Generalised Chazy

We follow the notation of Bureau [10]. The general ODE with the same dominant terms of Chazy equation is [8,11]:

$$-u''' + auu'' - \tfrac{3}{2}au'^2 + a_1u'' + c_1uu' + c_0u' + d_3u^3 + d_2u^2 + d_1u + d_0 = 0\ , \qquad (3.1a)$$

$$-w''' + (au+a_1)w'' + (c_1u-3au')w' + (au''+c_1u'+3d_3u^2+2d_2u+d_1)w = 0\ , \qquad (3.1b)$$

where a, a_1, c_1, d_1 are arbitrary analytic functions of x. We can first use the transformation $u(x)\longrightarrow\lambda(x)u(\varphi(x))+\mu(x)$ to simplify this equation, setting:

$$a = 2\ ,\ a_1 = 0\ ,\ d_2 = 0\ . \qquad (3.2)$$

There exist two families of moveable singularities:

(i) singularity order $\alpha=1$, with indices $-3,-2,-1$,

(ii) singularity order $\alpha=2$, with indices $-1,0$,

but the second family will not be considered here.

With only negative resonances, the Painlevé part of the analysis cannot provide any stability condition, and one needs only compute terms $j=0,1,2$ in expansions for $u^{(p)}$ and w, with $\alpha=1$, in order to obtain the expressions of the Fuchs stability conditions. The results of this Fuchs-Painlevé analysis are:

$$u_0 = -6,\ u_1 = \tfrac{1}{2}(-3c_1 - 18d_3)\ ,\ u_2 = \tfrac{1}{20}(2c_0 - 3c_1^2 - 72c_1d_3 - 324d_3^2 + 18c_1' + 108d_3' - 4S) \qquad (3.3a)$$

$$w_0,\ w_1,\ w_2 \text{ arbitrary}\ , \qquad (3.3b)$$

together with the following compatibility conditions arising at index -2 and -1 in the Fuchs part of the analysis:

$$d_3 = 0\ ,\ c_1 = 0\ ,\ c_0 = 0\ , \qquad (3.4)$$

leading to:

$$-u''' + 2uu'' - 3u'^2 + d_1u + d_0 = 0\ . \qquad (3.5)$$

If we go beyond the Fuchs-Painlevé test and use second order perturbation [22], we can show that d_1 and d_0 also vanish, so that only Chazy's equation (1.4) remains. Chazy's derivation of this is given without detail in [11].

Example 2 : Anharmonic Oscillator

We consider the equation:

$$u'' + g_1u' + g_2u + g_3u^n + g_4 = 0\ ,\ n \in \mathbb{R}\ , \qquad (3.6)$$

where the g_i's are given analytic functions. In the variable $u^{(n-1)/2}$ or u^{n-1} and under certain conditions on g_4 and n, this equation is in class (1.2).

There exist two families of moveable singularities:

(i) singularity order $\alpha = \frac{2}{n-1}$, with $u_0^{n-1} = -2\frac{n+1}{(n-1)^2}g_3$, indices -1, $2+\frac{4}{n-1}$,

(ii) singularity order $\alpha=-2$, $u_0 = -\frac{1}{2}g_4$, indices -2, -1 .

Two conditions on (n, g_i), $C_1=0$, $C_2=0$, are given in [23] to ensure the existence of a particular first integral of type:

$$I = f_1u'^2 + f_2\int_0^u u^2du + f_3uu' + f_4u^2 + f_5u\ . \qquad (3.7)$$

A Fuchs analysis on family (ii) gives no stability conditions at index -1. On the other hand, for family (i), whenever the index $r=2+\frac{4}{n-1}$ is an integer, both the pure Painlevé analysis (when $r>0$) and the Fuchs-Painlevé analysis (when $r<0$) provide stability conditions $Q_r=0$, which happen to be differential consequences of the condition $C_1=0$, $C_2=0$.

The details of C_i and Q_r are given in [23] for the case of positive integer r, using pure Painlevé. The correspondence between C_i and Q_r for negative integer r is new and required the Fuchs-Painlevé analysis.

4. Conclusions

We have explained the 'Fuchs-Painlevé method', recently introduced in [1], which enables us to analyse negative resonances by considering the given nonlinear equation and its linearisation. The power of this method is illustrated by analysing two equations which are beyond the scope of the standard Painlevé test. Further examples can be found in [1]. In particular, an equation of Bureau (example 3 of [1]) shows that 'secondary branches' (previously thought to contain <u>less</u> information than the principal branch) can contain <u>full</u> information, but that it is necessary to use the Fuchsian part of our analysis to extract this.

A further aspect considered in [1], but not here, is the relationship between negative resonances and lower order commuting flows, when considering hierarchies of integrable equations.

We have generalised the 'Fuchs-Painlevé' approach, which may be considered as analysing a nonlinear equation and its <u>first</u> perturbation, to an analysis of a nonlinear equation and higher order perturbations. For integrable equations with a principal branch this gives nothing new, but for equation (3.5) this gives rise to new constraints! Further work on this type of analysis is being carried out and will be presented elsewhere [22].

Acknowledgements

We thank Martin Kruskal for the many suggestions and remarks made during the workshop.

References

[1] A.P.Fordy and A.Pickering, Analysing negative resonances in the Painlevé test, Phys.Letts.A (in press).

[2] F.J.Bureau, Differential equations with fixed critical points, Annali di Matematica pura ed applicata <u>LXIV</u>, 229-364 (1964)

[3] P.Painlevé, Acta Math. <u>25</u>, 1-85 (1902).

[4] P.Painlevé, Leçons sur la théorie analytique des équations différentielles (Lecons de Stockholm, 1895), Hermann, Paris (1897).

[5] E.Hille, Ordinary differential equations in the complex domain, Wiley, NY, 1976.

[6] B.Gambier, Sur les équations différentielles du second ordre et du premier degré dont l'intégrale générale est à points critiques fixes, Thèse, Paris, 1909; Acta math., <u>33</u>, 1-55 (1910).

[7] F.J.Bureau, Sur la recherche des équations différentielles du second order dont l'intégrable générale est à points critiques fixes, Bulletin de la Classe des Sciences <u>XXV</u>, 51-68 (1939).

[8] J.Chazy, Sur les équations différentielles dont l'intégrale générale est uniforme et admet des singularités essentielles mobiles, C.R.Acad. Sc. Paris <u>149</u>, 563-5 (1909).

[9] R.Garnier, Annales scientifiques de l'École normale supérieure, <u>29</u>, 1-126 (1911).

[10] F.J.Bureau, Differential equations with fixed critical points, Annali di Matematica pura ed applicata, LXVI, 1-116, (1964).

[11] J.Chazy, Sur les équations différentielles du troisième ordre et d'ordre supérieur dont l'intégrale générale a ses points critiques fixes, Acta math. 34, 317-85 (1911).

[12] M.J.Ablowitz and H.Segur, Exact linearisation of a Painlevé transcendent, Phys.Rev.Lett. 38, 1103-6 (1977).

[13] M.J.Ablowitz, A.Ramani and H.Segur, Nonlinear evolution equations and ordinary differential equations of Painlevé type. Lett.Nuovo Cimento 23, 333-8 (1978).

[14] M.J.Ablowitz, A.Ramani and H.Segur, A connection between nonlinear evolution equations and ordinary differential equations of P-type. I, J.Math.Phys. 21, 715-21 (1980); II, J.Math.Phys. 21, 1006-15 (1980).

[15] S.Kowalevskaya, Acta.Math. 12, 177-232 (1889); 14, 81-3 (1889).

[16] J.Weiss, M.Tabor and G.Carnevale, The Painlevé property for partial differential equations, J.Math.Phys. 24, 522-6 (1983).

[17] M.Tabor, Painlevé property for partial diffrential equations, pp.427-46, Soliton theory: a survey of results, ed. A.P.Fordy, MUP, Manchester 1990.

[18] M.D.Kruskal, A.Ramani and B.Grammaticos, Singularity analysis and its relation to complete, partial and non-integrability, pp. 321-72, Partially Integrable Evolution Equations in Physics, (eds) R.Conte and N. Boccara, Kluwer, Dordrecht, 1990.

[19] M.D.Kruskal and P.A.Clarkson, The Painlevé-Kowalevski and Poly-Painlevé Tests for Integrability, preprint (1991).

[20] D.Levi and P.Winternitz (eds), Painlevé transcendents, their asymptotics and physical applications, Plenum, NY, 1991.

[21] R.Conte, Invariant Painlevé analysis of partial differential equations, Phys.Letts.A, 140, 383-90 (1989).

[22] R.Conte, A.P.Fordy and A.Pickering, A perturbative Painlevé approach to nonlinear partial differential equations, in preparation.

[23] R.Conte, Partial integrability of damped, forced, anharmonic oscillators, to be published in proceedings of "Nonlinear coherent structures in physics and biology", (eds) M.Peyrard and M.Remoissenet, Springer, Berlin, 1991.

Appendix

In this appendix we state some formulae from Conte's invariant analysis [21]. The functions χ, in this paper, and the usual WTC φ are related by:

$$\chi = (\varphi_x/\varphi - \varphi_{xx}/2\varphi_x)^{-1} . \tag{A1}$$

This χ satisfies the two Riccati equations:

$$\chi_x = 1 + \frac{1}{2}S\chi^2 , \quad \chi_t = -C+C_x\chi - \frac{1}{2}(C_{xx}+CS)\chi^2 , \tag{A2}$$

with integrability condition:

$$S_t+C_{xxx}+2C_xS+CS_x = 0 , \tag{A3}$$

where S and C are the unique differential invariants of the Möbius group, given by:

$$S = \varphi_{xxx}/\varphi_x - \frac{3}{2}(\varphi_{xx}/\varphi_x)^2 , \quad C = -\varphi_t/\varphi_x . \tag{A4}$$

The practical advantage of using this approach is to reduce the size of many expressions and thus simplify many of the calculations.

Modulation Equations for Nearly Integrable PDEs with Periodic Boundary Conditions

M.G. Forest

Department of Mathematics, Ohio State University,
Columbus, OH 43210, USA

This note, following my lecture, is a summary of the current state of practical application of periodic, N-phase modulation equations. First, I sketched interesting bifurcation phenomena that have been documented in [1] through extensive direct numerical investigations. These long-lived "attractors" of the damped, driven periodic sine-Gordon equation consist of various flows among coherent spatial structures. The temporal behavior ranges from periodic to quasiperiodic to chaotic, quantified in terms of several diagnostics. The spatial resolution during each flow is coherent and low-dimensional. In addition to Fourier projections, we use the sine-Gordon spectral transform to independently measure the "N-phase integrable mode approximations" of the perturbed field at each time step. The upshot is that the attractors are well-approximated, at least locally, by modulated N-phase periodic waves, where N remains small into the chaotic regime.

Next I simply wrote down results from [2] which yield a closed dynamical system for modulated, periodic N-phase sine-Gordon solutions in the presence of weak damping and driving. The way in which integrable theory can be exploited to explicitly diagonalize the perturbed energy-phase equations is both beautiful and remarkable, but beyond such a lecture. In particular, the numerical experiments in [1] indicate which class of N-phase waves to modulate in the various parameter regimes.

In order to "benchmark" the perturbed modulation theory, we first want to establish agreement with direct numerical ezperiments in the pre-chaotic regime. Results of two types were reported, both for even, periodic, two-phase breather trains. The first result is an averaging calculation from [3] which predicts frequency and phase locked breather trains, periodic in x and t. The energy and phase parameters of the integrable solution are uniquely specified in terms of the perturbation parameters to yield a leading order approximation to this space-and time-periodic, near integrable attractor. The modulation prediction is then compared with direct numerical simulation of the pde for initial data given by our prediction. There is agreement within the order of the perturbation theory. The final topic presented was a dynamical simulation of the perturbed energy-phase modulation equations for even, periodic breather trains. These results from [4] clarify the averaging predictions and determine robustness of these 2-phase attractors to variations in perturbation parameters and initial data. Our present focus is the dynamics of chaotic attractors in terms of near-integrable coördinates, which requires an additional phase and poses as a nontrivial next step.

REFERENCES

[1] Bishop, A. R., Forest, M. G., McLaughlin, D. W., and Overman, E. A., Physica 23D (1986) , 293-328; Physics Letters A 127(6,7) (1988), 335-340.

[2] Ercolani, N. M., Forest, M. G., and McLaughlin, D. W., "Fully nonlinear modal equations for nearly integrable pdes", to be submitted.

[3] Forest, M. G., Sheu, S.-P., and Sinha, A., "Frequency and phase locking of spatially periodic perturbed sine-Gordon breather trains", to appear, SIAM Journal of Applied Mathematics, in press.

[4] Forest, M. G. and Sinha, A., "A numerical study of nearly integrable modulation equations", submitted to Singular Limits of Dispersive Wave Equations, edited by Ercolani, Levermore, and Serre.

Spectrum of Domain Wall Excitations in YIG

A.V. Mikhailov[1,2] *and I.A. Shimokhin*[1]

[1]L.D. Landau Institute for Theoretical Physics,
ul. Kosygina 2, 117940 Moscow, Russia
[2]Institute for Scientific Interchange Foundation, Torino, Italy

Here we present results of our numerical and analytical study of excitation spectrum of 180° domain wall in yttrium iron garnets (YIG). Samples which are made from YIG are widely used in experiments since they have extremely low dissipation of spin waves and high Curie temperature. The effective anisotropy of YIG is quite small and therefore the dipole-dipole interaction becomes very important. There is a theory [1] which describes a dynamics of a domain wall if the anisotropy of a sample is strong enough (or, so called, the quality parameter $Q >> 1$). A straightforward extrapolation of results of that theory to the case of small anisotropy is obviously inconsistent and leads to a disagreement with experimental data, in particular, the experimentally found group velocity of excitations propagating along the domain wall proved to be unexpectedly high [2]. YIG have a complicated crystal structure and the theory which have been recently developed for uniaxial anisotropy [3] requires in modifications. Moreover, due to this structure new branches of excitation spectrum do appear. Any non-linear model which claims to describe a dynamics of domain walls should have a correct linear limit, i.e. demonstrate the true dispersion law corresponding to the excitation spectrum of a domain wall.

The crystallomagnetic structure of yttrium iron garnets can be approximated by a cubic and uniaxial anisotropy with the easy directions lie along of one of the cube diagonals. Thus we study the Landau-Lifshitz model in the magneto static approximation:

$$\frac{1}{\gamma}\dot{\mathbf{M}} = \mathbf{M} \times \frac{\delta H}{\delta \mathbf{M}}, \quad \mathrm{div}(\mathbf{H} + 4\pi\mathbf{M}) = 0, \quad \mathrm{rot}\mathbf{H} = 0, \tag{1}$$

where $\mathbf{M}$ is the magnetization vector, $\mathbf{H}$ is the magnetic field and γ is the gyromagnetic ratio. Here the Hamiltonian $H = \int(h_{ex} + h_a + h_{d-d})dv$ corresponds to the exchange, anisotropy and dipole-dipole interactions:

$$h_{ex} = \frac{\alpha}{2M_s^2}\sum_{i=1}^{3}\left(\frac{\partial \mathbf{M}}{\partial x_i}\right)^2, \quad h_{d-d} = -\frac{1}{2}(\mathbf{M}\cdot\mathbf{H})$$

$$h_a = \frac{K_1}{M_s^4}(M_1^2M_2^2 + M_2^2M_3^2 + M_3^2M_1^2) + \frac{\tau|K_1|}{12}\left(1 - \frac{(\mathbf{M}\cdot\mathbf{n})^2}{M_s^2}\right),$$

where M_s is the saturation magnetization ($\mathbf{M}^2 = M_s^2$), α is the exchange constant, K_1 is the cubic anisotropy constant and constant τ characterizes the uniaxial anisotropy along the unit vector $\mathbf{n} = (1,1,1)/\sqrt{3}$. More accurate accounting of the magneto-elastic interaction slightly modifies coefficients in h_a (see [4], [5], [6]) but does not affect much on the following consideration.

An exact solution of equations (1) which depends on one spatial coordinate y and corresponds to the Bloch-type domain wall has been found by Lilley [7]. It can be rewritten as[1]

$$\mathbf{S}^0 = (\frac{1}{\sqrt{1+\Delta(by)^2}}, 0, \frac{\Delta(by)}{\sqrt{1+\Delta(by)^2}}), \quad \mathbf{H}^0 = 0, \tag{2}$$

where

$$a = \frac{\sqrt{\tau^2+9\tau}}{8+\tau}, \; b = \sqrt{\frac{|K_1|(8+\tau)}{12\alpha}}, \; c = \frac{2\sqrt{2}}{8+\tau} \; \Delta(by) = a\sinh(by) + c.$$

For small τ the 180° domain wall (2) is a combination of two domain walls of $\sim 71°$ and $\sim 109°$ separated with a distance about

$$\delta y \approx \frac{1}{b}\sinh^{-1}(\frac{3\sqrt{2}}{4a}) \approx \sqrt{\frac{3\alpha}{2|K_1|}}\log\frac{1}{\tau}.$$

To study small excitations of the domain wall we linearize equations (1) near the exact solution (2) $\mathbf{S} = \mathbf{S}^0 + \mathbf{s}$, $|\mathbf{s}| \ll 1$, $\mathbf{H} = M_s \mathrm{div}\phi$, separate variables assuming that $\mathbf{s}$, ϕ are plane waves with the wave vector $\kappa = (\kappa_x, \kappa_z)$ and frequency Ω and introduce dimensionless variables:

$$\begin{aligned} s_x &= S_3^0\Phi(\xi)\cos(\kappa_x x + \kappa_z z - \Omega t), & s_y &= \Psi(\xi)\sin(\kappa_x x + \kappa_z z - \Omega t), \\ -s_z &= S_1^0\Phi(\xi)\cos(\kappa_x x + \kappa_z z - \Omega t), & \phi &= 2\alpha b f(\xi)\sin(\kappa_x x + \kappa_z z - \Omega t), \end{aligned}$$

Where $\Omega = 2\gamma\alpha b^2\omega/M_s$, $\kappa_x = \eta b\sin\vartheta$, $\kappa_z = \eta b\cos\vartheta$, $y = \xi/b$. Functions Φ and Ψ characterize the deviation of $\mathbf{S}$ in the plain of the domain wall and across it respectively. In these variables the linearized equations have the form:

$$\begin{aligned} D_1\Phi &= \omega\Psi - \eta T(\xi,\vartheta)f, & D_2\Psi &= \omega\Phi + f', \\ f'' - \eta^2 f &= (\Psi' + \eta T(\xi,\vartheta)\Phi)/Q, \end{aligned} \tag{3}$$

where $Q = (8+\tau)|K_1|/(24\pi M_s^2)$, $T(\xi,\vartheta) = (\cos\vartheta - \sin\vartheta\Delta(\xi))/\sqrt{1+\Delta(\xi)^2}$ and the differential operators D_1, D_2 are of the form

$$\begin{aligned} D_1 &= \frac{d^2}{d\xi^2} - [1 + \eta^2 - \frac{6(a^2+c^2-1) - 2 - 6c\Delta(\xi)}{1+\Delta(\xi)^2} \\ &+ \frac{16c\Delta(\xi) - 8(a^2+c^2-1)}{(1+\Delta(\xi)^2)^2}], \end{aligned}$$

[1] For convenience we introduce the unit length vector field $\mathbf{S}$ making an orthogonal change of coordinates and rescaling $M_1 = M_s(-2S_1+\sqrt{2}S_3)/\sqrt{6}$, $M_2 = M_s(S_1-\sqrt{3}S_2+\sqrt{2}S_3)/\sqrt{6}$, $M_3 = M_s(S_1+\sqrt{3}S_2+\sqrt{2}S_3)/\sqrt{6}$.

$$D_2 = \frac{d^2}{d\xi^2} - [1+\eta^2 - \frac{2(a^2+c^2-1)-2+6c\Delta(\xi)}{1+\Delta(\xi)^2} + \frac{6c\Delta(\xi)-3(a^2+c^2-1)}{(1+\Delta(\xi)^2)^2}].$$

Now the problem to find the excitation spectrum of the domain wall can be formulated in the following way: in three dimensional space with coordinates $(\eta, \vartheta, \omega)$ one has to find a set of points such that the system of equations (5) possesses a solution vanishing with $\xi \to \pm\infty$. This set is nothing else but the dispersion law of linear excitations of the domain wall. It depends on the parameters Q and τ.

The excitation spectrum of a domain wall represents a set of sheets located within a band forbidden for space volume spin waves:

$$\omega^2 < (1+\eta^2)(1+\eta^2+(\sin\vartheta)^2/Q).$$

The long wavelength asymptote of the low frequency branch can be found analytically. The method that we have used is rather similar to the one that we have developed for the case of the uniaxial anisotropy [3]. But the problem under consideration is much more complicated and we have to assume $Q \ll 1$ to obtain the main contribution in the long wavelength asymptotic of the low frequency branch of the spectrum. In this short note we bypath whole theory and present only the result in dimensionless variables

$$\omega = \sqrt{\eta}\frac{\sqrt{2}|\cos\vartheta|}{aQ^{1/2}(Q+(\sin\vartheta)^2)^{1/4}J_0^{1/2}} + O(\eta). \tag{4}$$

Where J_0 is a quadrature,

$$J_0 = \int_{-\infty}^{+\infty} \frac{\cosh(\xi)^2}{[1+(a\sinh(\xi)+c)^2]^2} d\xi.$$

that can be in principle, evaluated in elementary functions but we have not found a compact form for the result. If a is small (like in the case of yttrium iron garnets) one can expand J_0:

$$J_0 = \frac{1}{a^2}(1+c\tan^{-1}(c)) + O(1), \quad \text{if } a \to 0.$$

Like the uniaxial case the low frequency brunch of the spectrum has a square root singularity at $|\kappa| = \eta b \to 0$ for $\vartheta \neq \pm\pi/2$.

The result obtained perfectly matches with the one for the uniaxial case (see [3]) if $Q \to 0$ and $\tau \to \infty$.

Discussion of numerical results

We have been studying the spectrum of excitations numerically for the parameters $Q = 3.4 \cdot 10^{-2}$ and $\tau = 3.8 \cdot 10^{-2}$ that experimentalists usually assume

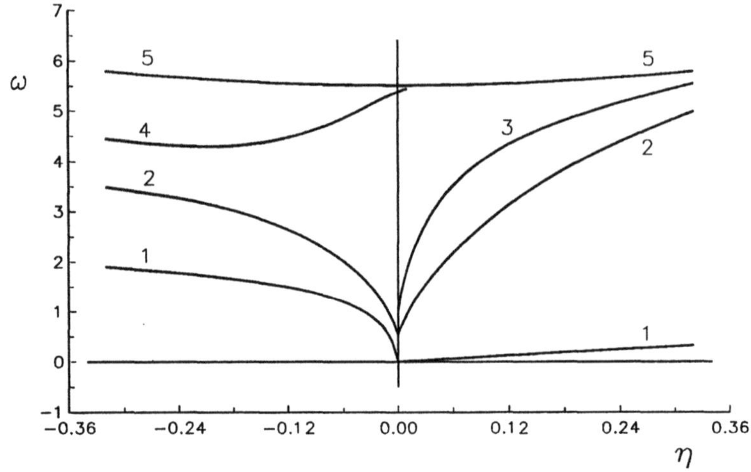

Figure 1 The excitation spectrum for $\vartheta = \pm\pi/2$. Curves "1"-"4" correspond to different branches and "5" is the bottom of the space volume spin wave zone.

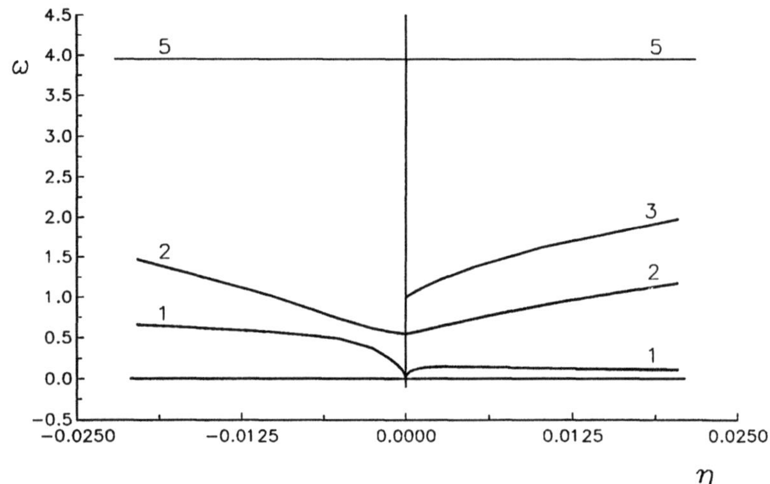

Figure 2 The excitation spectrum for $\vartheta = \pi/4,\ 5\pi/4$.

to describe the yttrium iron garnet [8]. In the most interesting long wavelength region we have found four branches of the spectrum (see Fig.1). We remind that in the case of uniaxial anisotropy there are two branches only, namely the low frequency branch and the Gilinskii branch [3].

The low frequency branch corresponds to a bend Goldstone mode (see Figs.1, 2, 3, branch "1"). Due to the dipole-dipole interaction it has a square root singularity at $\eta \to 0$ in all directions (except of $\vartheta = \pm\pi/2$ where it has a linear behavior). Our asymptotical result (4) perfectly coincides whit the computer

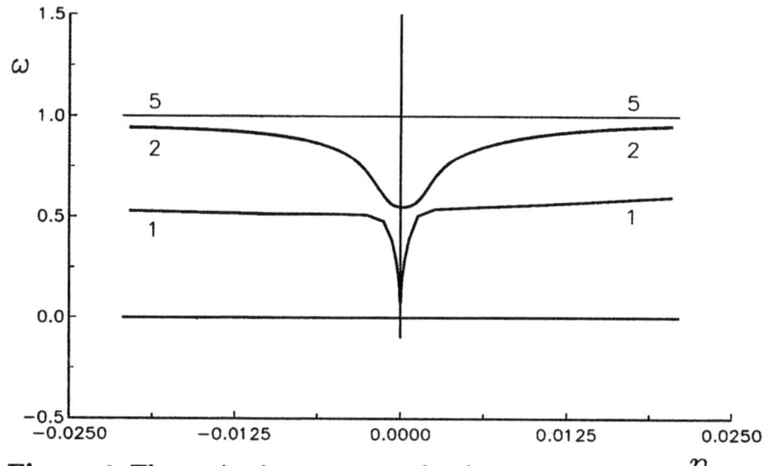

Figure 3 The excitation spectrum for $\vartheta = 0,\ \pi$.

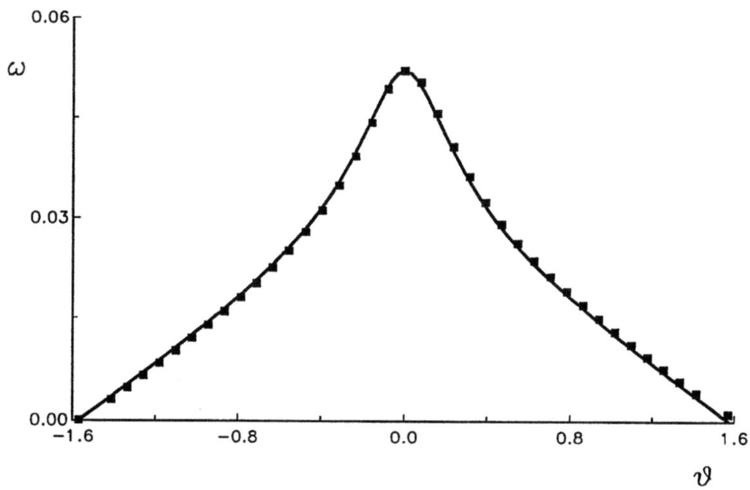

Figure 4 Angular dependence of the frequency in the Goldstone branch "1". Solid line corresponds to (24), blocks are numerical results. Here $Q = 3.4 \cdot 10^{-2}, \tau = 3.84 \cdot 10^{-2}, \eta = 10^{-5}$.

simulation at small η (see Fig. 4). In particular, this means that the group velocity $\mathbf{v}_{gr} = \mathrm{grad}_{\eta}\omega$ becomes very high at low values of η (and can in principle violate the applicability of the magneto static approximation). In the real experiments [2] an anomalously high velocity of propagation of pulses along a domain wall in the direction $\vartheta = 0$ has been observed.

The branch "2" on Figs. 1, 2, 3 has a gap and corresponds to oscillations of the width of a domain wall. This mode resembles the one that exists in the simple mechanical model of two weakly interacting strings. It does not exist in uniaxial ferromagnets. There was an attempt to find the gap of the spectrum at $\eta = 0$ theoretically [9]. Unfortunately the result obtained is not in a good agreement with our computer simulation. Our analysis has shown that the theoretical construction in [9] had been based on a wrong assumption. Namely, the author has neglected the contribution of the fast exponents $\exp(-|\xi|/\sqrt{Q})$, which become important for the finite gap modes (in contrary to the Goldstone one). Our computed eigenfunction does not fit the assumption which had been made in [9].

The branch "3" exists if $0 < \vartheta < \pi$. It corresponds to the Gilinskii branch in the uniaxial case [10], [3]. Gilinskii had found it at $\vartheta = \pi/2$, the angular dependency has been recently described in [11]. Here we do not see a big difference with the uniaxial case.

The last branch "4" exists into a narrow region of angles $\vartheta \sim \pi/2$ when the spin wave gap is high enough. It seems that this branch has been unknown.

References

[1] A. P. Malozemoff and J. C. Slonczewski, *Magnetic Domain Walls in Bubble Materials* (Academic Press. New York, London, Toronto, Sydney, San Francisco, 1979).

[2] V. S. Gornakov, L. M. Dedukh, V. I. Nikitenko, and V. T. Synogach, *Sov. Phys.-* **JETP 90**, 2090 (1986).

[3] A. V. Mikhailov, I. A. Shimokhin, *Zh. Eksp. Teor. Fiz. (Sovet Phys.-* **JETP***)* **97**, 1966 (1990).

[4] G. Rieder, *Abhandi. Braunschw. Wiss. Ges.* **11**, 20 (1959).

[5] W. Döring, *Handbuch d. Physik* Springer, Berlin, Heidelberg, New York, 1966, Bd. 18/2.

[6] A. Hubert, *Theorie Der Domanenwande in Geordneten Medien,* (Springer-Verlag, Berlin, Heidelberg, New York, 1974).

[7] B. A. Lilley, *Phil. Mag.* **41**, 792 (1950).

[8] L. M. Dedukh, V. I. Nikitenko and V. T. Synogach, *Sov. Phys.-* **JETP 94**, 312 (1988).

[9] H. E. Khodenkov, *Solid State Physics-* **31**, 226 (1989).

[10] I. A. Gilinskii, *Sov. Phys.-* **JETP 68**, 1032 (1975).

[11] I. A. Shimokhin, *Phys. Stat. Sol.* (b), **167**, 243, (1991).

Interaction of Defects in Nonlinear Dissipative Fields

L.M. Pismen, J. Rubinstein, A.A. Nepomnyashchy, and J.D. Rodriguez
Technion, Israel Institute of Technology, 32000 Haifa, Israel

Introduction. Singularities of the order parameter fields, commonly called *defects*, play a decisive role in statics and dynamics of various ordered media, such as superfluids and superconductors [17], liquid crystals [8, 14], ferromagnets [18], and dissipative structures arising as a result of chemical [16] or hydrodynamical [23] instabilities.

Defects are particularly important dynamically, since they are readily formed at the initial stages of evolution towards an ordered state in an extended initially disordered system. At the initial rapid stage, resulting in the formation of ordered domains, defects emerge due to the incompatibility of local ordering. The following stage of slow evolution is dominated by the motion and coalescence of defects that effectively serves to spread the order over ever larger domains, as long as it is compatible with both topological constraints in the order parameter space and constraints due to boundary conditions in the real space.

Once defects are present, they interact with each other and with the regular components of the field. Defects dominate the solutions to the field equations. In some systems they mediate order by an annihilation process [23]. In other cases, e.g. superfluids [11, 33], their entanglement creates turbulence. Spontaneous generation and annihilation of defects is seen as the principal mechanism of weak turbulence in dissipative wave patterns [6]. The optical and rheological properties of liquid crystals are greatly influenced by the presence of defects, and the internal structure of ferromagnets is almost completely determined by them. It is important therefore to develop a general formalism that would allow to derive equations of motion of defects in various cases of interest.

Particle–field approximation for potential systems. In order to illustrate our approach to ordered media with defects, we consider a dissipative system that evolves as to minimize the generalized Ginzburg-Landau (GL) Lagrangian

$$G(u) = \frac{1}{2}\int \left[g_{ab}\nabla u^a \cdot \nabla u^b + V(|u|)\right] d^n\mathbf{x}. \tag{1}$$

Here u^a is the order parameter defined in a p-dimensional space with a metric tensor g_{ab}, ∇ is the gradient operator in an n-dimensional real space, and $V(|u|)$ is a potential dependent on a suitably defined norm of the order parameter $|u|$. Suppose that the minimum of $V(|u|)$ is achieved at a $(p-m)$-dimensional symmetric manifold in the order parameter space (minimal stratum). The simplest example is the potential of the form $V = \frac{1}{2}(1-|u|^2)^2$ minimized on the $(p-1)$-sphere $|u| = 1$.

The respective dynamic equations follow the variational principle $\partial_t u^a = -g^{ab}\delta G/\delta u^b$, and, in a general case, can be expressed through Cristoffel symbols Γ^a_{bc} corresponding to the chosen metric of the order parameter space:

$$\partial_t u^a = \nabla^2 u^a + \Gamma^a_{bc} \nabla u^b \cdot \nabla u^c - g^{ab} \partial V/\partial u^b. \tag{2}$$

We shall denote any smooth parametrisation of a minimal stratum as *phase variables*, and call the order parameter field a *phase field* when it can be expressed through phase variables alone. This implies that the order parameter is confined everywhere to a vicinity of the minimal stratum. A codimension m defect is defined as a $(n-m)$-dimensional manifold of the real space where the *phase field* is singular. The existence of a defect implies that in its vicinity the phase description is inadequate, even though the full order parameter field may be nonsingular.

The precise classification of topological defects is given by homotopy groups π_m that are computed on the basis of the symmetry group of the minimal manifold [20]. A codimension m defect exists when the $(m-1)$-th homotopy group is nontrivial. The large-scale dynamics of a diluted gas of defects follows almost everywhere phase equations corresponding to the residual symmetry group of the underlying system. In the vicinity of a defect, full nonlinear field equations are in force, but the solution is close to a symmetric static form due to the weakness of interactions. The mechanism which sets the defects into motion is the incompatibility of the symmetries of the far (phase) and near field approximations. The defect's velocity is computed as an adjustable parameter in an asymptotic procedure matching the near and far field solutions. This determines equations of motion of defects that have to be further solved simultaneously with far field equations.

Unlike a more standard Peach-Köhler type approach [28], this method of derivation of equations of motion is not restricted to potential systems. It is also free of paradoxes that have lead in the past to erroneous conclusions in cases when the energy of the system diverges with its size.

Vortices in $U(1)$ vacuum: Ginzburg - Landau (GL) equation. The simplest soluble system is the GL equation for a scalar complex order parameter that comes in two very dissimilar forms: dissipative (or real)

$$\partial_t u = \nabla^2 u + u - |u|^2 u. \tag{3}$$

and conservative

$$-i\partial_t u = \nabla^2 u + u - |u|^2 u. \tag{4}$$

Eq.(3), applicable to certain kinds of non-equilibrium patterns [4], is potential. The conservative system (3) has been extensively studied in the context of superfluid dynamics [17, 11]. The symmetry group of the minimal stratum in both cases is $U(1)$, and the phase equations are linear. The defects are points in $2d$, and lines in $3d$.

In $2d$, the dynamical problem for the conservative system is almost trivial, and the equation of motion of a point defect (vortex), containing the well-known Magnus force [17], immediately follows from the fluid-mechanical analogy of the phase field. On the contrary, the dynamics of point defects in the dissipative system (3) had not been well understood before the equation of motion containing a logarithmic correction was obtained [4] using a modification of the Peach-Köhler method that took into account large distortions of the phase field introduced by an arbitrary small, but non-vanishing defect velocity. This result was later reproduced using the method of asymptotic matching [25].

It was found that the defect moves in the direction orthogonal to the imposed phase gradient in such a way that likely charged defects repel, and oppositely charged ones attract each other. The velocity is connected with the magnitude of the local gradient A by the mobility relationship

$$2A = v \ln \frac{v_0}{v}, \quad v_0 = 3.29. \tag{5}$$

When several defects are present, their combined influence in the neighborhood of a test defect can be approximated by a linear function. We then have

$$\mathbf{A} = \sum N_i \nabla \theta_i (q(t) - q_i(t)), \tag{6}$$

where N_i is the charge and q_i is the instantaneous position of the i-th defect. The phase field is defined by the heat equation

$$\partial_t \theta = \nabla^2 \theta. \tag{7}$$

Equations (7), (5) and (6) determine the evolution of a of a diluted gas of defects. An example of the numerical solution of this system is given in Ref.[30].

In $3d$, the motion of a line defect is dominated by its local geometry. Now the conservative problem is far more interesting than the dissipative one: in the latter case the line just tends to shorten [31], while in the former its dynamics is complex, fascinating, and still not sufficiently understood in spite of a long history of studies, in particular, in the context of hydrodynamics of ideal fluids [2]. The equation of motion of a line defect in the conservative GL equation was obtained [27] using the method of matched asymptotic expansions; remarkably, computations employed for estimating local contributions to the binormal motion of a line segment were identical to those leading to eq.(5), and gave the same numerical coefficient. Nonlocal contributions introduce also a normal component of motion, and thus lead to line stretching or shrinking. Similar results were independently obtained for thin vortex lines in the Navier–Stokes equation [13].

$U(1)$ field defects: dislocations in $2d$ patterns A a more complex situation, where already the far field equations are nonlinear, arises in pattern-forming systems, as described, for example, by the Newell–Whitehead–Segel (NWS) equation [22, 34]

$$\partial_t u = (2\partial_x - i\partial_y^2)^2 u + u - |u|^2 u, \tag{8}$$

defining the complex amplitude $u = \rho e^{i\theta}$ of the envelope of a basic pattern of rolls parallel to the y axis. The phase diffusion approximation corresponds to the case when the characteristic scale of phase distortions is small even on the extended NWS scale, *viz.* $\theta_x = O(\epsilon)$, $\theta_y = O(\sqrt{\epsilon})$. Then the real amplitude ρ follows the phase adiabatically. The phase equation is, however, nonlinear:

$$\theta_t = -\theta_{yyyy} + 4\theta_{xx} + 4\theta_x\theta_{yy} + 8\theta_{xy}\theta_y + 6\theta_y^2\theta_{yy}. \tag{9}$$

Meiron and Newell [19] showed that the stationary singular solution of eq.(9) corresponding to a a solitary quiescent defect in the pattern with the optimal wave number has a self-similar form, i.e. can be presented as a function of a single coordinate $\eta = y/\sqrt{2|x|}$. A remarkably simple analytical form of this solution was found in Ref.[21]:

$$\theta = -\text{sign}(x) \log \left[\frac{1+e^{-\pi}}{2} - \frac{1-e^{-\pi}}{2}\text{erf}(\eta)\right] \tag{10}$$

This solution is fundamental for pattern-forming systems characterized by a non-vanishing optimal wave number, playing the same role of a 'basic non-trivial state' as the solution $\theta = \phi$ (polar angle) of eq.(3). Other solutions containing arbitrarily spaced rows of defects can be obtained by logarithmic superposition [21].

The properties of the available analytic solutions indicate the inadequacy of earlier approaches to this problem [35, 36] based on a truncated linearized version of the phase equation (9). The defects interact in an overlap region where nonlinear terms are dominant, and the influence of the defect cores, apparently, cannot be discarded.

$U(1)$ **field defects: rotating spirals.** Another problem where the interaction of defects is mediated by a nonlinear far field is the dynamics of rotating spiral waves (vortices) in the complex Ginzburg–Landau equation (CGLE)

$$\partial_t u = (1 + i\eta)\nabla^2 u + u - (1 + i\nu)|u|^2 u. \tag{11}$$

This equation arises as the amplitude equation in the vicinity of the Hopf bifurcation in spatially extended systems[16], and therefore is generic for active media displaying wave patterns. Lacking either potential or Hamiltonian structure, this equation displays a particularly rich variety of solutions. One of these is a rotating spiral wave [12], which is one of the best known non-equilibrium structures, observed in numerous experiments with the Belousov–Zhabotinsky reaction[16, 7], and in other systems.

The problem of motion and interaction of spiral waves attracted recently wide interest. Numerical simulations [32, 5, 15] showed that the attractive interaction of oppositely charged vortices changes to repulsion at larger distances. Further simulations [1] indicated still another change of sign, that would cause formation of stable bound pairs. Formation of pairs was predicted also by analytical studies [9, 1] which, however, were based on incompatible approaches and lead to contradictory results.

The analytical theory of interaction of well-separated spiral waves can take as a ground state some superposition of Hagan's solutions. The *ansatz* of Refs. [9, 29], that reduces to a phase superposition in the far field, ignores the mutual screening of spiral waves, and grossly exaggerates the strength of their interaction. Aranson, Weber and Kramer [1] observed that the far field equation reduces to the Burgers equation in the limit of small asymptotic wavenumbers, and that a superposition principle holds for the linear equation obtained via the Hopf–Cole transformation. In this case, however, remains a more subtle obstacle: the topological circulation condition becomes nonlinear after the logarithmic transformation, which renders the superposition principle inapplicable.

The topological correction was taken into account in Ref. [26], where the asymptotic expansion scheme was used to derive a mobility relationship for spiral waves. The method is restricted to very small values of the asymptotic wavenumber, and cannot be compared to the numerical experiment, since the interesting effect of pair formation does not occur within the applicability limits of the analytical theory.

SO_3 vacuum: point defects in nematics The simplest nolinear medium possessing point defects in $3d$ is a nematic liquid crystal. When it is perfectly aligned, the order parameter can be defined as a directionless unit vector, and parametrized either by two spherical angles or by a complex number obtained by stereographic projection of the unit sphere on the complex plane. The generalized GL form (1) corresponds to the so-called '1-constant approximation'. The potential V vanishes when the system is isotropic. Assuming gradient dynamics, and using polar coordinates in both real $(\theta,\ \phi)$ and the order parameter $(\alpha,\ \beta)$ space yields the dynamic equations

$$\partial_t \alpha = \nabla^2 \alpha - \frac{1}{2} \sin 2\alpha |\nabla \beta|^2, \tag{12}$$

$$\partial_t \beta = \nabla^2 \beta + 2 \cot \alpha \nabla \alpha \cdot \nabla \beta. \tag{13}$$

A variety of exact analytic solutions corresponding to non-trivial equilibria of (12), (13) can be obtained exploiting the complex analytic structure of both the $2d$ stratum of the order parameter space and a $2d$ manifold of the real space wherein the solution sought for is defined. One such family of solutions is an ensemble of quiescent straight line vortices. Another family of 'hedgehog' solutions can be constructed presuming the orientation of the director to be independent of r (distance from a point defect).

The perfectly aligned nematic corresponds, indeed, to a simplest of chiral models that have been studied in connection to non-Abelian gauge field theories [3, 24]. The solutions belonging to both families are highly degenerate, and there are indications that the dynamical problem of interaction of point defects becomes meaningless without taking the core effects into account.

A full system of equations that is also valid in the core region takes also in account an imperfect molecular alignment [10]. It can be obtained from the general form 1 by specifying some symmetric potential $V(|u|)$, say, a quatric $\frac{1}{4}(1-|u|^2)^2$. Equations (12), (13) are seen then, in our terms, as far field equations. Unlike the $2d$ case, they are nonlinear, which is due to the fact that the symmetry group of the minimal stratum is non-Abelian.

A dynamical theory of codimension 3 defects is a major open problem. While there is little chance to come up with explicit solutions, one can hope that matching of near and far fields can be accomplished via an asymptotic method similar to that used in $2d$, with an added selection mechanism within the hedgehog and line solution families.

Acknowledgement. This research was supported in parts by the US–Israel Binational Science Foundation.

References

[1] I.S. Aranson, L. Kramer and A. Weber (1991) Physica D, in press.

[2] G.K. Batchelor (1967) An Introduction to Fluid Dynamics, Cambridge U.P.

[3] A.A. Belavin and A.M. Polyakov (1975) JETP Lett. **22**, 245.

[4] E. Bodenschatz, W. Pesch and L. Kramer (1988) Physica D **32**, 135.

[5] E. Bodenschatz, A. Weber and L. Kramer, in *Nonlinear Wave Processes in Excitable Media*, Eds. A.V. Holden, M. Markus and H.G Othmer, NATO ASI series v.244, Plenum Press, New York, 1991.

[6] P. Coullet, L. Gil and J. Lega (1989) Phys. Rev. Lett. **62**, 1619.

[7] R.J. Field and M. Burger *Oscillations and Travelling Waves in Chemical Systems*, J.Wiley, New York (1985).

[8] P. de Gennes (1974) The Physics of Liquid Crystals, Oxford University Press.

[9] C. Elphick and E. Meron (1989, 1991) Weizmann Institute preprints (unpublished).

[10] J. Ericksen (1991) Arch. Rat. Mech Anal. **113**, 97.

[11] W.I. Glaberson and R.J. Donnelly (1986) Progr. Low Temp. Phys., v. 9, Elsevier.

[12] P. Hagan (1982) SIAM J. Appl. Math. **42**, 762.

[13] R. Klein and A.J. Majda (1991) Physica D **49**, 323.

[14] M. Kleman (1983) Points, Lines and Walls, John Wiley.

[15] L. Kramer, E. Bodenschatz, M. Kaiser, W. Pesch, A. Weber and W. Zimmermann, in *New Trends in Nonlinear Dynamics and Pattern Forming Phenomena*, Eds. P. Coullet and P. Huerre, NATO ASI series, Plenum Press, New York, 1990.

[16] Y. Kuramoto (1984) Chemical Waves, Oscillations and Turbulence, Springer-Verlag

[17] L.D. Landau and E.M. Lifshitz (1980) Stastical Mechanics, Pergamon Press.

[18] A.P. Malozemoff and J.C. Slonczewski (1979) Magnetic Domain Walls in Bubble Meterial, Academic Press.

[19] D.Meiron and A.C.Newell, Phys. Lett., **113A** 289 (1985).

[20] N.D. Mermin (1979) Rev. Mod. Phys. **51**, 591.

[21] A.A. Nepomnyashchy and L.M. Pismen (1991) Phys. Lett. A, **153**, 427.

[22] A.C.Newell and J.A.Whitehead, J. Fluid Mech. **38** 279 (1969);

[23] A. Newell (1988) in Propagation in Systems Far From Equlibrium, Wesfried et. al. (eds.), Springer-Verlag, 122.

[24] A.M. Perelomov (1987) Phys. Rep. **146**, 135.

[25] L.M. Pismen and J.D. Rodriguez (1990) Phys. Rev. A **42**, 2471.

[26] L.M. Pismen and A.A. Nepomnyashchy (1991) Phys. Rev. A, in press.

[27] L.M. Pismen and J. Rubinstein (1990) Physica D **47**, 353.

[28] Y. Pomeau, S. Zaleski and P. Manneville (1983) Phys. Rev. A **27**, 2710.

[29] S. Rica and E. Tirapegui (1990) Phys. Rev. Lett. **64**, 878; (1991) Physica D **48**, 396.

[30] J.D. Rodriguez, L.M. Pismen and L. Sirovich (1991) Phys. Rev. A, in press.

[31] J. Rubinstein (1990) Self-induced motion of line defects, Quart. Appl. Math., in press.

[32] H. Sakaguchi (1989) Progr. Theor. Phys. **82**, 7.

[33] K.W. Schwarz (1985) Phys. Rev. B **31**, 5782; (1988) Phys. Rev. B **38**, 2398.

[34] L.A.Segel, J. Fluid Mech. **38** 203 (1969).

[35] E.Siggia and A.Zippelius, Phys. Rev. **A24** 1036 (1981);

[36] G.Tesauro and M.C.Cross, Phys. Rev. **A34** 1363 (1986).

On the Analytic Degenerate Dispersion Laws

E.I. Schulman and D.D. Tskhakaya

P.P. Shirshov Institute for Oceanology,
23 Krasikova St., 117218 Moscow, Russia

0.Introduction The concept of degenerate dispersion laws arose in the theory of Hamiltonian wave systems [1]. Their Hamiltonians in normal variables have the form

$$H = \sum_{\alpha=1}^{Q} \int \omega^{\alpha}(k) a_k^{\alpha} a_k^{\alpha*2} dk + H_{int}$$

The functions $\omega^{\alpha}(\vec{k}), \alpha = 1, ..., Q$ represent dispersion laws of linear modes available in the system. Consider the simplest case Q = 1. The dynamics of the solution is controlled by usual Hamiltonian equation

$$i\partial a_k/\partial t = \delta H/\delta a_k^* \tag{0.1}$$

It is shown in [1], (for review of further results see [2], [3]) that the following alternative is the necessary condition for system (0.1) to possess an additional motion invariant of the form

$$I = \int f_k a_k a_k^* dk + higher\ terms.$$

For any nonlinear process "n waves into m waves" in points of general position of correspondent resonant surface

$$\vec{k}_1 + ... + \vec{k}_n = \vec{k}_{n+1} + ... + \vec{k}_{n+m}$$

$$\omega(\vec{k}_1) + ... + \omega(\vec{k}_n) = \omega(\vec{k}_{n+1}) + ... + \omega(\vec{k}_{n+m}) \tag{0.2}$$

one of two possibilities should be true: or the amplitude of the process (0.2) equals zero,

$$\mathrm{W}_{1,...,n;n+1,...,n+m} = 0$$

or the function f_k satisfies on the surface (0.2) the same condition as frequency,

$$f(\vec{k}_1) + ... + f(\vec{k}_n) = f(\vec{k}_{n+1}) + ... + f(\vec{k}_{n+m}) \qquad (0.3).$$

In the latter case the function $\omega(k)$ is very special. For typical ω one will obtain

$$f_k = f_k^{triv} = (\vec{v}, \vec{k}) + C\omega(\vec{k}). \tag{0.4}$$

Here $\vec{v}$ and C are constant vector and scalar respectively. If (0.4) has nontrivial solution, the dispersion law is called degenerate to the process (0.2).

1. Analytic degenerate dispersion laws. We will consider here dispersion laws which are analytic in components of the wave vector $\vec{k}$ and degenerate to the "1 into 2" process

$$\omega(\vec{k_1}+\vec{k_2})=\omega(\vec{k_1})+\omega(\vec{k_2}) \tag{1.1}$$

The example of this kind was found recently by Balk, Nazarenko and Zakharov for two - dimensional case, $\vec{k}=(p,q)$ in [3]: $\omega=pq^2, f=q^3/p^2$ and related example

$\omega=q/(1+p^2); f=q^3/[(1+p^2)^3p^2]$. Before going further we have to emphasize that the property of degeneracy depends strongly on the functional class in which we search for the solution of equation

$$f(\vec{k}_1+\vec{k}_2)=f(\vec{k}_1)+f(\vec{k}_2) \tag{1.2}$$

For example if we search for analytic f_k any analytic dispersion law is nondegenerate. So we present the rigorous definition of degenerate dispersion law.

Definition

The function $\omega(k)$ is called degenerate to the process (1.1) in the class m of functions in the point P of (1.1) if there is a solution of (1.2), belonging to the class m in the vicinity of P and linearly independent of $\omega(k)$ and components of $\vec{k}$. In study of analytic dispersion laws we will use the following class m: let $\vec{k}=\vec{\xi}\tau$. Then $f(\vec{\xi}\tau)$ may be decomposed in a Laurent series in τ with finite number of pole terms. In what follows we will consider $\omega(0)=0$, so $k_2=0$ belong to (1.1). We also introduce the equivalence relation: f_1 is equivalent to f_2 if $Cf_1-f_2=f^{triv}$. The results are following:

(i.) If the dimensionality of space is three or more, there is no degenerate analytic dispersion law.

If the space is two-dimensional , $\vec{k}=(p,q)$, then

(ii.) Up to the equivalence relation ω should begin with cubic terms,

$$\omega=\omega_3+\sum_{s>3}C_s\omega_s;\ \omega_s(\lambda k)=\lambda^s\omega_s(k)$$

(iii.) Contrary to the known cases [1], [2] the corresponding $f(k)$ is unique

(iv.) At small p,q

$$f(k)=q\psi(\xi)+\sum_{j>L}q^{j+1}\psi^{(j)}(\xi);$$

$$\xi=q/p$$

(v.) If all $C_s=0$ then ω_3 is of the form found in [3] up to linear in p, q transformations.

Sketch of the proof. If ω is degenerate, it is degenerate in the vicinity of $\vec{k_2}=0$ so we may search for the parametrization of (1.1) of the form

$$p_2=\sum_{i>1}\kappa^i p_{2i}(p_1,q_1);$$

$$q_2=\sum_{i>1}\kappa^i q_{2i}(p_1,q_1) \tag{1.3}$$

Here $\kappa << 1, p_1, q_1$ are independent coordinates parametrizing (1.1) in the vicinity of $k_2 = 0$. The functions p_{2i}, q_{2i} are to be obtained. We may expand (1.1) (using (1.3)) in powers of κ and we need to collect terms at $\kappa, \kappa^2, \kappa^3$. As a result we obtain

$$p_{21} = \omega_q; \; q_{21} = -\omega_p \tag{1.4}$$

where $\omega_{p,q} = \partial\omega/\partial p, q$. Substituting(1.3) and (1.4) into (1.2) we find

$$\kappa[\omega_q f_p - \omega_p f_q] = f(\kappa\omega_q, -\kappa\omega_p) \tag{1.5}$$

If the r.h.s. of (1.5) is not linear in κ then $J(\omega, f) = 0$, $J(f,g) = f_p g_q - f_q g_p$, $f = F(\omega)$ and it is not easy to show that $f = f^{triv}$. So the leading term in f is homogeneous of degree 1 and iv. is proved. Further, omitting indices of p_1, q_1 we obtain

$$J(\omega, f) = \omega_p \psi(-\omega_p/\omega_q)$$

Let p,q be also small. Then we may expand ω in homogeneous polynomials $\omega = \sum_{n=N}^{\infty} \omega_{(n)}(k)$ and write down equations arising in the first three orders in κ. We obtain that N is odd and the polynomial $\Omega = \omega_{(N)}(p, \xi p)/p^N$ has at least one real root ξ_0. So by linear transformation we may reduce ω to the form

$$\omega_{(N)} = q^m p^{N-m} + \sum_{n=1}^{N-m} c_n q^{m+n} p^{N-m-n}$$

where c_m are constants and m is the multiplicity to ξ_0. If also $\psi = \xi^s + ...$ as ξ tends to ∞ we obtain the following equation

$$s\frac{n}{m} + \frac{N-m}{m} = (-1)^s[\frac{N-m}{m}]^s \tag{1.6}$$

and from the next two orders one may extract that

$$s = (m-N)/N; \; N > 3$$

and

$$s = 5(N-m) - 8; \; N = 3. \tag{1.7}$$

It is easy to see that (1.7) is comparible with (1.6) only for N = 3, so (ii.) is proved. It is interesting to notice that any two - dimensional cubic dispersion law is degenerate to (1.1). Really, by linear transformation we may reduce it to the form

$$\omega = \beta q p^2 - \gamma q^3, \; \beta > 0 \tag{1.9}$$

were $\gamma >, < or = 0$ depending on if all roots of ω are real, complex or coincide. One may see, that

$$f(k) = q^3/(p^2 + 3\gamma q^2)$$

satisfy (1.2) on (1.1). The parametrization of (1.1) in that case reads

$$q_2 = \lambda q_1; q_1 = [Q(p_1, p_2, \lambda)/3\gamma\lambda(1+\lambda)]^{1/2}$$

$$Q = {p_2}^2 + 2p_1p_2 + \lambda({p_1}^2 + 2p_1p_2)$$

Nevertheless if we search for f(k) in the class m the γ has to be equal to zero for cubic dispersion law may be added by higher order terms and still be degenerate. It may be proved using three orders of expansion of (1.1) in κ. So the drift wave dispersion law $\omega = q/(1+p^2+q^2)$ is nondegenerate in this class. But it starts to be degenerate if we allow ω to have logarithmic singularities [4]. It is interesting to notice that example founded in [3] is related to the dispersion laws of Kadomtsev - Petviashvili hierarchy. They may be parametrized as follows [1] : $p = \sigma_1 - \sigma_2$, $q = {\sigma_1}^2 - {\sigma_2}^2$, $\omega_n = {\sigma_1}^n - {\sigma_2}^n$.

For $n = 4$ we obtain $\omega_4 = 8[q^3/p^2 + qp^2]$. The resonant surface (1.1) for this dispersion law splits into three pieces .Two of them are common for all equations of KP hierarchy while the third is parametrized as

$q_1 = r({p_1}^2 + 2p_1p_2)$, $q_2 = -r({p_1}^2 + 2p_1p_2)$. As r tends to zero or to ∞ we obtain that the second or the first terms become leading. And both of them satisfy (1.1) independently. The amplitude of the process (1.1) for KP-4 is proportional to $(q_1+q_2)/(p_1+p_2) + q_1/p_1 + q_2/p_2$ and equals zero on this piece of the resonant surface.

References

[1] V.E. Zakharov, E.I. Schulman. Physica D, v.1, 2, 191 - 202 (1980)

[2] V.E. Zakharov, E.I. Schulman. Physica D, v.29, 3, 283 - 321 (1988) V.E. Zakharov, E.I. Schulman. Integrability of nonlinear systems and perturbation theory, in “What is integrability?”,ed. V.E. Zakharov, Springer-Verlag, 1991.

[3] A.M. Balk, S.V. Nazarenko, V.E. Zakharov. Phys. Lett., v. 152, 5-6, 278-280 (1990).

[4] A.M. Balk. Private communication.

Part II

Inverse Scattering Transforms

Breaking Solitons. Systems of Hydrodynamic Type

*O.I. Bogoyavlenskij**

The Fields Institute for Research in Mathematical Sciences,
Waterloo, Ontario, Canada
Steklov Mathematical Institute, Vavilov St. 42,
117966 Moscow, GSP-1, Russia

1. Introduction

A certain class of integrable $(n+1)$-dimensional equations was studied by F. Calogero and A. Degasperis in works [1,2,3] by using the generalized Wronskian relations. A general Lax type operator equation was proposed by V.E. Zakharov [4] for constructing $(n+1)$-dimensional integrable equations. These constructions were discussed also in the monograph by R.K. Dodd, J.C. Ellbeck, J.D. Gibbon and G.C. Morris [5].

The equations which are studied in this paper and in [6,7] are not integrable for the general initial data, but their N-soliton solutions may be found explicitly and they possess the breaking behavior. We consider the differential equations, which are equivalent to the following equation in space of linear operators L and A:

$$L_t = P(L) + \sum_{k=1}^{n} R_k(L, L_{y_k}) + [L, A], \tag{1.1}$$

where $P(L)$ and $R_k(L, L_{y_k})$ are certain meromorphic functions of operator L, functions of $R_k(L, L_{y_k})$ are linear with respect to L_{y_k}. We assume that operators L and A depend on the variables $t, y_1, \cdots, y_n$ and $L_{y_k} = \partial L/\partial y_k$. L and A are supposed to be $n \times n$ matrices or 1-dimensional differential operators (in the last case L is self-adjoint operator, A is skew-symmetric operator).

The coefficients of the meromorphic functions $P(L), R_k(L, L_{y_k})$ are assumed to depend on invariants of operator L and their derivatives with respect to variables $t, y_1 \cdots, y_n$, that is the coefficients do not change after the transformation $L \to QLQ^{-1}$.

Lemma 1 *In view of equation (1.1) the eigenvalues $f(t, y_1, \cdots, y_n)$ of the operator L satisfy the system of equations*

$$f_t = P(f) + \sum_{k=1}^{n} R_k(f, f_{y_k}). \tag{1.2}$$

If the coefficients of the functions $P(L), R_k(L, L_{k_y})$ are constant, the system (1.2) is split into noninteracting equations for each eigenvalue $f_j(t, y, \cdots, y_n)$:

*Supported by the Ministry of Colleges and Universities of Ontario and the Natural Sciences and Engineering Research Council of Canada

$$f_{j_t} = P(f_j) + \sum_{k=1}^{n} R_k(f_j, f_{j_{y_k}}). \tag{1.3}$$

A proof is given in [7].
In the case

$$P(L) = 0, \qquad R_k(L, L_{y_k}) = \sum_{0\le i\le m}^{\infty} c_{ik}^m L^{m-i} L_{y_k} L^i \tag{1.4}$$

the equation (1.3) is the conservation law and we get as a consequence

$$(f_j^p)_t = \sum_{k=1}^{n} \left[\sum_{0\le i\le m}^{\infty} \frac{p c_{ik}^m}{m+p} f_j^{m+p} \right] y_k. \tag{1.5}$$

Hence, assuming that the eigenvalues $f_j(t, y_1, \cdots, y_n)$ tend to zero rapidly enough for $|y_k| \to \infty$ and applying the Gauss-Ostrogradskij theorem, we obtain the conserved quantities

$$\frac{dJ_P}{dt} = 0, \qquad J_p = \int_{r^n} (f_j)^p dy_1 \cdots dy_n. \tag{1.6}$$

These properties distinguish essentially the equation (1.1) from Lax equation $L_t = [L, A]$. Equation (1.1) as $P(L) = 0, R_k(L, L_{y_k}) = 0$ has the attractors in the phase space, see [6].

2. Systems of hydrodynamic type, connected with the Toda lattice

Let us consider the operator equation

$$L_t = LL_y + L_y L + [L, A], \tag{2.1}$$

where the matrices L and A have the form

$$L = \begin{bmatrix} p_1 & a_1 & 0 & & \\ a_1 & p_2 & a_2 & & \\ 0 & a_2 & p_3 & & \\ & & & \ddots & a_{n-1} \\ 0 & & & a_{n-1} & p_n \end{bmatrix}, A = \begin{bmatrix} 0 & x_1 & & & 0 \\ -x_1 & 0 & x_2 & & \\ & -x_2 & 0 & & \\ & & & \ddots & x_{n-1} \\ 0 & & & -x_{n-1} & 0 \end{bmatrix} \tag{2.2}$$

After the substitution $a_i = \exp(q_{i+1} - q_i)$ the operator equation (2.1), (2.2) is reduced to the following system of equations

$$\begin{aligned} p_{i_t} &= 2p_i p_{i_y} + 4(q_{i+1_y} + \beta)e^{2(q_{i+1}-q_i)} - 4(q_{i-1_y} + \beta)e^{2(q_i - q_{i-1})} \\ q_{i_t} &= 2p_i q_{i_y} + p_{i_y} + 2\sum_{k=1}^{k=i-1} p_{k_y} + 2\beta p_i \end{aligned} \tag{2.3}$$

where β is an arbitrary function of t, y. For solutions independent of y system (2.3) turns into the famous Toda lattice [9,10,11,12,13], system (2.3) as $\beta = 0$ is a system of hydrodynamic type, following the terminology of [8,15].

According to Lemma 1, the eigenvalues $f_k(t,y)$ of the matrix L (2.2) due to the system (2.3) satisfy the equation

$$f_{k_t} = 2 f_k f_{k_y}. \tag{2.4}$$

Hence the eigenvalues $f_k(t,y)$ are the Riemann invariants for the system (2.3). Obviously they possess the breaking behaviour.

3. Systems of hydrodynamic type, connected with the Volterra model

I. Let us consider the operator equation

$$L_t = L L_y L + [L, A] \tag{3.1}$$

for matrices L and A of the following form

$$L = \begin{bmatrix} 0 & \sqrt{a_1} & & & 0 \\ \sqrt{a_1} & 0 & \sqrt{a_2} & & \\ & \sqrt{a_2} & 0 & & \\ & & & \ddots & \sqrt{a_{n-1}} \\ 0 & & & \sqrt{a_{n-1}} & 0 \end{bmatrix},$$

$$A = \begin{bmatrix} 0 & 0 & x_1 & & & 0 \\ 0 & 0 & 0 & x_2 & & \\ -x_1 & 0 & 0 & & \ddots & \\ & -x_2 & & \ddots & & \\ & & \ddots & & & x_{n-2} \\ & & & & & 0 \\ 0 & & & -x_{n-2} & 0 & 0 \end{bmatrix} \tag{3.2}$$

The operator equation (3.1), (3.2) is equivalent to the system of equations

$$a_{i_t} = a_i \left[\beta(a_{i+1} - a_{i-1}) + a_{i_y} + a_{(i+1)_y} + a_{i+1} \sum_{k=1}^{k=i} a_{k_y} a_k^{-1} - a_{i-1} \sum_{k=1}^{k=i-2} a_{k_y} a_k^{-1} \right], \tag{3.3}$$

where β is an arbitrary function of t, y. System (3.3) for solution independent of y turns into Volterra model (or Kac van Moerbeke equation [14]) system (3.3) as $\beta = 0$ is system of hydrodynamic type.

Due to Lemma 1 we get that eigenvalues $f_k(t,y)$ of the matrix L (3.2) in view of system (3.3) satisfy the equation

$$f_{k_t} = f_k^2 f_{k_y}. \tag{3.4}$$

Hence they are the Riemann invariants for the system (3.3) and have the breaking behaviour.

II. System (3.3) after transformation $a_i = \exp u_i$ turns into the system

$$u_{i_t} = \beta(e^{u_{i+1}} - e^{u_{i-1}}) + e^{u_i}u_{i_y} + e^{u_{i+1}} \sum_{k=1}^{k=i+1} u_{k_y} + e^{u_{i-1}} \sum_{k=1}^{k=i-2} u_{k_y}. \quad (3.5)$$

System (3.5) has the following form

$$u_{i_t} = \sum_{j=1}^{n-1} A^{ij} \frac{\partial H}{\partial u_j}, \qquad H = \frac{1}{2} TrL^2 = e^{u_1} + \cdots + e^{u_{n-1}}. \quad (3.6)$$

The operators A^{ij} are skew-symmetric and have the form

$$A^{ij} = g^{ij} \frac{\partial}{\partial y} + \sum_{k=1}^{n-1} b_k^{ij} u_{k_y} + \beta I^{ij}. \quad (3.7)$$

Here the coefficients g^{ij}, b_k^{ij}, K^{ij} are constants and non- zero only in the following cases

$$\begin{aligned} g^{ii} &= g^{i,i+1} = g^{i+1,i} = 1 \\ I^{i,i+1} &= -I^{i+1,i} = 1 \\ b_k^{i,i+1} &= -b_k^{i+1,i} = 1, \text{ as } 1 \le k \le i. \end{aligned} \quad (3.8)$$

As a consequence of (3.6) we get the conservation law

$$\int_{-\infty}^{\infty} H dy = \frac{1}{2} \int_{-\infty}^{\infty} TrL^2 dy = \text{const.} \quad (3.9)$$

The same conservation law (3.9) and representation of the form (3.6), (3.7) do exist also for the system (2.3).

III. Let us suppose that there exists a smooth function $v(t, x, y)$, such that

$$a_j(t, y) = 1 - \varepsilon^2 v(t, x_j, y), \qquad x_j = j\varepsilon, \qquad \beta = 3\beta_0 \varepsilon^{-1}. \quad (3.10)$$

System (3.3) after the substitution of $t\prime = -\epsilon^2 t, y\prime = y + 4t, x\prime = x + 6\beta_0 t$ and passing to the continuous limit $\varepsilon \to 0$ is transformed into the equation

$$v_t = 4vv_y + 2v_x \int_0^x v_y(t, \xi, y) d\xi - v_{xxy} + \beta_0(6vv_x - v_{xxx}). \quad (3.11)$$

This equation belongs to the class of equations studied by F. Calogero and A. Degasperis [1,2,3]. In the following section we study the concrete properties of the equation (3.11).

4. Equation of interaction of Riemann breaking wave with transversal KdV long waves

I. **Physical sense.** Equation (3.11) for functions $v = v(t, y)$ takes the form of the Riemann breaking wave equation

$$v_t = 4vv_y. \tag{4.1}$$

For the functions $v = v(t, z), z = x + cy$ equation (3.11) turns into the KdV equation

$$v_t = (c + \beta_0)6vv_z - v_{zzz}. \tag{4.2}$$

So equation (3.11) describes interaction between Riemann breaking waves (4.1), propagating in the y-direction and KdV long waves (4.2), traveling in transversal directions.

II. **Hamiltonian structure.** Equation (3.11) after substitution $v = u_{x\prime}, x\prime = x - \beta_0 y, y\prime = y$ has the Hamiltonian form

$$u_{tx} = 4u_x u_{xy} + 2u_y u_{xx} - u_{xxxy}. \tag{4.3}$$

This equation takes the form

$$u_t = \partial_x^{-1}\frac{\delta H}{\delta u}, \qquad H = \int\int_{-\infty}^{\infty}\left(\frac{1}{2}u_{xxx} - u_x^2\right)u_y dxdy. \tag{4.4}$$

III. **Operator representations.** Equation (4.3) is equivalent to the following operator equation

$$L_t = 2(LL_y + L_yL) + [L, A], \qquad L = -\partial_x^2 + u_x, \qquad A = -u_y\partial_x - \partial_x u_y. \tag{4.5}$$

So according to Lemma 1 the eigenvalues $f_k(t, y)$ of the Schroedinger operator $L = -\partial_x^2 + u_x$ in view of equation (4.3) satisfy the Riemann breaking wave equation

$$f_{k_t} = 4f_k f_{k_y}. \tag{4.6}$$

If one takes the operator A of the form

$$A = -u_y\partial_x - \partial_x u_y - 2F(t, y)\partial_x, \tag{4.7}$$

where $F(t, y)$ is and arbitrary function of t, y, then from equation (4.5) one gets

$$u_{xt} = 4u_x u_{xy} + 2(u_y + F(t, y))u_{xx} - u_{xxxy}. \tag{4.8}$$

Operator equation (4.5) may be written also in the Lax form

$$L_t = [L, A - L\partial_y - \partial_y L].$$

Equations (4.3), (4.8) possess also the operator representation analogous to the zero-curvature representation

$$
\begin{aligned}
U_t &- V_x + [U,V] = 4\lambda^2 U_y,\\
U &= -i\lambda \begin{bmatrix} 1 & 0 \\ 0 & -1 \end{bmatrix} + \begin{bmatrix} 0 & 1 \\ u_x & 0 \end{bmatrix},\\
V &= -2i\lambda \begin{bmatrix} u_y + F & 0 \\ -u_{xy} & -u_y - F \end{bmatrix}\\
&+ \begin{bmatrix} -u_{xy} & 2(u_y+F) \\ 2u_x(u_y+F) - u_{xxy} & u_{xy} \end{bmatrix}.
\end{aligned} \tag{4.9}
$$

Here λ is an arbitrary spectral parameter. Equation (4.9) is the compatibility condition for linear system of equations

$$
\psi_x = U\psi, \qquad \psi_t = 4\lambda^2\psi_y + V\psi \tag{4.10}
$$

and has the commutator form

$$
\left[\partial_x - U, \quad \partial_t - 4\lambda^2\partial_y - V\right] = 0 \tag{4.11}
$$

IV. **Evolution of scattering data.** We consider the one- dimensional scattering problem, associated with the one-dimensional Schroedinger operator $L = -\partial_x^2 + u_x$. We suppose that potential $u_x(t,x,y)$ tends to zero as $x \to \pm\infty$. The primitive function $u(t,x,y)$ has the following asymptotics:

$$
\begin{aligned}
u(t,x,y) &\to g(t,y), \qquad \text{as } x \to +\infty,\\
u(t,x,y) &\to h(t,y), \qquad \text{as } x \to -\infty.
\end{aligned} \tag{4.12}
$$

Evolution of the scattering data $a(k,t,y), b(k,t,y), f_k(t,y), b_d(t,y)$ due to the equation

$$
u_{tx} = 4u_xu_{xy} + 2(u_y + F(t,y))u_{xx} - u_{xxxy} + \gamma(6u_xu_{xx} - u_{xxxx}) \tag{4.13}
$$

is described by equations

$$
\begin{aligned}
a_t - 4K^2a_y &= 2ik(g_y - h_y)a,\\
b_t - 4k^2b_y &= (2ik(g_y + h_y + 2F(t,y)) + 8\gamma ik^3)b,\\
b_{k_t} + 4\lambda_k b_{k_y} &= (-2\lambda_k(g_y + h_y + 2F(t,y)) + 8\gamma\lambda_k^3)b_k,\\
f_{t_k} &= 4f_kf_{k_y},\\
f_k &= -\lambda_k^2.
\end{aligned} \tag{4.14}
$$

These equations as $g(t,y) \equiv 0, h(t,y) \equiv 0$ coincide with equations obtained by F. Calogero and A. Degasperis [1,2,3], and in this case are integrable linear equations.

V. **N-soliton solutions.** One-soliton solution of equation (4.13) was found in [1,2,3] and has the form

$$u_x = \frac{-2\lambda^2}{\cosh^2(\lambda x - \varphi)}, \qquad \lambda_t + 4\lambda^2\lambda_y = 0, \qquad \varphi_t + 4\lambda^2\varphi_y = 4\gamma\lambda^3, \qquad \lambda^2 = -f. \tag{4.15}$$

N-soliton solutions, found in [6], are determined in accordance with Hirota's method by the formulae

$$\begin{aligned}
u(t,x,y) &= -2\frac{d}{dx}lndetA(t,x,y) - 2\sum_{n=1}^{N}\lambda_n(t,y), \\
A_{kj}(t,x,y) &= \delta_{kj} + \frac{\beta_k(t,y)}{\lambda_k+\lambda_j}e^{-(\lambda_k+\lambda_j)x}, \\
\beta_n(t,y) &= \frac{b_n(t,y)}{ia\prime(i\lambda_n)}, \\
a(k,t,y) &= \prod_{n=1}^{N}\frac{k-i\lambda_n}{k+i\lambda_n}, \\
b_{n_t} + 4\lambda_n b_{n_y} &= 8\gamma\lambda_n^3 b_n, \\
\lambda_{n_t} + 4\lambda_n^2\lambda_{n_y} &= 0, \\
-\lambda_n^2 &= f_n
\end{aligned} \tag{4.16}$$

These formulae describe breaking N-soliton solutions. The breaking of the graph of the function $u(t,x,y)$ takes place simultaneously on all axis x with the breaking of the graph for one of the functions $\lambda_n(t,y)$. A single-valued branch of the function $u(t,x,y)$ corresponds to each choice of single-valued branches of the functions $\lambda_n(t,y)$. The derivative $u_x(t,x,y)$ has the form of the N-soliton solution of the KdV equation for each branch.

Solutions (4.16) are localized on the plane x,y if functions $\lambda_n(t,y)$ exponentially tend to zero as $|y| \to \infty$. The function $\lambda(t,y)$ in formula (4.15) may be taken as smooth solution of equation $\lambda_t + 4\lambda^2\lambda_y = 0$ identically equal zero as $|y| > C$; corresponding function $u(t,x,y) = 0$ as $|y| > C$. The main difference with localized multisoliton solutions of Davey-Stewartson equation [16,17,18,19] consists of phenomenon of breaking for solutions (4.16).

VI. **Shock N-soliton solutions.** Equation (4.16) for each eigenvalue $f_k(t,y)$ is the conservation law

$$f_{k_t} - (2f_k^2)_y = 0 \tag{4.17}$$

It is possible to consider discontinuous solutions of (4.17) with Rankine-Hugoniot condition

$$s\,[f_k] = \left[-2f_k^2\right], \qquad s = -2(f_{k_+} - f_{k_-}), \tag{4.18}$$

where s denotes the speed of propagation of line of discontinuity $y = y(t)$, that is $s = dy/dt$. Corresponding N-soliton solution (4.16) has N shock waves, traveling with different speeds.

VII. **Modified $2+1$-dimensional equation.** Equation (4.3) after Miura transformation

$$u_x = v^2 + \sigma v_x \tag{4.19}$$

acquires the modified form

$$v_t = 4v^2 v_y + 2v_x \int_0^x (v^2)_y(t, \xi, y)d\xi - s v_{xxy}, \qquad s = \sigma^2. \tag{4.20}$$

Modified equation (4.20) possesses the operator representation

$$L_t = \alpha(L_y L^2 + L^2 L_y) + [L, A] \tag{4.21}$$

with matrix operator L:

$$L = \begin{bmatrix} ip_1 & 0 \\ 0 & ip_2 \end{bmatrix} \partial_x + v \begin{bmatrix} 0 & 1 \\ 1 & 0 \end{bmatrix}, \tag{4.22}$$

where $\alpha = -(p_1 - p_2)^2/2p_1p_2$.

Breaking soliton of equation (4.20) as $s = -1$ has the form

$$v = \frac{\lambda}{\cosh(\lambda x - \varphi)}, \qquad \lambda_t = \lambda^2 \lambda_y, \qquad \varphi_t = \lambda^2 \varphi_y. \tag{4.23}$$

Equations for scattering data evolution are found in [6].

VIII. **Countable set of conservation laws.** Equation (4.3) after the Gardner transformation

$$u_x = w + \varepsilon w_x + \varepsilon^2 w^2 \tag{4.24}$$

turns into the equation

$$w_t = 2(w(u_y - \varepsilon w_y))_x + (w^2 - w_{xx})_y. \tag{4.25}$$

Substituting into this equation the formal power series

$$\begin{aligned} w(t, x, y, \varepsilon) &= \sum_{n=0}^{\infty} P_n(u_x)\varepsilon^n = u_x - u_{xx}\varepsilon + (u_{xxx} - u_x^2)\varepsilon^2 + \cdots, \\ 2w(u_y - \varepsilon w_y) &= \sum_{n=0}^{\infty} Q_n(u_x, u_y)\varepsilon^n, \qquad w^2 - w_{xx} = \sum_{n=0}^{\infty} R_n(u_x)\varepsilon^n, \end{aligned} \tag{4.26}$$

we get the conservation laws

$$\frac{\partial P_n(u_x)}{\partial t} = \frac{\partial Q_n(u_x, u_y)}{\partial x} + \frac{\partial R_n(u_x)}{\partial y}. \tag{4.27}$$

Here $P_n(u_x)$ are the same differential polynomials of u_x as in the theory of KdV equation. From (4.27) the relation follows

$$\frac{d}{dt} \int \int_{-\infty}^{\infty} P_n(u_x) dx dy = - \int_{-\infty}^{\infty} P_{n+2}(u_x) dx \Bigg|_{y=-\infty}^{y=+\infty}. \tag{4.28}$$

IX. Connection with the integrable Klein-Gordon equations.

Proposition 1 *Any solution of Klein-Gordon equation*

$$\varphi_{xy} = f(\varphi), \tag{4.29}$$

where the function $f(\varphi)$ satisfies the linear equation

$$sf''(\varphi) = f(\varphi), \tag{4.30}$$

defines the solution of the modified equation

$$v_t = 4v^2 v_y + 2v_x \partial_x^{-1}(v^2)_y - s v_{xxy} \tag{4.31}$$

by the formula

$$v(t, x, y) = \frac{1}{2}\varphi_x(x + c(t), y), \tag{4.32}$$

where $c(t)$ is an arbitrary function.

Obviously there are the following non-equivalent cases of equations (4.29), (4.30):

$$s = +1: \quad \varphi_{xy} = e^{\varphi}, \quad \varphi_{xy} = \sinh\varphi, \quad \varphi_{xy} = \cosh\varphi;$$
$$s = -1: \quad \varphi_{xy} = \sin\varphi.$$

Exact solutions of the Liouville equation $\varphi_{xy} = e^{\varphi}$ lead to exact solutions of equation (4.31) ($s = +1$)

$$v(t, x, y) = \frac{1}{2}\frac{a''(x + c(t))}{a'(x + c(t))} - \frac{a'(x + c(t))}{a(x + c(t) + b(y))}, \tag{4.33}$$

which depend on three arbitrary functions $a(x), b(y), c(t)$.

References

[1] F. Calogero and A. Degasperis, Nuovo Cim. B (11), **32** (1976), 201-242.

[2] F. Calogero and A. Degasperis, Nuovo Cim. B (11), **39** (1977), 1-54.

[3] F. Calogero and A. Degasperis, in: R.K. Bullough and P.J. Caudrey (eds.), Solitons, Springer, Berlin-New York 1980, pp. 301- 323.

[4] V.E. Zakharov, Appendix Russ. transl. of R.K. Bullough and P.J. Caudrey (eds.), Solitons, Mir, Moscow, 1983.

[5] R.K. Dodd, J.C. Ellbeck, J.D. Gibbon, and H.C. Morris, Solitons and non-linear wave equations, Academic Press, London, 1982.

[6] O.I. Bogoyavlenskij, Russ. Math. Surv. vol. **45**, N 4 (1990), 1-86.

[7] O.I. Bogoyavlenskij, Math. USSR Izv., vol. **55**, N 3 (1991).

[8] G.B. Whitham, Proc. Royal. Soc. A **283** (1965), 238- 261.

[9] M. Toda, Progr. Th. Phys. Suppl. **45** (1970), 174- 200.

[10] M. Henon, Phys. Rev. **B9** (1974), 1921-1923.

[11] H. Flaschka, Phys. Rev. **B9** (1974), 1924-1925.

[12] H. Flaschka, Progr. Th. Phys. **51** (1974), 703-716.

[13] S.V. Manakov, Zh. Eksp. Theor. Fiz. **67** (1974), 543-555.

[14] M. Kac and P. van Moerbeke, Adv. Math. **16** (1975), 160-169.

[15] B.A. Dubrovin and S.P. Novikov, Russ. Math. Surv. v. **44** N 6 (1989), 35-124.

[16] M. Boiti, J.J.-P. Leon, L. Martina, F. Pempinelli, Phys Lett. **A 132** (1988), 432-439.

[17] M. Boiti, J.J.-P. Leon, F. Pempinelli, Phys. Lett. **A 141** (1989), 101-107.

[18] A.S. Fokas, P.M. Santini, Phys. Rev. Lett. **63** (1989), 1329-1333.

[19] A.S. Fokas, P.M. Santini. Physica **D 44** (1990), 99- 130.

Real and Virtual Multidimensional Solitons

M. Boiti, L. Martina, O.K. Pashaev, and F. Pempinelli*

Dipartimento di Fisica dell'Università, Sezione INFN, I-73100 Lecce, Italy
*Permanent address: Joint Institute for Nuclear Research, Dubna, Russia

Recently it has been shown that in two spatial and one temporal dimensions (2+1) there exist localized solitons [1-4]. These coherent structures display a richer phenomenology than the one dimensional solitons. Different effects have been reported successively in a series of papers [5-8]. Some of them are due to the fact that the soliton solution is structurally unstable with respect to special choices of the parameters [6]. Also some quantum-like effects as the non conservation of the number of solitons have been discovered by using direct methods [9-10]. This report is dedicated to the study of the origin and generality of these new effects in the context of the Spectral Transform (ST) theory developed in [11-12].

By choosing more general boundaries than those used in the previous papers [11,7,8] we derive an N^2-soliton solution, which is parameterized by a point in a space of 4N(N+1) real parameters. Of these parameters 2N(N+2) are determined by the choice of the boundaries and fix the velocity and the possible location of the solitons in the plane at large times, while the remaining 2N govern the dynamics of the solitons during the interaction.

The total mass of solitons is conserved but, in general, the mass of the single soliton is not preserved by the interaction. The extreme cases in which the masses of one or more solitons are zero at $t = -\infty$ or/and $t = +\infty$ are also allowed. We call these solitons with asymptotic zero masses and, consequently, with asymptotic zero amplitudes virtual solitons. The total momentum of solitons is not conserved because the boundaries act as external forces. Solitons can simulate inelastic scattering processes of quantum particles including creation and annihilation of particles.

The solitons we are considering are solutions of the so-called Davey-Stewartson I (DSI) equation, which can be written in characteristic coordinates $u = x + y$ and $v = x - y$ as follows

$$iQ_t + \sigma_3(Q_{uu} + Qvv) + [A, Q] = 0 \tag{1}$$

where Q is a 2×2 off diagonal matrix field

$$Q = \begin{pmatrix} 0 & q(u, vt) \\ r(u, v, t) & 0 \end{pmatrix} \tag{2}$$

and the diagonal 2×2 matrix field A, so called auxiliary field, is chosen to have arbitrary boundary values a_1 and a_2 according to the formula

$$A = \frac{1}{2}\begin{pmatrix} -\int_{-\infty}^{u} du'(Q^2)_v + a_1(v,t) & 0 \\ 0 & \int_{-\infty}^{v} dv'(Q^2)_u + a_2(u,t) \end{pmatrix} \tag{3}$$

Of interest is the reduced case $r = -\epsilon\bar{q}$ (the overbar means complex conjugation and $\epsilon^2 = 1$) and specifically the so called focusing case $\epsilon = 1$. Then the DSI system is a model of the evolution of shallow water waves weakly nonlinear and nearly monochromatic, when the effects of the surface tension are important. The field q is the complex envelope of the free surface wave and the auxiliary field $\sigma_3 A$ is the real velocity for the mean motion generated by the surface wave. However, more generally, the DSI equation is obtained as a universal multiscale limit of nonlinear dispersive wave equations [13,14].

In order to linearize the initial value problem for the DSI equation one needs to give the ST of Q considered as potential in the Zakharov-Shabat spectral problem

$$(\partial_x + \sigma_3\partial_y + Q)\psi = 0 \tag{4}$$

and of the boundaries a_1 and a_2 considered as potentials in two different time-dependent Schrödinger equations

$$(i\partial_t + \partial_v^2 + a_1)\phi_1 = 0 \tag{5}$$

$$(i\partial_t + \partial_u^2 + a_2)\phi_2 = 0. \tag{6}$$

The Green operator G of the corresponding spectral problems can be chosen to be sectionally holomorphic in the spectral parameter k and the values of the eigenfunctions on the two sides $\pm\Im m\, k > 0$ of the real k axis are given by integral equations of the form $\Phi^{\pm} = \Phi_0^{\pm} + G_{\Phi}^{\pm}\Phi^{\pm}$ with $\Phi = \psi, \phi_1, \phi_2$, where $\Phi_0 = \psi_0, \phi_{10}, \phi_{20}$ are solutions of the corresponding spectral problem with identically zero potential. Different choices of Φ_0 result in different analytical properties of the eigenfunction Φ and, consequently, in different spectral transforms. In [11-12] it has been shown that it is possible to define a spectral transform for the DSI equation that satisfies the following fundamental requirements: i) for vanishing boundaries values a_1 and a_2 it reduces to the ST introduced by Fokas and Ablowitz [15]; ii) its time evolution can be explicitly integrated; iii) the discrete part of the spectrum corresponds to solitons and the continuous part to radiation. The integral equation for ψ is of Volterra type and therefore the singularities of ψ in the complex plane are those of ψ_0 and of the sectionally holomorphic Green function. Consequently, ψ has simple poles only if ψ_0 has simple poles. The matrix ψ_0 is related to the eigenfunctions of the time dependent Schrödinger equations (5) and (6) and has poles for potentials a_i which are solitons of the Kadomtsev-Petviashvili (KP) equation. We deduce that solutions Q of the DSI equation have a discrete ST only for these special boundaries. The wave soliton solutions of the KP have been first described in ref. [16] and the N-soliton solution of the DSI corresponding to these boundaries has been already derived in the context of a ST satisfying the requirements i) to iii) [11,6] (for the same soliton solution in the context of a different ST theory see ref. [8]). However, this soliton solution has not enough free parameters to describe all the dynamical richness of the processes in which the total number of solitons is not conserved (for some of these special processes obtained by using direct methods see [9,10]).

We need, therefore, to consider the most general discrete spectral transforms of a_1 and a_2 [17]

$$r_1(k,l) = 2\pi i \sum_{n,m} e^{-i\bar{\lambda}_m v_{om} + i\lambda_n v_{on} - i(k^2-l^2)t} D_{nm}(\bar{\lambda}_m - \lambda_m)\delta(l-\bar{\lambda}_m)\delta(k-\lambda_n) \quad (7)$$

$$r_2(k,l) = 2\pi i \sum_{n,m} e^{i\bar{\mu}_m u_{om} - i\mu_n u_{on} - i(k^2-l^2)t} C_{nm}(\bar{\mu}_m - \mu_m)\delta(l-\bar{\mu}_m)\delta(k-\mu_n) \quad (8)$$

The complex parameters λ_n and μ_n are the discrete eigenvalues, the real parameters v_{on} and u_{on} fix the position of the nth wave soliton in the corresponding plane, and the two matrices D_{nm}, C_{nm} satisfy

$$D_{nm}^2 = C_{nm}^2 = 1, \quad D_{nm}(\bar{\lambda}_m - \lambda_m) = \bar{D}_{mn}(\bar{\lambda}_n - \lambda_n),$$
$$C_{nm}(\bar{\mu}_m - \mu_m) = \bar{C}_{mn}(\bar{\mu}_n - \mu_n). \quad (9)$$

For definiteness we choose the same number N of λ's and μ's.

Let us now consider the following discrete spectral transform $R_d(k,l)$ of Q

$$R_d(k,l) = -i\pi \begin{pmatrix} 0 & -\tau_2 e^{-i(k^2+l^2)t} \\ \tau_1 e^{i(k^2+l^2)t} & 0 \end{pmatrix} \quad (10)$$

with

$$\tau_1 = \sum_{in} \rho_{in} e^{ilu_o(l)+ikv_o(k)} [(l-\mu_n)\delta(l-\mu_n) + \sum_m C_{nm}(\bar{\mu}_m - \mu_m)\delta(l-\bar{\mu}_m)]\delta(k-\lambda_i) \quad (11)$$

$$\tau_2 = \sum_{in} \eta_{in} e^{-ilv_o(l)-iku_o(k)} [(l-\lambda_n)\delta(l-\lambda_n) + \sum_m D_{nm}(\bar{\lambda}_m - \lambda_m)\delta(l-\bar{\lambda}_m)]\delta(k-\mu_i) \quad (12)$$

where ρ and η are arbitrary complex constant matrices and $v_o(\lambda_n) = v_o(\bar{\lambda}_n) = v_{on}$, $u_o(\mu_n) = v_o(\bar{\mu}_n) = u_{on}$. The distributions $\delta(l-l_o)$ and $(l-l_o)\delta(l-l_o)$ are defined as $\iint dl \wedge d\bar{l}\delta(l-l_o)f(l) = f(l_o)$ and $\iint dl \wedge d\bar{l}(l-l_o)\delta(l-l_o)f(l) = \mathrm{Res}(f,l_o)$.

One can get explicit algebraic formulae for Q and A (for details see ref.[19]). One can prove that the reduced case $r = -\epsilon\bar{q}$, $\epsilon^2 = 1$, the one we consider in the following, is obtained if and only if $\sum_m \rho_{nm} C_{ms}(\bar{\mu}_s - \mu_s) = -\epsilon \sum_m D_{nm}(\bar{\lambda}_m - \lambda_m)\bar{\eta}_{sm}$. In the so called focusing case $\epsilon = 1$ one can prove that if the hermitian matrices $iC_{nm}(\bar{\mu}_m - \mu_m)$ and $-iD_{nm}(\bar{\lambda}_m - \lambda_m)$ have positive eigenvalues than the N^2-soliton solution and the auxiliary field A are regular.

One can show that the solution q derived in the previous paragraph describes N^2 localized coherent structures interacting in a complicated way at finite times, but moving in the far past and in the far future with constant velocities $V_{in} = (2\lambda_{i\Re}, 2\mu_{n\Re})$ $(n, i = 1, 2, ..., N)$ ($\lambda_\Re$ and $\lambda_\Im$ mean real and imaginary part of λ). It is therefore natural to call this solution the N^2-soliton solution. One can also show that this N^2-soliton solution is structurally unstable at $t = -\infty$ and at $t = +\infty$ when any couple of discrete eigenvalues λ_n, λ_m or μ_n, μ_m have the same real part or are equal (see [6]).

A relevant information in the global dynamical behaviour of the solitons is furnished by the mass of the solution $M = \int\int | q |^2 dudv$ and by the masses of the solitons at $t = \pm\infty$ $M_{in}^{(\pm)} = \int\int | q_{in}^{(\pm)} |^2 dudv$ $(i, n = 1, 2, ..., N)$ where $q_{in}^{(\pm)}$ is the asymptotic behaviour of q at $t = \pm\infty$ computed in the rest reference frame of the (i, n) soliton. In general, it results that $M = \sum_{in} M_{in}^{(-)} = \sum_{in} M_{in}^{(+)}$. However, the mass of the single soliton is not conserved and in particular it can be zero at $t = +\infty$ or at $t = -\infty$. For a special choice of the parameters the mass of the (i, n) soliton can be zero at $t = \pm\infty$ also when the coefficients ρ_{in} and η_{in} are not both equal to zero. We call these solitons with zero mass virtual solitons and they generate peculiar effects.

Because the four soliton solution displays all the richness of the general case we examine it. Of special interest are the cases in which one or more masses are zero. For definiteness, we choose the masses $M_{ij}^{(\pm)}$ $(i \neq j)$ different from zero and we consider the case in which some $M_{ii}^{(\pm)}$ are zero. The following cases are possible. For $M_{22}^{(-)} = 0$ or $M_{11}^{(-)} = 0$ or $M_{22}^{(\pm)} = M_{11}^{(-)} = 0$ the solution describes the creation of a soliton, for $M_{22}^{(+)} = 0$ or $M_{11}^{(+)} = 0$ or $M_{22}^{(\pm)} = M_{11}^{(+)} = 0$ the annihilation of a soliton, for $M_{22}^{(-)} = M_{11}^{(-)} = 0$ the creation of a pair of solitons, for $M_{22}^{(+)} = M_{11}^{(+)} = 0$ the annihilation of a pair of solitons and for $M_{22}^{(-)} = M_{11}^{(+)} = 0$ a soliton changes its mass and velocity. For all masses $M_{ii}^{(\pm)} = 0$ one gets a solution describing two interacting solitons which is, however, different from the two soliton solution obtained by taking the same boundaries and off diagonal matrices ρ and η. All these dynamical behaviours can be obtained by choosing the same boundaries. Boundaries do not give any information on the dynamics of solitons, but fix only their asymptotic kinematic, i.e. their possible locations in the plane and their velocities at large times.

Of special interest is the case in which two spectral data are equal, say $\mu_1 = \mu_2$. In general, this solution describes the interaction of two solitons. The masses $M_{ii}^{(\pm)}$ $(i = 1, 2)$ in this degenerate case depend also on the initial position of the solitons (see ref. [18]). With a special choice of the parameters one can get a solution describing the fission of a soliton and the fusion of two solitons. If in addition one chooses $\det(\rho\eta) = 0$ the solution describes a single soliton that by the interaction with a virtual soliton is forced to change velocity.

REFERENCES

1. M. Boiti, J. Léon, L. Martina and F. Pempinelli, Phys. Lett. A 132, 432 (1988)

2. M. Boiti, J. Léon, L. Martina and F. Pempinelli, "Solitons in two dimensions" in "Integrable Systems and Applications", eds. M. Balabane et al., Lecture Notes in Physics n. 342 (Springer, 1989)

3. M. Boiti, J. Léon, L. Martina and F. Pempinelli, "Localized Solitons in the Plane" in "Nonlinear Evolution Equations, Integrability and Spectral Methods", eds. A. Degasperis et al. (Manchester University Press, 1989)

4. M. Boiti, J. Léon and F. Pempinelli, J. Math. Phys. 31, 2612 (1990)

5. A. S. Fokas and P. M. Santini, Phys. Rev. Lett. 63, 1329 (1989)

6. M. Boiti, J. Léon and F. Pempinelli, Inverse Problems 6, 715 (1990)

7. P. M. Santini, Physica D 41, 26 (1990)

8. A. S. Fokas and P. M. Santini, Physica D 44, 99 (1990)

9. J. Hietarinta and R. Hirota, Phys. Lett. A 145, 237 (1990)

10. R. Hernandez Heredero, L. Martinez Alonso and E. Medina Reus, "Fusion and fission of dromions in the Davey-Stewartson equation", Preprint, April 1990 (Madrid)

11. M. Boiti, J. Léon and F. Pempinelli, Phys. Lett. A 141, 96 and 101 (1989)

12. M. Boiti, J. Léon, F. Pempinelli and A. K. Pogrebkov, "Solitons and Spectral Transform for DSI and KPI equations" in "Solitons and Applications", Eds. V. G. Makhanhov, V. K. Fedyanin and O. K. Pashaev (World Scientific, Singapore, 1990)

13. V. E. Zakharov and E. A. Kuznetsov, Physica 18 D, 455 (1986)

14. F. Calogero and V. Eckhaus, Inverse Problems 3, 27 (1987); F. Calogero and A. Maccari, "Equations of nonlinear Schrödingher type in 1+1 and 2+1 dimensions, obtained from integrable PDEs in "Inverse Problems: An Interdisciplinary Study", ed. P. C. Sabatier (Academic Press, 1987)

15. A. S. Fokas, Phys. Rev. Lett. 51, 3 (1983), A. S. Fokas and M. J. Ablowitz, J. Math. Phys. 25, 2494 (1984)

16. S. V. Manakov, V. E. Zakharov, L. A. Bordag, A. R. Its and V. B. Matveev, Phys. Lett. A 63, 205 (1977)

17. B. A. Dubrovin, T. M. Malanyuk, I. M. Krichever and V. G. Makhankov, Sov. J. Part. Nucl. 19, 252 (1988)

18. M. Boiti, J. Léon, L. Martina, F. Pempinelli and D. Perrone. "Asymptotic bifurcation of multidimensional solitons" in Proceedings of the Workshop NEEDS '90 held in Dubna (1990) (Springer-Verlag)

19. M. Boiti, L. Martina, O.K. Pashaev and F. Pempinelli, "Dynamics of multidimensional solitons", preprint (December 1990, Lecce), to be published in Phys. Lett. A.

Resolvent Approach for the Nonstationary Schrödinger Equation with Line-Type Potential*

M. Boiti, F. Pempinelli, A.K. Pogrebkov†*, and M.C. Polivanov*†

Dipartimento di Fisica dell'Università, Sezione INFN, I-73100 Lecce, Italy
†Permanent address: Steklov Mathematical Institute,
Vavilov St. 42, 117966 Moscow, GSP-1, Russia

The Spectral Transform for the nonstationary Schrödinger equation is considered for the potentials approaching zero in every direction except a finite number, which corresponds to the physical situation of long waves mutually interacting in the plane. These potentials are of relevant interest for the Davey-Stewartson-I equation as they lead to localized soliton solutions [1, 2, 3, 4, 5]. In order to overcome the problems which arise for such potentials in the standard approach, we introduce a new technique based on the notion of the resolvent operator. We show, that the resolvent function even in this case of the line-type potentials is a well defined object with necessary analytic properties which enables us to get automatically the proper generalizations of Jost functions and Spectral Data. This talk is continuation of the works [6, 7, 8, 9], the complete version will be published elsewhere.

We give a description of the resolvent method having a general validity in two dimensional scattering. Here we apply it, as an example, to the nonstationary Schrödinger equation

$$(-i\partial_y + \partial_x^2 - u(x,y))\Phi = 0, \tag{1}$$

where $u(x,y)$ is a smooth real function of the line-type

$$u(x,y) = u_0(x,y) + \sum_{j=1}^{N} u_j(x - 2\lambda_j y) \tag{2}$$

with smooth decaying functions $u_0(x,y)$ and $u_j(x)$ resp. in two and one variables and parameters $\lambda_1 \cdots \lambda_N$. In the case of absence of one-dimensional potentials this problem is thoroughly studied (see e.g. [10, 11, 12]).

The resolvent operators $\gamma_\pm$ with kernels $\gamma_\pm(x,y,x',y')$ are defined by the equations $(\mathbf{I} - \mathbf{G}_\pm)\gamma_\pm = \mathbf{I}$, or their dual: $\gamma_\pm(\mathbf{I} - \mathbf{G}_\pm) = \mathbf{I}$. Here $\mathbf{I}$ is the identity integral operator (product of δ-functions), $\mathbf{G}_\pm$ are integral operators with retarded and advanced Green functions $G_\pm(x,y,x',y')$ of the eq. (1) as kernels (see, e.g. [10]).

More precisely we consider a resolvent function μ as given by the integral equation

*Work supported in part by M.U.R.S.T.

$$\mu(\mathbf{p},\mathbf{q},p',q') = \delta(p')\,\delta(q') + \frac{1}{\mathbf{q}-\mathbf{p}^2}\iint d\tilde{p}d\tilde{q}\,v(\tilde{p},\tilde{q})\,\mu(\mathbf{p}-\tilde{p},\mathbf{q}-\tilde{q},p'+\tilde{p},q'+\tilde{q}), \tag{3}$$

or its dual

$$\mu(\mathbf{p},\mathbf{q},p',q') = \delta(p')\,\delta(q') + \iint d\tilde{p}d\tilde{q}\,\mu(\mathbf{p},\mathbf{q},\tilde{p},\tilde{q})\,\frac{v(\tilde{p}-p',\tilde{q}-q')}{\mathbf{q}+\tilde{q}-(\mathbf{p}+\tilde{p})^2}. \tag{4}$$

Here v is the Fourier transform of the potential u and boldface letters means that the corresponding arguments are in complex domain. Then the Fourier transform of the functions $\gamma_\pm(x,y,x',y')$ are given as $\tilde{\gamma}_\pm(p,q,p',q') = \lim_{\mathbf{q}_\Im\to\pm 0}\mu(p,\mathbf{q},p'-p,q'-q)$, $q=\mathbf{q}_\Re$. The function $\mu(\mathbf{p},\mathbf{q},p',q')$ is not analytic in the two complex variables $\mathbf{p}$ and $\mathbf{q}$. However, there are two and only two cases, when this function is analytic in one complex variable $\mathbf{k}$ for $\mathbf{k}_\Im \neq 0$: $\mu(p,q+\mathbf{k},p',q')$ and $\mu(p+\mathbf{k},q+\mathbf{k}^2,p',q')$ with real p and q.

The resolvent function due to reality of the potential $u(x,y)$ obeys the relation

$$\frac{\mu(\mathbf{p}+p',\mathbf{q}+q',-p',-q')}{\mathbf{q}-\mathbf{p}^2} = \frac{\overline{\mu(\overline{\mathbf{p}},\overline{\mathbf{q}},p',q')}}{\mathbf{q}+q'-(\mathbf{p}+p')^2} \tag{5}$$

and due to the eqs. (3) or (4) the asymptotic expansion in $\mathbf{q}$

$$\mu(\mathbf{p},\mathbf{q},p',q') = \delta(p')\,\delta(q') + \frac{1}{\mathbf{q}}\,v(-p',-q') + o\Big(\frac{1}{\mathbf{q}}\Big). \tag{6}$$

In order to introduce the Jost solutions we notice that $\mu(\mathbf{p},\mathbf{q},p',q')$ has singularity at the point $\mathbf{q}=\mathbf{p}^2$ of the order 1. Let us consider the 'residuum' μ at this point:

$$B_\pm(p',q',\mathbf{k}) = \lim_{q\to\pm 0} q[\mu(\mathbf{k},q+\mathbf{k}^2,p',q') - \delta(p')\,\delta(q')]. \tag{7}$$

These functions satisfy the integral equation

$$B_\omega(p,q,\mathbf{k}) = v(-p,-q) + \iint\frac{d\tilde{p}d\tilde{q}}{\tilde{p}+i\omega\mathbf{k}_\Im 0}\,v(\tilde{p}-p,\tilde{q}-q)\left(\frac{\tilde{p}\,B_\omega(\tilde{p},\tilde{q},\mathbf{k})}{\tilde{q}-\tilde{p}(\tilde{p}+2\mathbf{k})}\right). \quad \omega=\pm, \tag{8}$$

The eq. (5) under limiting procedure (7) suggests to introduce the function

$$A_\pm(p',q',\mathbf{k}) = \lim_{q\to\pm 0}\mu(p'+\mathbf{k},q+q'+\mathbf{k}^2,-p',-q'), \tag{9}$$

which due to the eq. (5) results to be related to B_ω by the equation

$$A_\omega(p,q,\mathbf{k}) = \delta(p)\,\delta(q) + \frac{1}{p+i\omega\mathbf{k}_\Im 0}\left(\frac{p\,\overline{B_\omega(p,q,\overline{\mathbf{k}})}}{q-p(p+2\mathbf{k})}\right), \quad \omega=\pm. \tag{10}$$

This function can be considered to be defined by the integral equation

$$A_\omega(p,q,\mathbf{k}) = \delta(p)\delta(q) +$$

$$\frac{1}{p+i\omega\mathbf{k}_\Im 0}\left(\frac{p}{q-p(p+2\mathbf{k})}\iint d\tilde{p}d\tilde{q}\,v(\tilde{p},\tilde{q})A_\omega(p-\tilde{p},q-\tilde{q},\mathbf{k})\right), \tag{11}$$

obtained from (3) in the limit (9). By comparing (11) with (10) one gets the additional relation

$$\overline{B_\omega(p,q,\overline{\mathbf{k}})} = \iint d\tilde{p}d\tilde{q}\, v(\tilde{p},\tilde{q})\, A_\omega(p-\tilde{p},q-\tilde{q},\mathbf{k}). \tag{12}$$

In the case of decaying potentials $A_\omega(p,q,\mathbf{k})$ is independent of ω and after Fourier transform and multiplication by flat exponent coincides with the standard Jost solution of the eq. (1). In the case of line-potentials we have resp. two sets of generalized Jost solutions connected by the relation

$$A_-(p,q,\mathbf{k}) = \overline{T(\overline{\mathbf{k}})}\, A_+(p,q,\mathbf{k}), \tag{13}$$

where T is defined through

$$-\,2\pi i\,\mathrm{sign}\,\mathbf{k}_\Im \lim_{p\to 0} \frac{p\,B_+(p,q,\mathbf{k})}{q-p(p+2\mathbf{k})} = \left(\frac{1}{T(\mathbf{k})}-1\right)\delta(q). \tag{14}$$

Using the well known property of the resolvent that the solution of integral equation of the type (3) with some other inhomogeneous term is given as a convolution of the resolvent with this term, we have scalar products

$$\iint d\tilde{p}d\tilde{q}\,\overline{A_\omega(\tilde{p},\tilde{q},\overline{\mathbf{k}}+p)}\, A_{\omega'}(\tilde{p}+p,\tilde{q}+q,\mathbf{k})$$
$$= \delta(p)\,\delta(q)\left[\delta_{\omega,-\omega'} + \delta_{\omega,\omega'}\,(T(\mathbf{k}))^{-\omega'}\right], \qquad \omega,\omega' = \pm, \tag{15}$$

where $\delta_{\omega,\omega'}$ is the Kronecker symbol, and representation of the resolvent in terms of Jost solutions:

$$\mu(\mathbf{p},\mathbf{q},p',q') = \delta(p')\,\delta(q') + \iint d\tilde{p}d\tilde{q}\,T(\mathbf{p}-\tilde{p})\,\frac{A(\tilde{p},\tilde{q},\mathbf{p}-\tilde{p})\,\overline{B(p'+\tilde{p},q'+\tilde{q},\mathbf{p}-\tilde{p})}}{\mathbf{q}-\tilde{q}-(\mathbf{p}-\tilde{p})^2}. \tag{16}$$

Here only one set of Jost solutions corresponding to the sign '+' was used due to the eq. (13) and we omit the subscript '+' in what follows.

The last equation allows us to find the discontinuity of the function $A(p,q,\mathbf{k})$ analytic in the upper and bottom half-planes at the real axis. So let us introduce boundary values:

$$A^\sigma(p,q,k) = \lim_{\sigma\mathbf{k}_\Im\to+0} A(p,q,\mathbf{k}), \quad B^\sigma(p,q,k) = \lim_{\sigma\mathbf{k}_\Im\to+0} B(p,q,\mathbf{k}),$$
$$T^\sigma(k) = \lim_{\sigma\mathbf{k}_\Im\to+0} T(\mathbf{k}), \quad \sigma = \pm, \quad k = \mathbf{k}_\Re,$$
$$\tilde{\Phi}^\sigma(p,q,k) = A^\sigma(p-k,q-k^2,k), \quad \tilde{\Xi}^\sigma(p,q,k) = B^\sigma(p-k,q-k^2,k),$$

where $\tilde{\Phi}^\pm(p,q,k)$ are Fourier transforms of the Jost solutions $\Phi^\pm(x,y,k)$ of the eq. (1) and $\tilde{\Xi}^\pm(p,q,k)$ are Fourier transforms of the products $\overline{u(x,y)\Phi^\mp(x,y,k)}$.

The the discontinuity on the real axis can be written as

$$\mathbf{T}^\pm\,\Phi^\pm = \mathcal{F}^\mp\Phi^\mp, \tag{17}$$

where $\mathbf{T}^\pm$ and $\mathcal{F}$ are integral operators acting on the spectral parameter k. Their kernels are $T^\pm(k,p) = T^\pm(k)\,\delta(k-p)$ and

$$\mathcal{F}^{-\sigma} = \mathbf{R}^\sigma_\pm \mathbf{R}^{\sigma\,\dagger}_\pm. \tag{18}$$

Here the kernels of the integral operators $\mathbf{R}^{\sigma}_{\pm}$ are given by

$$
\begin{aligned}
R^{\sigma}_{\pm}(k,p) &= \frac{1}{2}\left(1+T^{\sigma}(k)\pm Z^{\sigma}(k)\right)\delta(k-p) \\
&\quad \mp\vartheta(\mp\sigma(k-p))\,r^{\sigma}(k,p), \quad \sigma=+,-, \\
Z^{\pm}(k)\,\delta(k-p) &= 2i\pi\left[T^{\pm}(k)\,\overline{\widetilde{\Xi}^{\mp}(p,p^2,k)} - \frac{T^{\pm}(k)}{k-p}\left(\overline{(k-p)\,\widetilde{\Xi}^{\mp}(p,p^2,k)}\right)\right], \\
r^{\pm}(k,p) &= -\frac{2i\pi\,T^{\pm}(k)}{k-p}\left(\overline{(k-p)\,\widetilde{\Xi}^{\mp}(p,p^2,k)}\right).
\end{aligned}
$$

The above equations concretize the heuristic formulae suggested in [6, 7]. These Spectral Data obey characterization equations $T^{\pm}(k)=\overline{T^{\mp}(k)}$, $\overline{Z^{\mp}(k)}=-Z^{\pm}(k)$ and

$$
\mathbf{R}^{\sigma}_{+}\,\mathbf{R}^{\sigma\,\dagger}_{+} = \mathbf{R}^{\sigma}_{-}\,\mathbf{R}^{\sigma\,\dagger}_{-}, \quad \mathbf{R}^{\sigma}_{\pm}\,\mathbf{R}^{-\sigma\,\dagger}_{\mp} = \mathbf{T}^{\sigma}, \quad \mathbf{R}^{-\sigma\,\dagger}_{\mp}\,\mathbf{T}^{\sigma\,-1}\,\mathbf{R}^{\sigma}_{\pm} = \mathbf{I}, \quad \sigma=+,-. \tag{19}
$$

It is easy to check that the operators $\mathcal{F}$ are characterized by the following equations

$$
\mathcal{F}^{\pm\,\dagger} = \mathcal{F}^{\pm}, \quad \mathcal{F}^{\pm}\,\mathbf{T}^{\pm\,-1}\,\mathcal{F}^{\mp} = \mathbf{T}^{\mp}. \tag{20}
$$

Evidently the above formulae need more detailed discussion. In particular $T^{\pm}(k)$ are given in terms of one dimensional Transition coefficients corresponding to the potentials u_j in the eq. (2). Obviously our formulae are generalizations for the line-type case of the ones known [10, 11, 12] for the standard case ($T^{\pm}(k)\equiv 1$, $Z^{\pm}(k)\equiv 0$). In this situation $r^{\pm}(k,k) = -2i\pi\,\overline{B^{\mp}(0,0,k)}$ is a generating functional of the integrals of motion of the KPI equation.

References

1. M. Boiti, J. Léon, L. Martina and F. Pempinelli, *Phys. Lett.* **A132** (1988) 432.

2. A. S. Fokas and P. M. Santini, *Phys. Rev. Lett.* **63** (1989) 1329; A. S. Fokas and P. M. Santini, *Physica* **D44** (1990) 99.

3. M. Boiti, J. Léon and F. Pempinelli, *Inverse Problems* **6** (1990) 715.

4. M. Boiti, L. Martina, O. K. Pashaev and F.Pempinelli, *"Dynamics of multidimensional solitons"*, *preprint Lecce*, December, 1990.

5. M. Boiti, J. Léon and F. Pempinelli, *Phys. Lett.* **A141** (1989) 96 and 101.

6. M. Boiti, J. Léon, F. Pempinelli and A. K. Pogrebkov, *"Solitons and spectral transform for DSI and KPI equations"*, in *"Solitons and applications"*, eds. V. G. Makhankov, V. K. Fedyanin and O. K. Pashaev, pp. 19 – 30, World Scientific Pub. Co., Singapore (1990).

7. F. Pempinelli, *"Solitons and spectral transform for the Kadomtsev - Petviashvili equation"*, in *"Inverse methods in action"*, ed. P. C. Sabatier, pp. 578-583, Springer-Verlag, Berlin (1990).

8. M. Boiti, F. Pempinelli, A. K. Pogrebkov and M. C. Polivanov, *Inverse Problems* **7** (1991) 43.

9. M. Boiti, F. Pempinelli, A. K. Pogrebkov and M. C. Polivanov, *"Resolvent approach for the nonstationary Schrödinger equation (standard case of rapidly decreasing potential)"*, in *Proceedings of the seventh Workshop on Nonlinear Evolution Equations and Dynamical Systems (NEEDS '91)*, World Scientific Pub. Co., Singapore (1992)

10. V. E. Zakharov, S. V. Manakov, *Sov. Sci. Rev. – Phys. Rev.* **1** (1979) 133; S. V. Manakov, *Physica* **D3** (1981) 420.

11. A. S. Fokas and M. J. Ablowitz, *Stud. Appl. Math.* **69** (1983) 211.

12. Xin Zhou, *Comm. Math. Phys.* **128** (1990) 551.

Lattice Construction of Quantum Integrable Systems

D.A. Coker

Institute for Theoretical Physics,
State University of New York at Stony Brook, USA

The construction of quantum integrable systems is performed using the Quantum Inverse Scattering Method. The R-matrix is preserved which produces a class of integrable models on the lattice whose continuum limit is the original continuum model. The Sine-Gordon model is used as an example.

1. Introduction

The Quantum Inverse Scattering Method (QISM)[1,2,6] has been successfully used to solve many quantum integrable systems (i.e. Heisenberg Magnet, Nonlinear Schrödinger Equation, Sine-Gordon model and Massive Thirring model). Their solution comes from the existence of an R-matrix which is a solution of the Yang-Baxter equation. A next step in the development of QISM was the construction of lattice versions[2,4] of these quantum models. For relativistic models, this is desirable since this provides a natural way of regulating ultraviolet divergences. There are several ways of putting an integrable model on the lattice. Unfortunately, not all of them preserve the integrability.

In this short overview, the method and results of Izergin and Korepin[4] for constructing the lattice version of an integrable quantum model will be reviewed. The key feature of their method is the preservation of the R-matrix. Thus the integrable structure of the model under consideration is kept which guarantees that critical exponents and scattering matrices are also preserved. Since the R-matrix is known, both the L-operator and the transfer matrix can be obtained. At certain values of the spectral parameter, the L-operator becomes a projector or inverse projector. This allows one to construct the integrable lattice Hamiltonian by means of trace identities[3] and express it as a function of the transfer matrix at these special values of the spectral parameter. For classical integrable systems, the lattice Hamiltonian is local and the integrals of motion coincide with the continuum model since there is no difference between the projector and the inverse projector.

Unfortunately, the quantum lattice Hamiltonian is generally quasilocal since the projector and inverse projector differ. Another feature of this method is that the L-operator becomes an explicit function of the lattice spacing Δ. Through analytic continuation in Δ, L-operators for other models having the same R-matrix can be also constructed. In the continuum limit, $\Delta \to 0$, the continuum Hamiltonian is regained. Hence the lattice formulation provides a nice way of defining the continuum model.

This method will be demonstrated by putting the integrable quantum Sine-Gordon (SG) model on the lattice following[4]. This is an integrable relativistic model which exhibits an ultraviolet divergence that is successfully regularized by the lattice. The lattice integrable Hamiltonian constructed from trace identities will be seen to be quasilocal rather than local. It should be noted, however, that the classical Sine-Gordon model remains local on the lattice.

2. Quantum Sine-Gordon Model

The Hamiltonian of the quantum Sine-Gordon model is given by

$$H = \int dx \left\{ \frac{p^2}{2} + \frac{(\partial_x u)^2}{2} + \frac{m^2}{\beta^2}(1 - \cos\beta u) \right\} \tag{1}$$

where β is the coupling constant $(0 < \gamma = \beta^2/8 < \pi)$ and the canonical fields u and p are quantum bosonic operators satisfying $[u(x), p(y)] = i\delta(x - y)$.

In putting the Hamiltonian (1) on the lattice, we wish to maintain integrability even though Lorentz invariance is lost. This is equivalent to leaving the R-matrix unchanged. Directly discretizing the above Hamiltonian produces a nonintegrable Hamiltonian which is undesirable. Thus the lattice quantum SG model will be constructed in such a way that the R-matrix is unchanged. Since the SG model is UV divergent, we also wish to construct the model in such a way that the lattice spacing regulates this divergence.

To define the model on the lattice, the local quantum fields $u_n(\simeq u)$ and $p_n(\simeq \Delta p)$ will be introduced where Δ is the lattice spacing. These fields obey the following commutation relation, $[u_n, p_m] = \delta_{nm}$. The R-matrix for the lattice SG model is chosen to be the same as in the continuum

$$R(\lambda, \mu) = \begin{pmatrix} \sinh(\alpha + i\gamma) & 0 & 0 & 0 \\ 0 & i\sin\gamma & \sinh\alpha & 0 \\ 0 & \sinh\alpha & i\sin\gamma & 0 \\ 0 & 0 & 0 & \sinh(\alpha + i\gamma) \end{pmatrix} \tag{2}$$

with $\exp(\alpha) = \lambda/\mu$ where λ and μ are spectral variables. An L-operator satisfying the intertwining relation with the R-matrix

$$R(\lambda, \mu) \left[L(n|\lambda) \bigotimes L(n|\mu) \right] = \left[L(n|\mu) \bigotimes L(n|\lambda) \right] R(\lambda, \mu) \tag{3}$$

is given by

$$L(n|\lambda) = \begin{pmatrix} e^{-i\beta p_n/8} \rho_n e^{-i\beta p_n/8} & \frac{m\Delta}{2} \sinh(\lambda - i\beta u_n/2) \\ -\frac{m\Delta}{2} \sinh(\lambda + i\beta u_n/2) & e^{i\beta p_n/8} \rho_n e^{i\beta p_n/8} \end{pmatrix} \tag{4}$$

where

$$\rho_n = \sqrt{1 + 2s\cos(\beta u_n)}, \qquad s = (\frac{m\Delta}{4})^2. \tag{5}$$

When $\Delta \to 0$, this L-operator agrees with the infinitesimal L-operator[2,6]. The symmetry of this L-operator is expressed as

$$\sigma_y L^*(\lambda^*)\sigma_y = L(\lambda) \tag{6}$$

where σ_y is a standard Pauli matrix. The quantum determinant of the L-operator is given by

$$d_q(\lambda) \equiv \det_q L(n|\lambda) = 1 + s(\lambda^2 e^{-i\gamma} + \lambda^{-2} e^{i\gamma}). \tag{7}$$

Now that the L-operator is known, the monodromy matrix can be constructed using

$$T(\lambda) = L(N|\lambda) \ldots L(2|\lambda)L(1|\lambda). \tag{8}$$

The transfer matrix, τ, (the trace of the monodromy matrix T) can now be used to generate the integrals of motion using trace identities.

The next step is to define a lattice Hamiltonian that in the continuum limit goes to the original Hamiltonian (1). This is accomplished by noting that $L(n|\lambda)$ becomes a projector at two points

$$\lambda^2 = \kappa_\pm \quad (d_q(\kappa_\pm) = 0) \tag{9}$$

and an inverse projector at two points

$$\lambda^2 = \nu_\pm \quad (d_q(\nu_\pm) = 0) \tag{10}$$

with

$$\kappa_\pm = \nu_\pm^* \quad \nu_\pm = -b^{\pm 1} e^{i\gamma} \quad b = 2s\left[1 + \sqrt{1 - 4s^2}\right]^{-1}. \tag{11}$$

The fact that these four values do not coincide implies that the lattice Hamiltonian will become quasilocal. The Hamiltonian is expressed in terms of the transfer matrix at these four values (9)(10)(11)

$$H_L = \frac{m^2\Delta}{16\sin\gamma} \sum_\pm \left(\frac{\partial}{\partial \lambda^{\pm 2}} F(\lambda^2 = \kappa_\pm) + \frac{\partial}{\partial \lambda^{\pm 2}} G(\lambda^2 = \nu_\pm) \right)$$

$$F(\lambda^2) = \ln\left[d_q^{-N/2}(\lambda)\tau(\lambda)\right] \quad G(\lambda^2) = \ln\left[d_q^{-N/2}(\lambda^{-1})\tau(\lambda)\right]. \tag{12}$$

N is the number of lattice sites. This Hamiltonian, both Hermitian and quasilocal, is integrable. Each site of the lattice interacts with all other sites with the strength decreasing as

$$\left(\frac{\sin\gamma}{2}\right)^{|n-m|} \tag{13}$$

where $|n-m|$ is the distance between sites.

In the quasiclassical limit, locality is restored because there is no difference between the projector and the inverse projector in the classical lattice SG model. In the limit $\Delta \to 0$, it can be easily shown that H_L (12) goes to H (1). However, it is not necessary to take the continuum limit to analyze this quantum system as H_L is still integrable.

The lattice SG model can be solved by means of QISM. The eigenvectors of the Hamiltonian H_L (and consequently the transfer matrix) can be constructed using the Algebraic Bethe Ansatz whereby the vacuum takes on a two site form. The spectral parameters are then found to obey the following set of Bethe equations

$$\left[\frac{d_q(\lambda_l)}{d_q(\lambda_l^{-1})}\right]^{N/2} = \prod_{k \neq l}^{m} \left(\frac{\lambda_k^2 e^{-i\gamma} - \lambda_l^2 e^{i\gamma}}{\lambda_k^2 e^{i\gamma} - \lambda_l^2 e^{-i\gamma}}\right) \tag{14}$$

where the λ_l are all different. The fact that $d_q(\lambda)/d_q(\lambda^{-1})$ depends on the lattice spacing, Δ, again demonstrates how different models with the same continuum limit can be constructed from the same L-operator.

In the continuum limit, the scattering matrix for particle-particle, particle-string, string-string and soliton-soliton have been computed explicitly[5]. Since the scattering matrix depends on the R-matrix, the scattering matrix on the lattice should be similar to that in the continuum.

3. Conclusion

The lattice quantum Sine-Gordon model has been constructed preserving integrability. This is of primary importance if the discrete system is to be analyzed. However, the resulting Hamiltonian was not local. This feature of lattice models has been overcome in a formal manner[7], but not explicitly for the SG model.

It should also be mentioned that the R-matrix (2) can also be used to generate a class of spin models on the lattice. These spin models are local with only nearest neighbor interactions. However, in the continuum limit, these models do not reduce to the SG model.

I would like to thank V.E. Korepin and I. Ivanov for useful suggestions.

References

[1] L.D. Faddeev, *Sov. Sci. Rev. Math. Phys.*, **C1**, 107, (1981).
[2] L.D. Faddeev, in *Les Houches 1982*, 561–608, eds. J.B. Zuber and R. Stora.
[3] A.G. Izergin, V.E. Korepin and F.A. Smirnov, *Teor. Mat. Fiz.* , **48**, 319, (1981).
[4] A.G. Izergin and V.E. Korepin, *Nucl. Phys.* **B 205**, 401–413, (1982).
[5] V.E. Korepin, *Teor. Mat. Fiz.* **41**, 953–966, (1979).
[6] V.E. Korepin, A.G. Izergin and N.M. Bogoliubov, *Quantum Inverse Scattering Method and Correlation Functions. Algebraic Bethe Ansätz* (Cambridge, to appear in 1982).
[7] V.O. Tarasov, L.A. Takhtajan and L.D. Faddeev, *Teor. Mat. Fiz.* **57**, 1059–1073, (1983).

Combining Dressing and Nonlocality

A. Degasperis

Dipartimento di Fisica, Università di Roma "La Sapienza",
P. le Aldo Moro 2, I-00185 Roma, Italy

The known hierarchies ILW_k of integro-differential evolution equations, which generalize the Intermediate Long Wave equation, are derived within the dressing method. The nonlocality due to the integral terms leads us to introduce formal expansions in powers of an operator-valued spectral parameter which does not commute with the space-variable x. Results obtained by D.Lebedev, M.Olshanetsky, S.Pakuliak, A.Perelomov, P.M.Santini and the author[1,2] are shortly reported here.

1. Introduction

By nonlocal evolution equations we understand here equations which are differential in the time variable, but not so in the space variable with respect to which they are integro-differential (or just purely integral). Known examples are the Benjamin-Ono (BO) equation

$$u_t = Hu_{xx} + 2uu_x \ , \qquad u = u(x,t) \ , \tag{1}$$

the Intermediate Long Wave (ILW) equation

$$u_t = Tu_{xx} + 2uu_x \ , \qquad u = u(x,t) \ , \tag{2}$$

and the Sine-Hilbert (SH) equation

$$Hu_t = sinu \ , \qquad u = u(x,t) \ , \tag{3}$$

where the convolution integral operators H and T are defined by the formulae

$$Tf(x) = (1/2\hbar)P.V. \int_{-\infty}^{+\infty} dy\, coth[\pi(y-x)/2\hbar] f(y) \ , \tag{4}$$

$$Hf(x) = \frac{1}{\pi} P.V. \int_{-\infty}^{+\infty} dy \, \frac{f(y)}{y-x} \tag{5}$$

Leaving aside the applicative relevance of eqs.(1) and (2) in fluid dynamics[3,8], these three equations are examples of nonlocal integrable equations; since they are, like the well-known Kortewed-de Vries equation[9], the compatibility condition for a pair of linear equations with a spectral parameter[10,13].

In order for the evolution equation to be nonlocal rather than partial differential, the associated spectral problem turns out to be a Riemann-Hilbert (RH) type problem rather than a differential equation. Thus, for instance, if L is the differential operator of order k

$$L = \sum_{n=0}^{k} u_n(x,t)\partial_x^n \ , \qquad u_k = 1 \ , \qquad k = 1,2,\ldots, \tag{6}$$

the spectral problem is that of finding a solution $\psi(z, \lambda, t)$, analytic in the strip $-2\hbar < Imz < 0(\hbar > 0)$, whose boundary values

$$\psi^+(x,\lambda,t) = \lim_{\varepsilon\to 0^+} \psi(x - i\varepsilon, \lambda, t) \ , \qquad x\epsilon\mathbf{R} \tag{7a}$$

$$\psi^-(x,\lambda,t) = \lim_{\varepsilon\to 0^+} \psi(x - 2i\hbar + i\varepsilon, \lambda, t) \ , \qquad x\epsilon\mathbf{R} \tag{7b}$$

satisfy the differential relation

$$L\psi^+ = \lambda\psi^- \ , \qquad \psi^\pm = \psi^\pm(x,\lambda,t) \ , \qquad x\epsilon\mathbf{R} \ . \tag{8}$$

Here $\hbar$ is a positive parameter, which may be considered as measuring the extent of the nonlocality (in the $\hbar = 0$ limit, indeed, the RH problem (8) reduces to an ordinary differential equation since $\psi^+ = \psi^-$). In particular the ILW equation (2) obtains with $k = 1$ (say L is first order, see (6)) and the BO equation is its limit for $\hbar \to \infty$ (while the KdV equation is its local limit for $\hbar \to 0$). More generally, by ILW$_k$ (i.e. generalized Intermediate Long Wave) equations we understand the hierarchy of evolution equations corresponding to the spectral problem (8), with (6). In the following we shortly show how the dressing method, as applied to the ILW$_2$ equations (with obvious generalization to ILW$_k$), naturally leads to introduce formal power expansions in the operator valued spectral variable

$$\hat{\lambda} \equiv \lambda exp(-2i\hbar\,\partial_x) \ , \tag{9}$$

whose commutator with the space variable x is nonvanishing, as it reads $[x, \hat{\lambda}] = 2i\hbar\hat{\lambda}$.

Dressing

In analogy with the local case, the dressing method in the λ-dependent version[14] requires the second-order differential relation (to simplify the notation, we set $\psi^+ \equiv \psi$)

$$\psi_{xx} + u_1(x,t)\psi_x + u_0(x,t)\psi = \hat{\lambda}\psi \ , \qquad \psi = \psi(x,\lambda,t) \tag{10}$$

to be rewritten in the 2×2 matrix first-order form

$$\mathcal{L}(\hat{\lambda})\begin{pmatrix}\psi_x\\ \psi\end{pmatrix} = 0 \ , \qquad \mathcal{L}(\hat{\lambda}) \equiv \partial_x - \begin{pmatrix}0 & \hat{\lambda}\\ 1 & 0\end{pmatrix} + \begin{pmatrix}u_1 & u_0\\ 0 & 0\end{pmatrix} \ . \tag{11}$$

Note that, because of the analyticity of $\psi(x,\lambda,t)$, we make use of the shift operator to write $\psi^-(x,\lambda,t) = exp(-2i\hbar\partial_x)\psi^+(x,\lambda,t)$. Solutions of equation (11) are obtained by dressing the "naked" equation

$$\mathcal{L}_0\begin{pmatrix}ik\\ 1\end{pmatrix} exp(ikx) = 0 \ , \qquad \lambda = -k^2 exp(-2\hbar k) \ , \tag{12a}$$

$$\mathcal{L}_0 \equiv \partial_x - \hat{\Lambda} \ , \hat{\Lambda} \equiv \begin{pmatrix}0 & \hat{\lambda}\\ 1 & 0\end{pmatrix} \ , \tag{12b}$$

by the dressing transformation

$$\begin{pmatrix} \psi_x \\ \psi \end{pmatrix} = G \begin{pmatrix} ik \\ 1 \end{pmatrix} e^{ikx} \tag{13}$$

which implies

$$G\mathcal{L}_0 G^{-1} = \mathcal{L}(\hat{\lambda}) , \tag{14}$$

where G is a 2×2 operator-valued matrix. Indeed, we prove that G has the formal power expansion

$$G = \sum_{n=0}^{\infty} \begin{pmatrix} K_n & K_{nx} \\ 0 & K_n \end{pmatrix} \hat{\Lambda}^{-n} \tag{15}$$

where the functions $K_n(x,t)$ are the boundary values at $z = x - i0$ of the expansion coefficients of the analytic pseudo-differential operator

$$K(z,t,\partial_z) = 1 + \sum_{j\geq 1} K_j(z,t)\partial_z^{-j} \tag{16}$$

which dresses the scalar version of spectral problem, i.e.

$$K^- \partial_x^2 (K^+)^{-1} = \partial_x^2 + u_1\partial_x + u_0 \tag{17}$$

with $K^+ = K(x - i0, t, \partial_x), K^- = K(x - 2i\hbar + i0, t, \partial_x)$.

The dependence on time is obtained by dressing the (trivial) commutator

$$[\partial_t - \hat{\Lambda}^s, \mathcal{L}_0] = 0 \tag{18}$$

which becomes

$$[\partial_t - G_t G^{-1} - G\hat{\Lambda}^s G^{-1}, \mathcal{L}(\hat{\lambda})] = 0 , \tag{19}$$

and by requiring that the coefficients of the negative powers of $\hat{\lambda}$ of the expansion of $G_t G^{-1} - G\hat{\Lambda}^s G^{-1}$ vanish, namely

$$[G_t G^{-1} + G\hat{\Lambda}^s G^{-1}]_- = 0 . \tag{20}$$

In this respect, we prove the following interesting Lemma: if the dressing operator G satisfies the evolution equation (20), then the scalar dressing operator K solves the KP hierarchy equation (here $(\cdot)_-$ means the expansion contribution from negative powers of ∂_x)

$$K_t K^{-1} + (K\partial_x^2 K^{-1})_- = 0 \tag{21}$$

and viceversa. As implied by (20), the operator

$$G_t G^{-1} + G\hat{\Lambda}^s G^{-1} = V(x,t,\hat{\lambda}) = V_0(x,t) + V_1(x,t)\hat{\lambda} + \ldots \tag{22}$$

is polynomial in the noncommutative spectral parameter $\hat{\lambda}$, and therefore (19) yields the lifting of the ILW_2 class of equations to the zero curvature representation with the spectral parameter

$$[\partial_t - V, \mathcal{L}(\hat{\lambda})] = 0 . \tag{23}$$

Moreover, we prove that this equation at $\lambda = 0$ does indeed coincide with the evolution equation constructed in the hamiltonian formalism[15–17], namely (in that notation) that

$$V_0(x,t) = Grad_q\, \tilde{\ell}_x \tag{24}$$

and, therefore,

$$\mathcal{L}_t(0) = [Grad_q \tilde{\ell}_x, \mathcal{L}(0)] \ , \tag{25}$$

where, of course, $\mathcal{L}(0) = \partial_x + \begin{pmatrix} u_1 & u_0 \\ -1 & 0 \end{pmatrix}$.

While we refer the reader to reference 1. for details and proofs, we conclude with the following three remarks:
1) the connection with the KP hierarchy (21) allows one to construct solutions of the ILW equation by applying appropriate conditions on solutions of the KP equation[18]. 2) The present investigation of nonlocal equations can be extended to include the class ILW_k for any integer k, and the nonlocal partners of the generalized MKdV equations and the two-dimensional Toda lattice[2]. 3) Going from the local equations to their nonlocal parteners, the algebra in which the motion takes place is deformed according to the nonvanishing of the parameter $\hbar$. For instance, in the simplest case of the ILW_1 class, we obtain the sinh-algebra commutation relations

$$[T(\vec{n}), T(\vec{m})] = sinh[\hbar(\vec{n} \times \vec{m})]T(\vec{n}+\vec{m}) \ , \tag{26}$$

where $\vec{n}$ and $\vec{m}$ are two-dimensional vectors with integer components, $\vec{n} \times \vec{m} = n_1 m_2 - n_2 m_1$ and

$$T(\vec{n}) = \frac{1}{2} exp(\hbar n_1 n_2) exp(i n_2 x) \hat{\lambda}^{n_1} \tag{27}$$

are the generators. This is an example of the so-called cross product Lie algebras[19].

References

1. Degasperis A., Lebedev D., Olshanetsky M., Pakuliak S., Perelomov A. and Santini P.M., *"Generalized intermediate long wave hierarchy in zero-curvature representation with noncommutative spectral parameter"* Preprint MPI/91-63, Max-Planck-Institute für Mathematik, Bonn.
2. Degasperis A., Lebedev D., Olshanetsky M., Pakuliak S., Perelomov A. and Santini P.M., *Commun.Math.Phys.* to be published.
3. Joseph R.I., *J.Phys.A.* **101**, 1225 (1977).
4. Kubota T., Ko D.R.S. and Dobbs D., *Hydronaut.* **12**, 157 (1978).
5. Joseph R.I. and Egri R., *J.Phys.A.* **11**, L97 (1978).
6. Chen H.H. and Lee Y.C., *Phys.Rev.Lett.* **43**, 264 (1979).
7. Benjamin T.B., *J.Fluid.Mech.* **29**, 559 (1967).
8. Ono H., *J.Phys.Soc. Japan* **39**, 1082 (1975).
9. Gardner C.S., Green J.M., Kruskal M.D. and Miura R.M., *Phys.Rev.Lett.* **19**, 1095 (1967).
10. Satsuma J., Ablowitz M.J. and Kodama Y., *Phys.Lett. A* **73**, 283 (1979).
11. Kodama Y., Ablowitz M.J. and Satsuma J., *J.Math.Phys.* **23**, 564 (1982).
12. Fokas A.S. and Ablowitz M.J., *Stud.Appl.Math.* **68**, 1 (1983).
13. Degasperis A., Santini P.M. and Ablowitz M.J., *J.Math.Phys.* **26**, 2469 (1985).
14. Zakharov V.E. and Shabat A.B., *Funct.Anal.* **13**, 13 (1979).
15. Lebedev D.R. and Radul A.O., *Commun.Math.Phys.* **91**, 543 (1983).

16. Drinfeld V. and Sokolov V., *Journal of Soviet Math.* **30**, 1975 (1984).
17. Lebedev D. and Pakuliak S., Preprint GEF-Th-9/1990.
18. Krichever I.M. *"The averaging procedure for the soliton-like solutions of integrable systems"*
in "Mechanics, Analysis and Geometry:200 years after Lagrange", M. Francaviglia (Editor) Elsevier Science Publishers, 1991 pp.99-125."
19. Saveliev M.V. and Vershik A.M., *Phys.Lett.* **143A**, 121 (1990).

Initial Boundary-Value Problems for Soliton Equations

A.S. Fokas

Department of Mathematics and Computer Science and
Institute of Nonlinear Studies, Clarkson University,
Potsdam, NY 13699-5815, USA

A method is presented for linearizing initial-boundary value problems for integrable nonlinear evolution equations with the spatial variable on an half-infinite line. This method yields the solution of a nonlinear equation in terms of the solution of two linear integral equations, whose analysis for large t, shows how the boundary conditions can generate solitons.

1 Introduction

Perhaps the most important open analytical problem associated with the soliton theory in one space variable, is the solution of nonlinear evolution equations on a finite domain. A first step in this direction would be the solution of equations on the half-infinite line; however, in spite of intense scrutiny by many researchers (see for example [1]), even this problem remains essentially open (the case of very particular homogeneous boundary conditions is studied in [2]-[5]).

A method is presented here for studying initial-boundary value problems on the semi-infinite line. This method is illustrated for the nonlinear Schrödinger (NLS) equation; also its application to the KdV equation and to the N-wave interaction equations, is briefly summarized. It was shown in [6], that initial-boundary value problems on the semi-infinite line, are similar to certain forced problems, where the forcing is of a distribution type. It is therefore not surprising that this method can also be used for the linearization of such forced problems.

Let us summarize the main steps of the method as applied to the NLS equation

$$iq_t + q_{xx} - 2\lambda|q|^2 q = 0, \quad x, t \epsilon [0, \infty); \lambda = \pm 1, \tag{1}$$

where $q(x,0)$ and $q(0,t)$ are given, they decay for large x and t, and they satisfy the necessary compatibility condition to ensure the existence of solution at $x = 0$, $t = 0$. The cases $\lambda = 1$, and $\lambda = -1$ will be referred to as the defocusing, and focusing cases, respectively. The NLS equation is the compatibility condition [7] of the Lax pair [8]

$$w_x + ikJw = Qw, \quad w_t + Uw = 0; U \doteqdot 2ik^2 J + i\lambda|q|^2 J - 2kQ + iQ_x J, \tag{2}$$

where $J = diag(1, -1)$ and Q is an off-diagonal matrix with 12 and 21 entries given by q and $\lambda\bar{q}$ respectively (throughout this paper bar denotes complex conjugate).

The first step of the method involves solving a direct and an inverse problem associated with the x-part of the Lax pair, i.e. equation (2.a). Let the 2×2 matrices $\varphi(x,t,k)$ and $\psi(x,t,k)$ be the solutions of equation (2.a) specified by the boundary conditions $\varphi(0,t,k) = I$ and $\lim_{x\to\infty} \exp(ikxJ)\ \psi(x,t,k) = I$ respectively. Writing Volterra integral equations for these eigenfunctions, and letting $\Phi = \varphi \exp(ikxJ)$, $\Psi = \psi \exp(ikxJ)$, it follows that Φ and Ψ have analytic extensions in the complex k-plane, given by $\Phi = (\Phi^+, \Phi^-)$ and $\Psi = (\Psi^-, \Psi^+)$. This notation means that the first column of the matrix Φ is analytic for $k_I \geq 0$ (where $k = k_R + ik_I$); similarly the second column of Φ is analytic for $k_I \leq 0$. These matrices are related by the scattering equation $\Phi = \Psi \exp(-ikx\hat{J})$ $(\psi(0,t,k))^{-1}$, where $\hat{J}f$ denotes the commutator of J with f and hence $(\exp \hat{J})f = (\exp J) f \exp(-J)$. The first column of this equation is

$$\frac{\Phi^+(x,t,k)}{\psi_2^+(0,t,k)} - \Psi^-(x,t,k) = -\frac{\psi_2^-(0,t,k)}{\psi_2^+(0,t,k)} e^{2ikx}\Psi^+(x,t,k), \quad k\epsilon\mathbf{R}. \tag{3}$$

The second column yields a similar equation, which is however redundant in lieu of the symmetry relations $\overline{\Phi_1^+(k)} = \Phi_2^-(\bar{k})$, $\overline{\Phi_2^+(k)} = \lambda\Phi_1^-(\bar{k})$, $\overline{\Psi_1^-(k)} = \Psi_2^+(\bar{k})$, $\overline{\Psi_2^-(k)} = \lambda\Psi_1^+(\bar{k})$. Equation (3), together with the above symmetry conditions, defines a Riemann-Hilbert (RH) problem, which, because of the underline symmetry, is always solvable (i.e. there exist a vanishing lemma [9]). To solve this RH problem one needs $\rho(t,k) = \psi_2^-(0,t,k)/\psi_2^+(0,t,k)$ for $k\epsilon\mathbf{R}$, the zeros of $\psi_2^+(0,t,k)$ for $k_I > 0$, and the k-derivative of ψ_2^+ at these zeros. In the defocusing case ($\lambda = 1$), ψ_2^+ cannot have zeros, and the knowledge of $\rho(t,k)$ is sufficient for the solution of the above RH problem. Having these data, which are usually called the scattering data, equation (3) can be solved via a linear integral equation. This problem will be referred to as the x-inverse problem.

The second step of the method involves finding $\psi(0,t,k)$. The eigenfunction $\psi(x,t,k)$ satisfies $\psi_t + U\psi = \psi c(t)$. Evaluating this equation as $x \to \infty$, it follows that $c(t) = 2ik^2 J$. Hence $\psi_t(0,t,k)$ solves

$$\psi_t(0,t,k) + 2ik^2[J, \psi(0,t,k)] = \hat{Q}(0,t,k)\psi(0,t,k), \tag{4}$$

where $\hat{Q}(0,t,k) = 2kQ(0,t) - i\lambda|q|^2(0,t)J - iQ_x(0,t)J$. The main difficulty associated with the existence of nontrivial boundary conditions appears at this stage. Since $Q(0,t)$ and $Q_x(0,t)$ are non zero, the time evolution of the scattering data is non trivial. Furthermore, $Q(0,t)$ is given but $Q_x(0,t)$ is unknown. A possible approach to this problem is the following: Eliminating $Q_x(0,t)$ from equation (4), one obtains two **nonlinear** equations for the four entries of $\psi(0,t,k)$. One then uses the fact that ψ has analyticity properties in k to supplement these two equations with two more equations involving Hilbert transforms. In this way one obtains two equations for $\psi_2^+(0,t,k)$ and $\psi_2^-(0,t,k)$, but these equations are nonlinear integrodifferential equations. Here, we follow a different approach: The main idea of our method is to solve equation (4) by formulating for this equation an inverse scattering problem, which will be referred to as the t-inverse problem. Let $\hat{\Phi}(t,k)$, $\hat{\Psi}(t,k)$ satisfy the same differential equation as $\psi(0,t,k)$ (i.e. equation (4)), and let them be specified by the boundary conditions $\hat{\Phi}(0,k) = I$ and

$\lim_{t\to\infty} \exp(2ik^2tJ)\hat{\Psi}(t,k) = I$, respectively. (We use the notation $\wedge$, to help the reader appreciate the analogy between Φ, Ψ and $\hat{\Phi}$, $\hat{\Psi}$). Writing Volterra integral equations for these eigenfunctions, it follows that $\hat{\Phi} = (\hat{\Phi}^+, \hat{\Phi}^-)$ and $\hat{\Psi} = (\hat{\Psi}^-, \hat{\Psi}^+)$, where when $+$ occurs with $\wedge$ it denotes analyticity in the first and third quadrant of the complex k-plane (similarly when $-$ occurs with $\wedge$ it denotes analyticity in the second and fourth quadrants). The first column of the matrix equation $\hat{\Phi} = \hat{\Psi}\exp(-2ik^2t\hat{J})\hat{\Psi}(0,k)^{-1}$, is

$$\frac{\hat{\Phi}^+(t,k)}{\hat{\Psi}_2^+(0,k)} - \hat{\Psi}^-(t,k) = -\frac{\hat{\Psi}_2^-(0,k)}{\hat{\Psi}_2^+(0,k)} e^{4ik^2t}\hat{\Psi}^+(t,k), \quad k^2 \epsilon \mathbf{R}. \tag{5}$$

The second column is again redundant since $\hat{\Phi}$ and $\hat{\Psi}$ satisfy the same symmetry relations as Φ and Ψ. Equation (5) expresses the solution of the t-inverse problem in terms of a RH problem. To solve this problem one needs $\hat{\rho}(k) \doteqdot \hat{\Psi}_2^-(0,k)/\hat{\Psi}_2^+(0,k)$ for $k^2 \epsilon \mathbf{R}$, the zeros of $\hat{\Psi}_2^+(0,k)$ in the first and third quadrant, and the k-derivative of $\hat{\Psi}_2^+(0,k)$ at these zeros. It turns out that some of these data are uniquely determined from $\psi(0,0,k)$, i.e. from $q(x,0)$. Indeed, $\psi(0,t,k)$ and $\hat{\Psi}^+(t,k)$ satisfy the same ODE (equation (4)), thus, in the domains of their definition, are simply related. For example, $\psi^+(0,t,k) = \alpha\hat{\Psi}^+(t,k)$ $+\beta\hat{\Psi}^-(t,k)\exp(-4ik^2t)$. Evaluating this relationship at $t=0$, it follows that

$$\psi^+(0,t,k) = \psi_2^+(0,0,k)\frac{\hat{\Psi}^+(t,k)}{\hat{\Psi}_2^+(0,k)}, \quad k_R \geq 0, \quad k_I \geq 0. \tag{6.a}$$

Similarly,

$$\psi^+(0,t,k) = \psi_2^+(0,0,k)\hat{\Phi}^-(t,k) + \psi_1^+(0,0,k)\hat{\Phi}^+(t,k)exp(-4ik^2t), \quad k_R \leq 0, \quad k_I \geq 0. \tag{6.b}$$

Equation (6.a) and its complex conjugate evaluated at $t=0$, imply that some of the jump conditions of the RH problem (5) in the first and fourth quadrant can be evaluated in terms of $\psi^+(0,0,k)$, i.e. in terms of the initial data. The rest of the jump conditions can in principle be specified from $q(0,t)$.

For brevity of presentation we concentrate in the case of $q(x,0) = 0$, which implies $\psi(0,0,k) = I$. We assume that $q(0,t)$ gives rise to N bound states k_j, $j = 1,\ldots,N$, i.e. $\hat{\Psi}_2^+(0,k_j) = 0$. Since $\hat{\Psi}_2^+(0,k) = 1$, for k in the first quadrant, it follows that k_j can occur only in the third quadrant, i.e. both k_{j_R} and k_{j_I} are negative. Equation (5) yields

$$\hat{\Psi}^-(t,k) = \begin{pmatrix} 1 \\ 0 \end{pmatrix} + \sum_{j=1}^{N} \frac{c_j}{k-k_j} e^{4ik_j^2t}\hat{\Psi}^+(t,k_j) - \frac{1}{2i\pi}\int dk' \frac{\hat{\Psi}_2^-(0,k')\hat{\Psi}^+(t,k')e^{4ik'^2t}}{k'-k}, \tag{7}$$

where the integral is along the positive imaginary axis and along the negative real axis. We note that, since $\hat{\Psi}^+(0,k') = (0,1)^T$, the evaluation of the second component of the integral equation (7) at $t=0$ implies an identity for $\hat{\Psi}_2^-(0,k)$, while the evaluation of the first component implies a relationship between $c_j\hat{\Psi}_1^+(0,k_j)$ and the k_l's. In what follows we discuss separately the focusing and defocusing cases.

(i) If $\lambda = 1$, all the c_j's are zero. For any given function $\hat{\Psi}_2^-(0,k')$, equation (7) implies $\hat{\Psi}^-(t,k)$. Then equations (6) imply $\psi^+(0,t,k)$ and hence the RH problem (3) can be solved for $\Psi^+(x,t,k)$. Finally, the solution of the NLS equation (1) follows from $q(x,t) = 2i \lim_{k\to\infty} \Psi_1^+(x,t,k)$. As $t \to \infty$, a stationary phase analysis of (7) implies that $\hat{\Psi}_2^-(t,k) = O(1/\sqrt{t})$, and hence $q(x,t)$ disperses away. This discussion shows that the general solution of the boundary value problem for the defocusing NLS equation can be expressed in terms of two linear integral equations uniquely defined in terms of $\hat{\Psi}_2^-(0,k')$. Given $q(0,t)$, $\hat{\Psi}^-(0,k')$ can in principle be found from the requirement that $q(0,t) = 2i \lim_{k\to\infty} \hat{\Psi}_1(t,k)$.
(ii) If $\lambda = -1$, the c_j's can be nonzero, depending on $q(0,t)$. For any given function $\hat{\Psi}_2^-(0,k')$, and constants c_j and k_j ($k_{j_R} < 0$, $k_{j_I} < 0$), equation (7) implies $\hat{\Psi}^-(t,k)$. Then the jump conditions for the RH problem (3) are known. Furthermore, using equations (6), $\psi_2^+(0,t,k)$ is known for $k_I > 0$ and its possible zeros can be determined. Then, the solution of the RH problem (3) yields $\Psi^+(x,t,k)$, and $q(x,t)$ follows as before. As $t \to \infty$, equation (7) implies that the main contribution comes from the discrete spectrum. Furthermore, it can be shown that if k_j is a zero of $\hat{\Psi}_2^+(0,k)$, then as $t \to \infty$, $\bar{k}_j$ is a zero of $\psi_2^+(0,t,k)$. Therefore, if $q(0,t)$ supports N discrete eigenvalues $k_1, ..., k_N$ then $q(x,t)$ will decompose into N solitons for large t, generated by $\bar{k}_1, ..., \bar{k}_N$. Since $k_{j_R} < 0$, these solitons travel away from the boundary.

As an example, let $j = 1$. Then equation (7) yields for large t,

$$\hat{\Psi}_2^-(t,k) \sim \frac{c_1}{k-k_1}\frac{e^{4ik_1^2 t}}{(1+A)}, \quad \hat{\Psi}_2^+(t,k) \sim 1 - \frac{2ik_{1_I}}{k-\bar{k}_1}\frac{A}{(1+A)},$$

$$\hat{\Phi}_2^-(t,k) \sim \frac{k-\bar{k}_1}{k-k_1}\hat{\Psi}_2^+(t,k), \tag{8}$$

where $A = |c_1|^2 \exp[4i(k_1^2 - \bar{k}_1^2)t]/4(k_{1_I})^2$. Also $\psi_2^+(0,t,k) = \hat{\Psi}_2^+(t,k)$ for k in the first quadrant, and since the zero of $\hat{\Psi}_2^+(t,k)$ occurs at $\nu \doteq k_R - ik_I(1-A)/(1+A)$ which belongs to the second quadrant, $\psi_2^+(0,t,k)$ does not have a zero in the first quadrant. $\psi_2^+(0,t,k)$ is $\hat{\Phi}_2^-(t,k)$ for k in the second quadrant and it has a zero at $k = \nu$; as $t \to \infty$, $A \to 0$ and $\nu \to \bar{k}_1$. Furthermore, as $t \to \infty$, $\hat{\Psi}_2^-/\hat{\Psi}_2^+$ is proportional to $\exp(-4i\bar{k}_j^2 t)/(k - \bar{k}_j)$. Using this expression in equation (3) we find a soliton solution moving away from $x = 0$.

We conclude with some remarks:

1. It is straightforward to make the above exact analysis rigorous: The direct problem satisfies Volterra integral equations, thus for $q(x,0)$ and $q(0,t)$ in $L^1(\mathbf{R}^+)$, these equations are always solvable. Also, both the x and the t inverse problems are formulated via RH problems which satisfy a certain Schwartz reflection symmetry, therefore if the jump functions are in $H^1(\mathbf{R})$ these RH problems are always solvable [9]. To make the above asymptotic analysis rigorous [10] one should use the recent work of [11].

2. For convenience of presentation we have presented the inverse problems as RH problems; this formulation demands strong decay for large x and large t. One can improve upon these results by using a formulation in terms of a Gel'fand-

Levitan integral equation [12]. (This equation can be obtained by taking the Fourier transform of the integral equation associated with the RH problem.)
3. There exist conceptual similarities between the problem studied here and the generation of dromions for the Davey-Stewartson I equation [13]: In both problems, it is the spectrum of the t-part of the Lax pair that determines the emergence of coherent structures.
4. The application of the method to the solution of the N-wave interactions on the half-infinite line is as follows. Let

$$Q_t = \hat{c}\hat{J}^{-1}Q_x + [\hat{c}\hat{J}^{-1}Q, Q],$$

where Q is an $N \times N$ off-diagonal matrix, and J, c are diagonal matrix with real, distinct entries. The x-part of the Lax pair associated with this equation is $w_x = ikJw + Qw$ and leads to an x-inverse problem which can be solved via the linear integral equation

$$\mu(x,t,k) = I + P^+\mu(x,t,k)e^{ikx\hat{J}}(\Psi^-(0,t,k)-I) - P^-\mu(x,t,k)e^{ikx\hat{J}}(\Psi^+(0,t,k)-I),$$

where $P^\pm$ denote the usual projectors in the k-complex plane (we assume for simplicity no bound states). This integral equation is uniquely defined in terms of $\Psi^\pm(0,t,k)$, which are in turn defined in terms of $T(t,k)$ through the equations $T_\pm(t,k) = T(t,k)\Psi^\pm(0,t,k)$, and the requirement that T_+, Ψ^-, and T_-, Ψ^+ are upper and lower triangular matrices respectively, and $\Psi^\pm$ have diagonal parts equal to unity. The evolution of T is given by $T_t = ik\hat{c}T - T\hat{c}\hat{J}^{-1}Q(0,t)$. In this case, in contract to the NLS equation, the evolution of T is completely determined, so one does not need to consider a t-inverse problem in order to find T. However, if one wants to analyze the behavior of the solution $Q(x,t)$, one still needs a spectral analysis of the above evolution equation in t. This situation is precisely analogous to the one for the Davey-Stewartson I equation.
5. The application of this method to the solution of the KdV equation is as follows. Let s, v, ψ be the solutions of the underline x-part of the Lax pair, i.e. of $w_{xx} + w(q+k^2)w = 0$, specified by the boundary conditions $s(0,t,k) = 0$, $s_x(0,t,k) = -2ik$; $v(0,t,k) = 2$, $v_x(0,t,k) = 0$; $\lim_{x\to\infty}\exp(-ikx)\psi = 1$. by writing Volterra integral equations for these eigenfunctions it follows that: (i) s is odd in k, and v is even in k; (ii) $s \sim e^{-ikx}F(t,k) - e^{ikx}F(t,-k)$ as $x \to \infty$, and v satisfies a similar expression with F replaced by G and $F(t,-k)$ by $-G(t,-k)$; (iii) $S \doteqdot se^{ikx}$, $\Psi \doteqdot \psi e^{-ikx}$, F, are analytic for $k_I \geq 0$; (iv) $F(t,k) = \psi(0,t,k)$, $ikG(t,k) = \psi_x(0,t,k)$, $F(t,k)G(t,-k) + F(t,-k)G(t,k) = 2$. Using these facts it follows that the x-inverse problem can be formulated as the RH problem

$$\frac{S(x,t,k)}{F(t,k)} - \Psi(x,t,-k) = -\frac{F(t,-k)}{F(t,k)}\Psi(x,t,k)e^{2ikx}.$$

For real q, $F(t,-k) = \overline{F(t,k)}$, hence, the proper scattering data are $\theta(t,k) = \arg F(t,k)$, the zeros of $F(t,k)$ for $k_I > 0$, and the k-derivative of F at these zeros. To find $F(t,k)$, one again formulates a t-inverse problem: $T_t - 4ik^3\hat{J}T = QT$, where T is a 2×2 matrix with 11, 21, 12, and 22 entries given by $\chi = F + G$, $\Omega = F - G$, $\bar{\Omega}$, and $\bar{\chi}$, respectively, and Q is a 2×2 matrix with entries

$$Q_{11} = -\frac{i(q_{xx}(0,t) + 2q^2(0,t))}{2k}, \quad Q_{21} = q_x(0,t) - 2ikq(0,t) + ikQ_{11},$$

$$Q_{21} = \bar{Q}_{12}, \quad Q_{22} = \bar{Q}_{22}.$$

Acknowledgements

This is part of a large program of study with several colleagues, including J. Bona, S. Brenner, A. Its, L. Sung and S. Venakides. The author is grateful to I.M. Gel'fand for his constant encouragement. This work was partially supported by the National Science Foundation under Grant Number DMS-8803471 and by the Air Force Office of Scientific Research under Grant Number 87-0310.

References

[1] C.K. Chu and R.L. Chou, Advances in Applied Mechanics, **27** 283-302 (1990); D.J. Kaup and P. Wycoff, Stud. Appl. Math. **81** 7-20 (1989).

[2] A.S. Fokas, Physics D **35**, 167-185 (1989).

[3] M.J. Ablowitz and H. Segur, J. Math. Phys. **16** 1054 (1975).

[4] E.K. Sklyanin, Funk. Analiz. (Func. Anal. Appl.) **21** (2), 86-7 (1987).

[5] R.F. Bikbaev and A.R. Its, Matem. Zametki (Math. Notes) **45** (3) 3-10 (1989); V.O. Tarasov, Inverse Problems **7** 435-449 (1991).

[6] A.S. Fokas and M.J. Ablowitz, Stud. in Appl. Math. **80**, 253-272 (1989).

[7] V.E. Zakharov and P.B. Shabat, Sov. Phys. JETP **34**, 62 (1972).

[8] P.D. Lax, Comm. Pure Appl. Math., **21** 467 (1968).

[9] X. Zhou, SIAM J. Math. Anal., **20** 966-986 (1989).

[10] S. Brenner, A.S. Fokas, A.R. Its, and L.Y. Sung, Analytical, Asymptotic, and Numerical Study of the NLS Equation on the Half-Infinite Line, (in preparation).

[11] P. Deift and X. Zhou, to appear in Annals of Math. (1991).

[12] I.M. Gel'fand and B.M. Levitan, Amer. Math. Soc. Transl. Ser. 2, **1** 253 (1955).

[13] A.S. Fokas and P.M. Santini, Phys. Rev. Lett. **63** 1329 (1989); Physica D **44** 99 (1990).

Non-Perturbative Two-Dimensional Quantum Gravity and the Isomonodromy Method

A.S. Fokas[1], *A.R. Its*[†], *and A.V. Kitaev*[2]

[1]Department of Mathematics and Computer Science and Institute of Nonlinear Studies, Clarkson University, Potsdam, NY 13699-5815, USA

[2]Department of Mathematics, Institute of Aircraft Instrument Engineering, Gertzen 67, 190000 St. Petersburg, Russia

[†]On leave of absence from University of St. Petersburg, Russia

1 Introduction

One of the main analytical achievements of the inverse spectral method is the implementation of global asymptotic analysis of the integrable nonlinear equations. This includes the explicit calculation of the asymptotics for the solution of the Cauchy problem, the explicit description of the connection formulae for the asymptotics in different regions, etc. The efficiency of the inverse spectral method for studying these questions was first demonstrated in the work of V.E. Zakharov and S.V. Manakov [1].

One of the versions [2,3] of the asymptotic analysis of integrable nonlinear equations can be represented as follows:

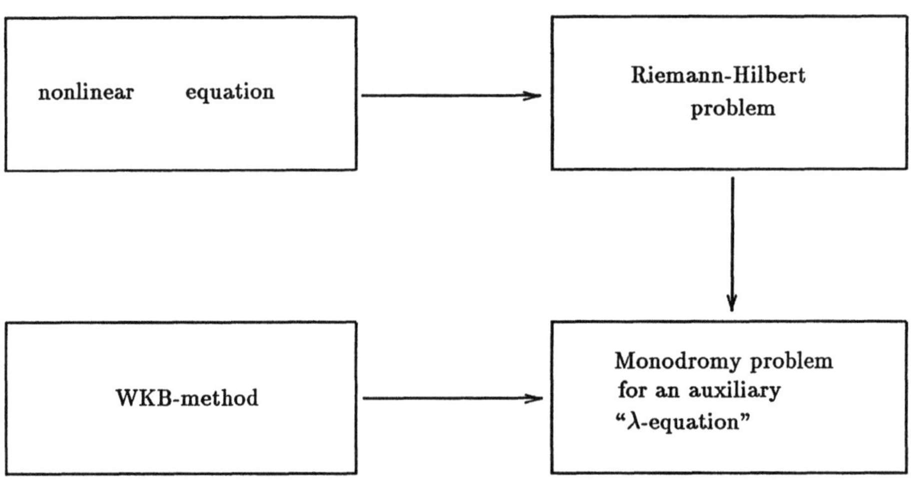

(1)

During the last ten years several well known integrable equations have been investigated by this approach.

In this note we present the results of the application of scheme (1) to a new situation that arose in connection with the two-dimensional quantum gravity.

We begin with the basic definitions of the classical theory of the orthogonal polynomials. Let us consider the orthogonal polynomial $Q_n(\lambda)$ with respect to the measure $\exp(-U(\lambda))d\lambda$:

$$\delta_{nm} = \int_{-\infty}^{\infty} Q_n(\lambda)Q_m(\lambda)e^{-U(\lambda)}d\lambda$$

where

$$U(\lambda) = \sum_{j=1}^{N} t_j\lambda^{2j}, \quad t_N > 0.$$

It is well known that $Q_n(\lambda)$ satisfy the linear recurrence relations

$$(L \cdot Q)_n \equiv \sum L_{nm}Q_m = \lambda Q_n$$

$$L_{nm} = \frac{1}{2}w_m^{\frac{1}{2}}\delta_{n+1,m} + \frac{1}{2}w_n^{\frac{1}{2}}\delta_{n-1,m}. \tag{2}$$

The coefficients w_n are functions of t_j, $w_n \equiv w_n(t_1,...,t_N)$, and satisfy [4-6] the nonlinear recurrence relation

$$n = \frac{1}{2}w_n^{\frac{1}{2}}(U'(L))_{n,n-1}, \quad n > 0, \tag{3}$$

supplemented with the initial conditions

$$w_0 = 0, w_1 = \frac{4\int_{-\infty}^{\infty}\lambda^2 e^{-U(\lambda)}d\lambda}{\int_{-\infty}^{\infty}e^{-U(\lambda)}d\lambda},$$

$$w_j = \frac{4h_j}{h_{j-1}}, j = 1,...,N-1, Q_j = \frac{1}{\sqrt{h_k}}\lambda^j + ... \tag{4}$$

Also, it has been known [9] that $w_n(t_1,...,t_N)$, as function of both n and $\{t_j\}$, satisfies the hierarchy of the Volterra equations:

$$\partial_{t_j}\ln w_n = L^{2j}_{n-1,n-1} - L^{2j}_{n,n}. \tag{5}$$

Both equations (3) and (5) admit the Lax-representations ((3) – [10-13]; (5) – [7,8])

$$[L, U^1_-(L)] = 1 \tag{3'}$$

$$L_{t_j} = [L^{2j}_- + \frac{1}{2}L^{2j}_0, L] \tag{5'}$$

where M_0 means the diagonal part of the operator M, and M_- means the lower diagonal part.

To summarize: the recurrence coefficients $w_n(t_1,...,t_N)$ are the special solutions of the Volterra hierarchy (5) satisfying the additional constraints (3), (4). Let us denote as $\tau_n(t_1,...,t_N)$ the corresponding τ-function. During the last two years this τ-function has been the object of investigation of a large number of papers. The reason is the works [5], [6] where the following two facts were discovered:

1. Let

$$t_j = \beta q_j, q_1 = \frac{1}{2}$$

and consider the following double-scaling limit of (3):

$$\begin{aligned} \beta &= C_1 h^{-5} \\ & \qquad\qquad h \to 0. \\ \tfrac{n}{\beta} &= C_2 + C_1^{-1} h^4 x \end{aligned} \tag{6}$$

It turns out that it is possible to choose $C_2 = C_2(q_2, ..., q_N)$ in such a way that

$$\tau_n \to Z(x)$$

where $Z(x)$ is the partition function for the two-dimensional quantum gravity. The variable x plays the role of the renormalized string coupling.

2. Assume that under the limit (6)

$$w_n \cong \rho(1 - 2h^2 u(x)). \tag{7}$$

Then, it is possible to choose $\rho = \rho(q_2, ..., q_N)$ in such a way that the ansatz (7) maps the equations (3) into the PI equation for the function $u(x)$:

$$u_{xx} = 6u^2 + x \tag{8}$$

(note, that $Z_{xx} = u$). Furthermore, the limiting equation (8) is the same for all of the $q_2, ..., q_N$ in the general position. At the same time, there exists a sequence of sets $M_m = \{q_2, ..., q_N\}$, $codim M_m = m - 2$, such that the limit (6,7) does not exist even on the formal level. For $(q_2, ..., q_N) \epsilon M_m$ one must consider the asymptotic ansatz

$$\beta = C_1 h^{-4-2/m}, \quad \frac{n}{\beta} = C_2 + C_1^{-1} h^4 x, w_n \cong \rho(1 - 2h^{4/m} u(x)),$$

and replace the equation (8) by

$$[H, A_m] = 1 \tag{9}$$

where $H = -\frac{d^2}{dx^2} + u(x)$ and A_m is A-operator of m-th KdV equation. In this case, τ_n tends to the partition function of the two-dimensional quantum gravity coupling with the m-th minimal conformal model.

In connection with these exciting results the following analytical problems arise:

1. To give the rigorous confirmation of the ansatz (7).
2. To calculate the parameters of the limiting solution $u(x)$.

It turns out [13-16], that using the general scheme (1) of the asymptotic analysis of the integrable systems, one can obtain complete answers for both these questions.

Regarding the first question, the answer is negative. Actually, this has already been known indirectly form [17]. Nevertheless, if one replaces in (4)

$$\int_{-\infty}^{\infty} d\lambda \text{ by } s_1 \int_{-\infty e^{\frac{i\pi}{N}}}^{\infty e^{\frac{i\pi}{N}}} d\lambda - s_2 \int_{-\infty e^{\frac{-i\pi}{N}}}^{\infty e^{\frac{-i\pi}{N}}} d\lambda, \quad s_1 \neq s_2,$$

then the answer becomes positive. Let's suppose that such modification has been accomplished. Then, the limiting solution $u(x)$ of PI equation can be uniquely characterized by the following asymptotic behavior on the ray $\arg x = \pi - \frac{2\pi}{5}$:

$$u(x) = e^{-\frac{i\pi}{5}} \sqrt{\frac{|x|}{6}} + \gamma_0 |x|^{-\frac{1}{8}} e^{-\frac{8i}{5}(\frac{3}{2})^{1/4}|x|^{5/4}} + o(|x|)^{-\frac{1}{8}}$$

$$\gamma_0 = -\frac{i}{\sqrt{8\pi}} e^{\frac{\pi i}{20}} (\frac{2}{3})^{\frac{1}{8}} \frac{p}{1+p}, \quad p = -\frac{s_2}{s_1}, \quad p \neq -1.$$

In [6] it was shown that $u(x)$ belongs to the one-parameter family of solutions of PI characterized by the following asymptotic behavior as $x \to -\infty$:

$$u(x) = \sqrt{\frac{-x}{6}} + \sum_{l=1}^{\infty} c_l (-x)^{\frac{1}{2} - \frac{5l}{2}} + \gamma (-x)^{-\frac{1}{8}} e^{-\frac{8}{5}(\frac{3}{2})^{1/4}(-x)^{5/4}} (1 + o(1)).$$

One of our main results is the calculation of the parameter γ,

$$\gamma = \frac{i}{2} \frac{1}{\sqrt{8\pi}} (\frac{2}{3})^{\frac{1}{8}} \frac{1-p}{1+p}.$$

This parameter describes the nonperturbative effect. We were informed by A. Orlov, that for the particular case $N = 2$, $p = 0$, this formula was also obtained in [18]. It should be mentioned, that the case $p = 0$ corresponds to the so-called "triply truncted solution" of PI. This solution has infinitely many poles only in the sector $\frac{7}{5}\pi < \arg x < \frac{9}{5}\pi$. Also, note that in order for $u(x)$ to be real, one needs $|p| = 1$.

To get the formulae for parameters γ_0, γ we made essential use of the asymptotic analysis of the PI equation developed in [24].

Remark: The theory of the general string equations (9) has been treated via the isomonodromy approach in [19,20]. The nonperturbative effect for $m = 3$ has been calculated in [21]. The original approach to the string equations (9) based on the methods of algebraic geometry was proposed in [22], [23]. The nice idea of considering equations (9) as the quantization of finite-gap potentials was put forward in [22] and [20].

References

[1] V.E. Zakharov and S.V. Manakov, Asymptotical Behavior of Nonlinear Wave Systems Integrable by the Inverse Scattering Method, Zh. Exp. Teor. Fiz. **71**, 203-215 (1976) (in Russian).

[2] A.R. Its, Asymptotics of Solutions of the Nonlinear Schrodinger Equation and Isomonodromic Deformations of Systems of Linear Differential Equations, Sov. Math. Dokl. **24**, 452-456 (1981).

[3] A.R. Its and V.Yu Novokshenov, The Isomonodromic Deformation Method in the Theory of Painlevé Equations, Springer Lect. Notes Math. **1191**, (1986).

[4] C. Itzykson and J.-B. Zuber, The Planar Approximation II, J. Math. Phys. **21** (3) (1980).

[5] M. Douglas and S. Shenker, Strings in Less than One Dimension, Rutgers preprint RU-89-34.

[6] D. Gross and A. Migdal, A Nonperturbative Treatment of Two-Dimensional Quantum Gravity, Princeton preprint PUPT-1159 (1989).

[7] S.V. Manakov, On Complete Integrability and Stochastization in the Discrete Dynamical Systems, Zh. Exp. Teor. Fiz., **67**, 2, 543-555 (1974).

[8] H. Flaschka, The Toda Lattice II. Inverse Scattering Solution, Prog. Theor. Phys. **51**, 3, 703-716 (1974).

[9] M. Kac and P. von Moerbeke, Advances in Math. **16**, 160-164, (1975).

[10] M. Douglas, String in Less Than One Dimension and the Generalized KdV Hierarchies, Rutgers University preprint.

[11] E. Witten, Two Dimensional Gravity and Intersection Theory on Modulli Space, IAS preprint, IASSNS-HEP-90/45.

[12] Emil J. Martinec, On the Origin of Integrability in Matrix Models, preprint EFI-90-67.

[13] A.R. Its, A.V. Kitaev and A.S. Fokas, Isomonodromic Approach in the Theory of Two-Dimensional Quantum Gravity, Usp. Matem. Nauk, **45**, 6 (276), 135-136, (1990) (in Russian).

[14] A.R. Its and A.V. Kitaev, Mathematical Aspects of 2D Quantum Gravity, MPLA, **5**, 25, 2079 (1990).

[15] A.S. Fokas, A.R. Its, and A.V. Kitaev, Discrete Painlevé Equations and Their Appearance in Quantum Gravity, Clarkson preprint, INS #164 (1990).

[16] A.R. Its, A.V. Kitaev, and A.S. Fokas, The Matrix Models of the Two-Dimensional Quantum Gravity and Isomonodromic Solutions of the Discrete Painlevé Equations; A.V. Kitaev, Calculations of Nonperturbation Parameter in Matrix Model. Φ^4, A.R. Its and A.V. Kitaev, Continuous Limit for Hermitian Matrix Model Φ^6, in the book: Zap. Nauch. Semin. LOMI, **187**, Differential Geometry, Li Groups and Mechanics, **12** (1991).

[17] F. David, Loop Equations and Non-Perturbative Effects in Two-Dimesnional Quantum Gravity, MPLA, **5**, 13, 1019-1029 (1990).

[18] P.G. Silvestrov and A.S. Yelkhovsky, Two-Dimensional Gravity as Analytical Continuation of the Random Matrix Model, INP preprint 90-81, Novosibirsk, (1990).

[19] G. Moore, Geometry of the String Equations, Commun. Math. Phys. **133**, 261-304 (1990).

[20] G. Moore, Matrix Models of 2D Gravity and Isomonodromic Deformation, YCTP-P17-90, RU-90-53.

[21] A.A. Kapaev, Weak Nonlinear Solutions of the P_1^2 Equation, Zap Nauch. Semin. LOMI, **187**, Differential Geometry, Li Groups, and Mechanics, **12**, p. 88 (1991).

[22] S.P. Novikov, Quantization of Finite-Gap Potentials and Nonlinear Quasiclassical Approximation in Nonperturbative String Theory, Funkt. Analiz i Ego Prilozh., **24**, 4, 43-53 (1990).

[23] I.M. Krichever, On Heisenberg Relations for the Ordinary Linear Differential Operators, ETH preprint (1990).

[24] A.A. Kapaev, Asymptotics of Solutions of the Painleve Equation of the First Kind, Diff. Equa., **24**, 10, 1684 (1988) [in Russian].

The Action of the Virasoro Nonisospectral KdV Symmetries of the Whitham Equations

P.G. Grinevich

L.D. Landau Institute for Theoretical Physics,
ul. Kosygina 2, 117940 Moscow, GSP-1, Russia

Abstract. A new set of symmetries of the Whitham equations (averaged Korteweg-de-Vries equations) is constructed. These symmetries form a noncommutative algebra (isomorphic to the algebra of polynomial vector fields on the complex plane), explicitly depend upon the coordinates and time and act nontrivialy at the constant Whitham solutions.

0. Introduction. Let us have a partial differential equation (in our paper we consider the Korteweg-de Vries equation (KdV)):

$$u_t = \frac{1}{4}u_{xxx} - \frac{3}{2}uu_x, \tag{0.1}$$

and a finite-parametric set of its periodic solutions

$$u(x,t) = V(\vec{U}x + \vec{W}t|c_1, \ldots, c_N). \tag{0.2}$$

If we look for the WKB-type solutions of (0.1)

$$u(x,t) = V(\vec{U}(X,T)x, \vec{W}(X,T)t|c_1(X,T), \ldots, c_N(X,T)),\ X = \epsilon x,\ T = \epsilon t, \tag{0.3}$$

then the coefficients $c_1(X,T)$, $c_N(X,T)$ satisfy some new hydrodynamical-type system

$$\partial c_i(X,T)/\partial T = v_i^j(c)\partial c_j(X,T)/\partial X \tag{0.4}$$

which are called Whitham equation (for KdV they where derived by Whitham [11] for the one-phase solutions and by Flaschka-Forest-McLaughlin [3], Lax-Levermore [8] in the multiphase situation).

If the original equation has some additional structure then we can hope that the structure is inherited in the averaged equation, but this connection can be very nontrivial (important examples - hamiltonian structure of the Whitham equations, constructed by Dubrovin and Novikov [2], integration procedure, based on the hodograph transform - Tzarev [10], integration by methods of algebraic geometry - Krichever [6].

It is well-known that (0.1) has an infinite set of symmetries (flows, commuting with (0.1)), corresponding to the Kruskal integrals (see, for example [12]). But it also has a second infinite family of the so-called nonisospectral symmetries (these symmetries change the spectrum of the auxiliary linear problem and form a noncommutative algebra). In our paper we construct the corresponding symmetries for (0.4).

1. Periodic (quasiperiodic) solutions of the KdV equation. Consider the Lax representation for KdV:

$$\partial L/\partial t = [L, A],\ L = -\partial_x^2 + u(x,t),\ A = \partial_x^3 - \frac{3}{4}(u\partial_x + \partial_x u) \tag{1.1}$$

For the periodic potential $u(x+\Pi, t) = u(x,t)$ the following spectral problems for L are considered (L is treated as a one-dimensional operator, depending on the parameter t):

a) Main problem $L\psi = E\psi$, ψ is bounded in x.

b) Auxiliary problem $L\psi_a = E\psi_a$, $\psi_a(0) = \psi_a(\Pi) = 0$.

The spectrum of the main problem consists of a set of intervals $[E_1, E_2]$, $[E_3, E_4], \ldots, [E_{2n-1}, E_{2n}], \ldots$ where $E_1 < E_2 \leq E_3 < E_4 \leq E_5 \ldots$, $E_{2n+1} - E_{2n} \to 0$ as $n \to \infty$.

The spectrun of the auxiliary problem consists of an infinite number of points $d_1 < d_2 < d_3 < \ldots$ located in the gaps $d_1 \in [E_2, E_3]$, $d_2 \in [E_4, E_5]$, $d_3 \in [E_6, E_7], \ldots$.

The Bloch eigenfunction $\psi(x, E)$ normalized by the conditions $\psi(x+\Pi, E) = \exp(\Pi i p(E))\ \psi(x, E)$ and $\psi(0, E) = 1$ is meromorphic on a two-sheeted Riemann surface Γ over the E-plane with branch points $E_1, E_2, \ldots, \infty$, and has simple poles in the points $\gamma_1, \gamma_2, \ldots$, such that the projection of γ_n to the E-plane coincides with d_n.

Explicit formulas for the periodic (quasiperiodic) solutions of KdV in terms of the Riemann Θ-functions can be obtained in the so-called finite-gap case when $E_{2n} = E_{2n+1}$ for all $n > g$. Then the data $E_1, E_2, \ldots, E_{2g+1}, \gamma_1, \gamma_2, \ldots, \gamma_g$ uniquely determines the potential $u(x)$ (see [12]).

The ends of gaps $E_1, \ldots, E_{2g+1}$ do not depend on t, the poles $\gamma_1, \ldots, \gamma_g$ plays the role of phases. The potential $u(x,t)$ is expressed via Riemann Θ -function (A.R. Its–V.B. Matveev formula).

$$u(x,t) = -2\partial_x^2 \log \Theta(\vec{U}x + \vec{W}t | B_{ij}) + const \tag{1.2}$$

2. Whitham equations for the finite-gap KdV solutions. In the [3] theory of averaged KdV the ends of gaps E_k are functions of slow variables $X = \epsilon x$, $T = \epsilon t$ and obey the following hydrodynamical-type equations

$$\partial E_k/\partial T = v_k(E_1, \ldots, E_{2g+1})\partial E_k/\partial X,\ v_k(E_1, \ldots, E_{2g+1}) = d\Omega_3(E_k)/d\Omega_1(E_k), \tag{2.1}$$

where $d\Omega_k$ is the abelian differential of the second kind with the unique pole in the point $E = \infty$, such that

$$d\Omega_k(E) \sim d(\sqrt{-E})^k,\ \int_{E_{2k}}^{E_{2k+1}} d\Omega_k = 0. \tag{2.2}$$

(For even $k = 2l$ $d\Omega_k = d(-E)^l$ and $v_k = 0$). Equation (2.1) means that E_k are Riemann invariants for the averaged KdV (i.e. the matrix v_i^j in (0.4) is diagonal).

Let $\lambda = \sqrt{-E}$ and Q_{km} be the expansion coefficients:

$$d\Omega_k = d(\lambda^k) + \sum_{m\geq 1} Q_{km} d\lambda/\lambda^{m+1} \quad as \quad \lambda \to \infty. \tag{2.3}$$

From the Riemann bilinear relations we have $Q_{km} = Q_{mk}$, $Q_{km} = 0$ if k or m is even.

3. Isospectral and nonisospectral symmetries of the KdV. The KdV equation possesses at least two infinite series of symmetries. The isospectral symmetries correspond to the Kruskal integrals and can be written in the following form:

$$\frac{\partial u}{\partial t_{2n+1}} = \frac{d}{dx}\Big(-\frac{\Lambda}{4}\Big)^n u = -2\frac{d}{dx}\Big(-\frac{\Lambda}{4}\Big)^{n+1} \cdot 1. \tag{3.1}$$

where $\Lambda = -\partial_x^2 + 2\partial_x^{-1} u\partial + 2u$ - recursion operator, $x = t_1$, $t = t_3$, they do not move E_k and have the Lax representation $[L, \partial_{t_{2n+1}} - A_{2n+1}] = 0$ with $A_{2n+1} = \partial_x^{2n+1} + \ldots$. The average flows have the hydrodynamical form

$$\partial E_k/\partial T_{2n+1} = w_k^{2n+1}(E_1, \ldots, E_{2g+1})\partial E_k/\partial X,$$

$$w_k^{2n+1}(E_1, \ldots, E_{2g+1}) = d\Omega_{2n+1}(E_k)/d\Omega_1(E_k), \tag{3.2}$$

The nonisospectral symmetries can be defined as:

$$\frac{\partial u}{\partial \beta_{2n}} = -2\frac{d}{dx}\Big(-\frac{\Lambda}{4}\Big)^{n+1} \Big(\sum_{k=0}^{\infty}((2k+1)t_{2k+1}\Big(-\frac{\Lambda}{4}\Big)^k\Big) \cdot 1. \tag{3.3}$$

(Ibragimov-Shabat formula, see for references [9], [4], [5]). For the calculations the following Orlov-Schulman formula [9] is more convenient:

$$\frac{\partial u}{\partial_{t_n}} = -2\frac{d}{dx}\, res\, |_{\lambda=\infty}\, \lambda^n \Psi(\lambda, \vec{t})\Psi^+(\lambda, \vec{t}),$$

$$\frac{\partial u}{\partial \beta_n} = -2\frac{d}{dx}\, res\, |_{\lambda=\infty}\, (\lambda^{n+1}\partial_\lambda \Psi(\lambda, \vec{t}))\Psi^+(\lambda, \vec{t}), \tag{3.4}$$

$$L\Psi = E\Psi,\ L\Psi^+ = E\Psi^+,\ (\partial_{t_{2n+1}} - A_{2n+1})\Psi = 0,\ (\partial_{t_{2n+1}} - A_{2n+1}^+)\Psi^+ = 0, \tag{3.5}$$

where $\lambda^2 = -E$, $\vec{t} = (t_1, t_3, t_5, \ldots)$, A_{2n+1}^+ is the formally conjugate operator to A_{2n+1}.

The nonisospectral flows do not commute with each other. The algebra of these flows for KdV is isomorphic to the algebra of polynomial vector fields in the complex plane.

In the finite-gap case the functions $\Psi(\lambda, \vec{t})$ and the holomorphic 1-differential $\Psi(\lambda, \vec{t})$ are uniquely determined by the following properties:

1a) $\Psi(\lambda, \vec{t})$ is meromorphic in λ on $\Gamma\backslash\infty$ with poles in $\gamma_1, \ldots, \gamma_g$,

1b) $\Psi^+(\lambda, \vec{t})$ is holomorphic in λ on $\Gamma\backslash\infty$ with zeroes in $\gamma_1, \ldots, \gamma_g$.

2a) $\Psi(\lambda, \vec{t}) = exp(\lambda x + \lambda^3 t + \lambda^5 t_5 + \ldots)[1 + \xi_1(\vec{t})/\lambda + \xi_2(\vec{t})/\lambda^2 + \ldots]$ as $\lambda \to \infty$

2b) $\Psi^+(\lambda, \vec{t}) = exp(-\lambda x - \lambda^3 t - \lambda^5 t_5 + \ldots)[1 + \xi_1^+(\vec{t})/\lambda + \xi_2^+(\vec{t})/\lambda^2 + \ldots]d\lambda$ as $\lambda \to \infty$

To write equation (3.3) in more convenient form we shall do the following. All the coefficients $\xi_k(\vec{t})$, $\xi_k^+(\vec{t})$ can be expressed via $u(\vec{t})$ from (3.4) (maybe in nonlocal form). All the exponents in $\Psi\Psi^+$ are reduced and the residue can be explicitly calculated.

4. Action of the nonisospectral symmetries on the finite-gap KdV solutions. This action was calculated in the papers [4], [5] for KdV and KP equations. For the KdV the result reads:

$$\partial E_s/\partial\beta_{2n} = 2(-1)^n E_s^{n+1}, \ \partial\gamma_k/\partial\beta_{2n} = 2(-1)^n\gamma_k^{n+1}. \tag{4.1}$$

So we have the shift of all the spectral data via the vector field $2(-1)^n E^{n+1}\partial/\partial E$.

5. Averaged nonisospectral symmetries. A general formula for the action of the nonisospectral symmetries on the Whitham solutions for KP was obtained by Krichever ([7]) in the terms of the averaged τ -function. But the methods of [4], [5] allow us to get the answer in more explicit form:

$$\frac{\partial E_s}{\partial\beta_{2n}} = 2(-1)^n E_s^{n+1} + \left\{\sum_{k\geq 0}(2k+1)t_{2k+1}w_s^{2k+2n+1} + \sum_{0\leq k<n} Q_{2k+1}w_s^{2n-2k-1}\right\}\frac{\partial E_s}{\partial X}. \tag{5.1}$$

where

$$Q_j = \int_{[0,0,0,\dots]}^{[T_1,T_3,T_5,\dots]} Q_{j,1}dT_1 + Q_{j,3}dT_3 + Q_{j,5}dT_5 + \dots \tag{5.2}$$

Krichever and Dubrovin ([6], [1]) have proved that for Whitham hierarchy solutions there exists a function $V_{\Omega\Omega}$ such that $Q_{i,j} = \partial_{T_i}\partial_{T_j}V_{\Omega\Omega}$ so Q_j are well defined.

The proof that the flows (5.1) commute with the flows (2.1), (3.2) is based upon a direct calculation, using the formulas (2.1), (3.2) (the Flaschka-Forest-McLaughlin expressions for velocities), (4.1) and the following Lemma.

Lemma. If we shift the branch points of Γ by the flow (4.1), then the variations of the differentials $d\Omega_k$ reads:

$$\partial d\Omega_k/\partial\beta_n = kd\Omega_{n+k} + \sum_{m=1}^{n-1} Q_{km}d\Omega_{n-m}. \tag{5.3}$$

The formula (5.3) was used, for example in [4], Appendix B.

The author is very grateful to B.A.Dubrovin, I.M.Krichevr and S.P.Tzarev for useful discussions.

References.

[1] B.A. Dubrovin. Hamiltonian formalism of Whitham-type hierarchies and topological Landau-Ginzburg models. Preprint 1991.

[2] B.A. Dubrovin, S.P. Novikov. Sov. Math. Doklady 27, p.665 (1983); Russian Math. Surveys 44:6 (1989), p.35.

[3] H. Flaschka, G. Forest, D.W. McLaughlin. Commun. Pure Appl. Math. 33, no. 6 (1980).

[4] P.G. Grinevich, A.Yu. Orlov. Virasoro action on Riemann surfaces, Grassmannians, det $\bar{\partial}_j$ and Segal-Vilson τ function. in "Problems of modern quantum field theory", ed. A.A. Belavin, A.U. Klimyk, A.B. Zamolodchikov, Springer 1989.

[5] P.G.Grinevich, A.Yu.Orlov. Funct. Anal. Appl. 24 no. 1, (1990); Higher symmetries of KP equation and the Virasoro action on Riemann surfaces. in "Nonlinear evolution equations and dynamical systems", ed. S. Carillo, O. Ragnisco, Springer 1990.

[6] I.M. Krichever. Funct. Anal. Appl. 11 (1988), p.15.

[7] I.M. Krichever. "The dispersionless Lax equations and topological minimal models". Preprint 1991 Torino; Lecture at the Sakharov Memorial Congress (1991).

[8] Peter D. Lax, C. David Levermore. Commun. Pure Appl. Math., 36, pp. 253-290, 571-593, 809-830 (1983)

[9] A.Yu. Orlov, E.I. Schulman. Additional symmetries of the integrable one-dimensional equations and the conformal algebra representations. Preprint IA and E. Novosibirsk (1984); Lett. Math. Phys. 12 (1986); A.Yu. Orlov in Proceedings Int. Workshop "Plasma theory and ..." in Kiev 1987, World Scientific (1988) p.116.

[10] S.P. Tzarev. Soviet. Math. Dokl. 31, pp. 488-491 (1985). Math. USSR Izvestiya, 54, no. 5 (1990).

[11] G.B. Whitham. Linear and nonlinear waves. N.Y., Wiley, 1974.

[12] V.E. Zakharov, S.V. Manakov, S.P. Novikov, L.P. Pitaevsky. Soliton theory. Plenum, New York, 1984.

A Generalized Sato's Equation of the KP Theory and Weyl Algebra

Y. Kodama

Department of Mathematics, Ohio State University,
Columbus, OH 43210, USA

We give a generalization of the Sato equation of the KP theory, and show that the corresponding flows form the Weyl algebra of infinite dimension. Giving several realizations of this algebra on the infinite dimensional Grassmann manifold, we discuss some connections to the string theory in 2-D gravity [1].

1. S. Aoyama and Y. Kodama, "A Generalized Sato Equation and W_{infinity} Algebra" to appear in Phys. Lett. B (1992).

Additional references for this subject

2. E. Brezin and V. Kazakov, Phys. Lett. B236 (1990) 144;
M. Douglas and S. Shenker, Nucl. Phys. B335 (1990) 635;
4. D. J. Gross and A. Migdal, Phys. Rev. Lett. 64 (1990) 127; Nucl. Phys. B349 (1990) 333.
5. M. Douglas, Phys. Lett. B238 (1990) 176.
6. T. Banks et al, Phys. Lett. B238 (1990) 279.
7. M. Fukuma et al, Intern. J. Mod. Phys. A6 (1990) 1385.
8. R. Dijkgraaf et al, Nucl. Phys. B348 (1991) 435.
9. A. Yu. Orlov and E. I. Schulman, Lett. Math. Phys. 12 (1986) 171.
10. S. Nishigaki and T. Yoneya, Nucl. Phys. B348 (1991) 787; Phys. Lett. B268 (1991) 35.
11. C. Pop et al, Phys. Lett. B242 (1990) 401.

Quasirelativistic Analogs of Lax Equations

B.A. Kupershmidt

The University of Tennessee Space Institute, Tullahoma, TN 37388-8897, USA

0. Integrable systems are rare; they are instances of well–organized structures. Even rarer are nonrelativistic mechanical systems which have (special) relativistic analogs viewed mathematically as regular deformations with the parameter $\epsilon = c^{-2}, c$ being the speed of light.

1. Until 1986 there was no hint that these two rare classes have common elements, when Ruijsenaars and Schneider found a relativistic version of the Calogero–Moser system and then later on Ruijsenaars discovered a relativistic Toda lattice:

$$\dot{a}_i = a_i[b_{i+1} - b_i + \epsilon(a_{i+1} - a_{i-1})], \quad \dot{b}_i = (1 + \epsilon b_i)(a_i - a_{i-1}), \tag{1}$$

in familiar mathematical notation. Bruschi, Calogero, Ragnisco, S. Alber, and others, have analyzed and elucidated these systems (physical background and up–to–date references can be found in [1]), and Gibbons and myself [2] found a general ansatz for all scalar Lax equations which, when applied to the Lax representation of the Toda lattice, produces a Lax representation for the relativistic Toda lattice. This ansatz is:

$$\dot{\mathcal{L}} = [\mathcal{P}_+ + S(\mathcal{P}_-), \mathcal{L}] = [\mathcal{L}, \mathcal{P}_- - S(\mathcal{P}_-)]. \tag{2}$$

$$\mathcal{L} = (1 - \epsilon\xi)^{-1} L, \quad L = \sum_{\rho}^{n} \xi^i q_i \ , \quad n \in \mathbf{N}, \quad q_n = \text{const}, \tag{3}$$

$\xi = \partial/\partial x$, $\rho = 0$ or $-\infty$ in the continuos case,
$\xi = exp(\partial/\partial x)$, $-\rho \in \mathbf{N} \cup \{\infty\}$ in the discrete case,

$\mathcal{P} \in Z(\mathcal{L})$, i.e. $[\mathcal{P}, \mathcal{L}] = 0$, and the relativistic symbol S is:

$$S(\Sigma f_i \xi^i) = \Sigma f_i \epsilon^{-i} \tag{4}$$

These deformations have the following properties: (a) All the flows are well defined, despite their apparent overdeterminacy; (b) All these flows commute and have a common infinite set of integrals $\{\text{Res}(\mathcal{P}) \mid \mathcal{P} \in Z(\mathcal{L})\}$; (c) More can be found in [2]. Here I report new developments, especially on the Hamiltonian aspects of the theory [the standard methods do not work for the deformed form (2)]. The word 'quasi' in the title is used as an accommodating sop for an occasional raging pedant who insists on using the adjective 'relativistic' only for familiar situations where the generators of the Lorentz group are exhibited explicitly.

2. The singular in ϵ constraint [with $a = q_1$, $b = q_0$ in (1)]

$$\{q_0 = -1/\epsilon\} \tag{5}$$

is invariant for all the discrete flows (2) with $n = 1$. For the Toda case (1), this results in a linear combination of (higher) Volterra (= Kac - van Moerbeke, Manakov) flows, together with a new representation (not of Lax or zero-curvature type) for them, with

$$\tilde{\mathcal{L}} = -\epsilon^{-1} + (1 - \epsilon\xi)^{-1}\xi^{-1}q_1 \tag{6}$$

3. For the matrix continuos case, formulae (3), (4) change into

$$\mathcal{L} = (1 - \epsilon q_n \xi)^{-1} L, \quad S(\Sigma f_i \xi^i) = \Sigma f_i (q_n \epsilon)^{-i}. \tag{7}$$

This change makes the flows (2) well–defined. Everything else hinges on the description of $Z(\mathcal{L})$ where q_n is a constant invertible diagonalizable, but, contrary to the classical theory [4], $q_{n-1} \notin Im(adq_n)$.

4. The relativistic flows are not isomorphic to the nonrelativistic ones. A proof is simple in the continuous case: the first nontrivial flow is not a x-shift anymore.

5 For the discrete case with $n = 1$, the (first) Hamiltonian structure is:

$$B_{ij}(\epsilon) = q_{i+j}\Delta^i - \Delta^{-j}q_{i+j} - \epsilon(q_{i+j+1}\Delta^{i+1} - \Delta^{-j-1}q_{i+j+1}). \tag{8}$$

The corresponding Lie algebra $\mathcal{G}(\epsilon)$ has the commutator

$$[\widehat{X}, \widehat{Y}] = \widehat{X}(1 - \epsilon\xi)\widehat{Y} - \widehat{Y}(1 - \epsilon\xi)\widehat{X}, \quad \widehat{X} = \sum_{i \geq 0} X_i \xi^i \;, \; \widehat{Y} = \dots \;, \tag{9}$$

so that $\mathcal{G}(\epsilon) \approx \mathcal{G}(0)$ and $B(\epsilon) \approx B(0)$. Thus, all the Hamiltonian maps from the standard theory [5] apply; in particular, one has the modified equations (not of the form (2) anymore!)

6. In the nonrelativistic case, one constructs τ-functions by the Cherednik - Flaschka method [6,5] on the basis of the equalities

$$\partial_{\mathcal{P}}(\text{Res } Q) = \partial_{Q}(\text{Res } \mathcal{P}), \quad \mathcal{P}, Q \in Z(\mathcal{L}). \tag{10}$$

Something similar should exist in the relativistic case.

7. The relativistic KP hierarchy, with $n = 1$, $\rho = -\infty$, $q_0 = 0$, and

$$\mathcal{L} = (1 - \epsilon\xi)^{-1}(\xi + \sum_{0}^{\infty} \xi^{-i-1} q_i) \tag{11}$$

has the first Hamiltonian matrix

$$B_{ij}(\epsilon) = B_{ij}^{(0)} - \epsilon B_{ij}^{(1)}, \; B_{ij}^{(r)} = \binom{i+r}{s} q_{i+j-s+r}\partial^r - (-\partial)^r \binom{j+r}{s} q_{i+j-s+r}. \tag{12}$$

Again, the corresponding Lie algebra $\mathcal{G}(\epsilon)$ has the commutator (9); hence $\mathcal{G}(\epsilon) \approx \mathcal{G}(0)$, $B(\epsilon) \approx B(0)$, so that (Hamiltonically) we are back to the nonrelativistic KP and can use the Hamiltonian maps from [7] to construct relativistic Hamiltonian mKP and DWW.

8. In the quasiclassical limit, the nonrelativistic KP hierarchy becomes the Benney hierarchy [8]. In the same limit, the RKP hierarchy (11) becomes the RBenney hierarchy; it has a $2+1-d$ free–surface representation, the Hamiltonian structure

$$B_{ij}(\epsilon) = iq_{i+j-1}\partial + \partial j q_{i+j-1} - \epsilon[(i+1)q_{i+j}\partial + \partial(j+1)q_{i+j}], \qquad (13)$$

and the Poisson representation

$$\dot{\mathcal{L}} = \{\mathcal{P}_+ + S(\mathcal{P}_-), \mathcal{L}\} = \{\mathcal{L}, \mathcal{P}_- - S(\mathcal{P}_-)\} \qquad (14)$$

The nonrelativistic moment representation $q_i = \int_0^h u^j dy$ when reduced to the case $\{u_y = 0\} \Leftrightarrow q_i = hu^j$, has a relativistic analog

$$q_j = hu^j(1-\epsilon u)^{-1}. \qquad (15)$$

This reduces the RBenney hierarchy to the R elativistic $1+1-d$ long surface waves hierarchy, with the fluid (second) flow

$$\begin{aligned} -\dot{u} &= \{(1-\epsilon u)^{-2}[(u+2\epsilon h)^2/2 + h(1+\epsilon^2 h)]\}_x, \\ -\dot{h} &= [(1-\epsilon u)^{-3} h(1+\epsilon^2 h)(u+2\epsilon h)]_x. \end{aligned} \qquad (16)$$

This hierarchy, as its nonrelativistic counterpart, is tri–Hamiltonian. The Hamiltonians $H_n = n^{-1}$ Res $(\mathcal{L}^n), \mathcal{L}$ in (11),(15), are:

$$\begin{aligned} &H_1 = h, \ H_2 = (1-\epsilon u)^{-1}(hu + \epsilon h^2), \\ &H_3 = (1-\epsilon u)^{-1}h^2 + (1-\epsilon u)^{-2}(hu^2 + 2\epsilon^2 h^3) - 3\epsilon u(1-\epsilon u)^{-4}, \ldots \end{aligned} \qquad (17)$$

There is a new effect here, similar to the separation of atomic spectra in magnetic field: although in the nonrelativistic case, for each $m \in \mathbf{N}$, there exists a unique (up to a multiplicative constant) polynomial integral of deg $= m$ (with deg $u = 1$, deg $h = 2$), for the relativistic case there are two such for each $m > 2$:

$$\overline{H}_2 = hu, \quad \overline{H}_3 = \epsilon^{-1}(H_2 - \overline{H}_2) = (1-\epsilon u)^{-1}(h^2 + hu^2), \ldots \qquad (18)$$

9. Like in the nonrelativistic case [7], the variables q_i with $i > 0$ and $i < 0$ (continuous case), and $i > 1$ and $i \leq 0$ (discrete case) split up in the Hamiltonian form, with the negative i's having the KP–type Hamiltonian form (12) or (8). It is more complicated for the positive i's. I've managed to check only the KdV case $L = \xi^2 + \xi v + u$, with the resulting matrix

$$B^I = \begin{pmatrix} 0 & -\epsilon\partial \\ -\epsilon\partial & 2\partial \end{pmatrix}. \qquad (19)$$

If the second Hamiltonian form B^{II} can be found, its localization will provide a Relativistic Virasoro algebra.

References

[1] S.N.M. Ruijsenaars, Comm. Math. Phys. (1990) 217; [3] (1990) 165.

[2] J. Gibbons and B. A. Kupershmidt, [3] (1990) 207.

[3] B. A. Kupershmidt, ed., "Integrable and Superintegrable Systems", World Scientific, Singapore, 1990.

[4] G. Wilson, Math. Proc. Cambr. Phil. Soc. 86 (1979) 131.

[5] B. A. Kupershmidt, "Discrete Lax Equations and Differential–Difference Calculus", Asterisque, Paris, 1985.

[6] H. Flaschka, Quart. J. Math. Oxford (2) 34 (1983) 61.

[7] B. A. Kupershmidt, Comm. Math. Phys. 99 (1985) 51.

[8] D. R. Lebedev and Yu. I. Manin, Phys. Lett. 74A (1979) 154.

[9] B. A. Kupershmidt and Yu. I. Manin, Funct. Anal. Appl. 11 (1977) 188; 12 (1978) 20.

$n \times n$ Zakharov-Shabat System of the Form $(d\psi/dx)(z^2 - 1/z^2)J\psi + (zQ + P + R/z)\psi$

J.-H. Lee

Institute of Mathematics, Academia Sinica,
Taipei 11529, Taiwan, P.R. China

Abstract. We consider the scattering and inverse scattering problem of the form $\frac{d\psi}{dx} = (z^2 - 1/z^2)J\psi + (zQ + P + R/z)\psi$, $J = \mathrm{diag}(id_1, id_2, \ldots, id_n)$, Q is off-diagonal and the diagonal part of P equals the diagonal part of $Q(adJ)^{-1}Q$, $adJ(A) = [J, A]$. We define a scattering transform following the set-up of Beals-Coifman[2]. We also derive the associate nonlinear evolution equations.For the special case n=2, we solve an evolution equation which can be transformed explicitly to the massive Thirring model in laboratory coordinates considered by Kuznetsov and Mikhailov[10]. Some analytic results are stated here and the proofs will be published elsewhere.

1. Forward Problem

Let $J = diag(id_1, id_2, ..., id_n)$, $d_1 < d_2 < \ldots d_n$, $i = \sqrt{-1}$. $\sum = \{z : Im(z^2 - 1/z^2) = 0\} = \{z : Im(z^2) = 0\} = R \cup iR$. Let Q, P, R be $M_n(\mathbf{C})$-valued functions and Q, $dQ/dx, P, R, dR/dx \in L^1$ and Q,R be off-diagonal ,the diagonal part of P equal the diagonal part of $Q(adJ)^{-1}Q$, $adJ(A) = [J, A]$. We consider the following system:

$$\frac{dM}{dx} = (z^2 - 1/z^2)[J, M] + (zQ + P + R/z)M, \quad z \notin \sum. \tag{1.1}$$

$M(., z)$ is bounded and absolutely continuous,$M(x, z) \to I$ as $x \to -\infty$. This constraint on the potentials is crucial to have M normalized at $z = \infty$ and we need this to pose the inverse problem. We may use the technique in [11], [13] to control z-dependence when $z \to 0$ and $z \to \infty$. Let $\Omega_+ = \{z : Im(z^2 - 1/z^2) > 0\}$, $\Omega_- = \{z : Im(z^2 - 1/z^2) < 0\}$. For the generic potential (Q, P, R), M has the following properties:

(i) for any $z \in \sum/\{0\}$, there is a unique matrix $v(z)$ such that

$$M^+(x, z) = M^-(x, z)exp(x(z^2 - 1/z^2)J)v(z)exp(-x(z^2 - 1/z^2)J), \tag{1.2}$$

where $M^{\pm}(x, z)$ =limit of M on $\sum$ from $\Omega_\pm$;

(ii) $M(x, .)$ has finite number of poles at $D = \{z_1, z_2, ...z_N\}$; for any z_j ,there is a matrix $v(z_j)$ such that $Res(M(x, .), z_j) = N(x, z_j)exp(x(z_j^2 - 1/z_j^2)J)v(z_j)exp(-x(z_j^2 - 1/z_j^2)J)$, where $N(x, z)$ is the regular part of $M(x, z)$ near $z = z_j$;

(iii) The map $(Q, P, R) \to (v(z); z_1, z_2, \ldots z_N; v(z_1), v(z_2), ...v(z_N))$ is injective.

If the potential has the symmetric conditions $Q^* = -Q$,$P^* = -P$,$R^* = -R$, by uniqueness $M^*(x, \bar{z}) = M^{-1}(x, z)$. It follows that $v^*(z) = v(z)$, $z \in \sum$, $v^*(\bar{z}_j) = -v(z_j)$,for $j = 1, 2, \ldots N$. Furthermore,if Q, P, R satisfy $Q^\rho = -Q$,$P^\rho = P$,$R^\rho = -R$, $J^\rho = J$,where ρ is an automorphism in $M_n(\mathbf{C})$,by uniqueness, $M^\rho(x, -z) = M(x, z)$. It follows that $v^\rho(-z) = v(z)$, $z \in \sum$, $v^\rho(-z_j) = -v(z_j)$, for $j = 1, 2, \ldots N$.

2. Inverse Problem

Given scattering data v satisfying the constraints as described in [16,p.328],the inverse problem amounts to solving an analytic factorization problem (Rieman-Hilbert problem) with one parameter x, i.e. we want to find $M(x, .)$ that is meromorphic on $\mathbf{C}/\sum$ such that

(2.1) $M^+(x,z) = M^-(x,z)exp(x(z^2-1/z^2)J)v(z)exp(-x(z^2-1/z^2)J), z \in \sum;$

$Res(M(x,.),z_j) = N(x,z_j)exp(x(z_j^2-1/z_j^2)J)v(z_j)exp(-x(z_j^2-1/z_j^2)J), \quad j=1,2,\dots N$ and $M \to I$ as $z \to \infty$. As in [11][13], we may use the augument of Beals and Coifman[2] and the property of weighted-boundedness of Hilbert transform to show that the inverse problem is solvable if $v(z)-I$ is small. If $v(z)-I$ is not small we may look for M of the form $rM^\sharp exp(x(z^2-1/z^2)J)u\, exp(-x(z^2-1/z^2)J)$,where u is a piecewise rational function,$M, M^\sharp$ is the eigenfunction associated with $v^\sharp$ and $v^\sharp - I$ is small and r is rational in z. The argument in Beals and Coifman establishes a Fredholm property[2]. We may use this to show that if the scattering data satisfies the symmetric conditions :$v^*(z) = v(z)$ $z \in \sum$, $v^*(\bar{z}_j) = -v(z_j)$, then the inverse problem is solvable and the associated M satisfies

$$\frac{dM}{dx} = (z^2 - 1/z^2)[J,M] + (zQ + P + R/z)M.$$

3. Evolution Equations

Suppose the scattering data evolves as

$$\text{(3.1)} \quad \begin{cases} \frac{dv(z,t)}{dt} = (az^L + bz^{-L})[J, v(z,t)], & z \in \sum,\ a,\ b \text{ are constants;} \\ \frac{dv(z_j,t)}{dt} = (az_j^L + bz_j^{-L})[J, v(z_j,t)], & z_j \text{ is fixed for each } j,\ j = 1,2,\dots,N. \end{cases}$$

Then the associated eigenfunction M evolves in t as

$$\text{(3.2)} \quad \begin{aligned} \frac{dM}{dt} = (az^L + bz^{-L})[J,M] + (a(G_1 z^{L-1} + G_2 z^{L-2} + \dots + G_L) \\ + b(F_{L-1}/z + \dots + F_1/z^L))M, \end{aligned}$$

where $MJM^{-1} \sim G_0 + G_1/z + G_2/z^2 + \dots$ as $z \to \infty$ and $MJM^{-1} \sim F_0 + F_1 z + F_2 z^2 + \dots$ as $z \to 0$. On the other hand $dM/dx = (z^2 - 1/z^2)[J,M] + (zQ + P + R/z)M$.Then from the compatibility condition we obtain the evolution equation

$$\text{(3.3)} \quad U_t - V_x + [U,V] = 0,$$

where $U = (z^2 - 1/z^2)J + (zQ + P + R/z)$; $V = (az^L + bz^{-L})J + a(G_1 z^{L-1} + G_2 z^{L-2} + \dots\dots G_L) + b(F_{L-1}/z + \dots . F_1/z^L)$.

4. Example

Massive Thirring Model(MTM) in laboratory coordinates is of the form :

$$\text{(4.1)} \quad \begin{aligned} -i\frac{\partial\psi_1}{\partial t} - i\frac{\partial\psi_1}{\partial x} + 2\psi_2 + 2|\psi_2|^2\psi_1 = 0 \\ -i\frac{\partial\psi_2}{\partial t} + i\frac{\partial\psi_2}{\partial x} + 2\psi_1 + 2|\psi_1|^2\psi_2 = 0, \end{aligned}$$

which was considered by Kuznetsov and Mikhailov[10]. Let $\tilde{Q} = \begin{pmatrix} 0 & i\psi_2^* \\ i\psi_2 & 0 \end{pmatrix}$, $J = \begin{pmatrix} -i/2 & 0 \\ 0 & i/2 \end{pmatrix}$, $\tilde{R} = \begin{pmatrix} 0 & i\psi_1^* \\ i\psi_1 & 0 \end{pmatrix}$, $\tilde{P}_1 = (|\psi_1|^2 - |\psi_2|^2)J = (1/2)\tilde{Q}(adJ)^{-1}\tilde{Q} - (1/2)\tilde{R}(adJ)^{-1}\tilde{R}$, $\tilde{P}_2 = (1/2)\tilde{Q}(adJ)^{-1}\tilde{Q} + (1/2)\tilde{R}(adJ)^{-1}\tilde{R}$. Then (4.1) can be written as

$$\tilde{Q}_t - \tilde{Q}_x + [\tilde{Q}, \tilde{P}_2 - \tilde{P}_1] - 2[J, \tilde{R}] = 0$$

(4.2)

$$\tilde{R}_t + \tilde{R}_x + [\tilde{R}, \tilde{P}_2 + \tilde{P}_1] - 2[J, \tilde{Q}] = 0.$$

(4.2) is the compatibility condition of the following system :

$$\frac{d\psi}{dx} = (z^2 - 1/z^2)J\psi + (z\tilde{Q} + \tilde{P}_1 + \tilde{R}/z)\psi$$

(4.3)

$$\frac{d\psi}{dx} = (z^2 + 1/z^2)J\psi + (z\tilde{Q} + \tilde{P}_2 - \tilde{R}/z)\psi.$$

A 2×2 special case of the system we considered is of the form: $dM/dx = (z^2 - 1/z^2)[J, M] + (zQ + P + R/z)$, with $P = Q(adJ)^{-1}Q$, $Q^* = -Q$,$R^* = -R$, Q and R are off-diagonal.The symmetric conditions implies the associated inverse problem is solvable.The associated evolution equation is of the form

$$Q_t - Q_x - 2[J, R] = 0$$

(4.4)

$$R_t + R_x + 2[R, P] - 2[J, Q] = 0$$

Let $U = exp(-\int_{-\infty}^{x}(1/2)(Q(adJ)^{-1}Q + R(adJ)^{-1}R)$, $\tilde{Q} = UQU^{-1}$, $\tilde{R} = URU^{-1}$. We observe that $(\tilde{Q}(adJ)^{-1}\tilde{Q})_t + (\tilde{R}(adJ)^{-1}\tilde{R})_t = (Q(adJ)^{-1}Q))_x - (R(adJ)^{-1}R)_x$. Then we can show that Q, R satisfy (4.2) i.e. the massive Thirring model in laboratory coordinates.

Remarks: Recently the author has noticed the results of Xin Chou who considerd non generic potentials and had a different set-up for the inverse problem of the generalized Zakharov-Shabat system based on a modification of the set-up of Beals-Coifman. [see SIAM J. MATH ANAL.,Vol.20, No.4, p.966-986]

Acknowledgement. This work was partially supported by the National Science Council of the R.O.C..

References

1. R. Beals and R.R. Coifman, Scattering, transformation spectrales,et equation d'evolution non lineaires, Seminaire Goulaouic-Meyer-Schwartz, 1980-81, exp.22, Ecole Polytechnique, Palaiseau.
2. R. Beals and R. R. Coifman, Scattering and inverse scattering for the first order system, Comm. Pure Appl. Math., Vol.37 (1984), p.39-90.
3. R. Beals and R. R. Coifman, Inverse scattering and evolution equations, Comm. Pure Appl. Math.,Vol. 38 (1985), p.29-42.
4. R. Beals and R.R. Coifman, Scattering and inverse scattering for the first order system: II, Inverse Problem 3 (1987), p.577-593.
5. R. Beals, P. Deift, C. Tomei, Direct and inverse scattering on the line, Math Surveys and Monographs, No. 28 (1988), AMS.
6. R.K. Bullough, P.J. Caudrey (eds), Solitons, Topics in Current Physics No.17, Springer-Verlag, 1980.
7. V.S. Gerzhikov et al, Quadratic bundle and nonlinear equations, Theoret. and Math, Phys. 44, No. 3 (1980), p.784-795.
8. D. J. Kaup and A.C. Newell, An exact solution for a derivative nonlinear Schrodinger equation, J. Math. Phys. 19 (4), 1978, p.789-801.
9. D.J. Kaup and A.C. Newell, On the Coleman correspondence and the solution of the Massive Thirring Model, Lettere Al Nuovo Cimento, Vol. 20, N.9 (1977) p.325-331.

10. E.A. Kuznetsov and A.B. Mikhailov, On the complete integrabity of the two - dimensional classical Thirring Model, Theoret. Math. Physics, Vol.30, No.3 (1977), p.193-200.
11. J-H. Lee, Analytic properties of Zakharov-Shabat inverse scattering problem with polynomial spectral dependence of degree 1 in the potential, Ph.D. Dissertation, Yale University, May 1983.
12. J-H. Lee, A Zakharov-Shabt inverse scattering problem and the associated evolution equations, Chinese J. of Mathematics, Vol.12, No.4, 1984, p.223-233.
13. J-H. Lee,Analytic properties of a Zakharov-Shabat inverse scattering problem(I), Chinese J. of Mathematics, Vol.14, No.4., Dec 1986, p.225-248.
14. J-H. Lee,Analytic properties of a Zakharov-Shabat inverse scattering Problem(II), Chinese J. of Mathematics, Vol.16, No.2, p.81-110.
15. J-H. Lee, Global solvability of the derivative nonlinear Schrodinger equation, Transaction of A.M.S., Vol.314, No.1 (1989), p.107-118.
16. J-H. Lee,On the dissipative evolution equations associated with the Zakharov-Shabat system with a quadratic spectral parameter, Transaction of A.M.S., Vol.316, No.1 (1989), p.327-336.
17. J-H. Lee, A NxN Zakharov-Shabat system with a quadratic spectral parameter, Proc. of the 5th workshop on the nonlinear evolution equations and dynamical systems, Kolymbari, Crete, July 2-16, Springer-Verlag.
18. V.E. Zakharov and A.B. Shabat, A refine theory of two-dimensional self-focusing and one-dimensional self-modulation of waves in non-linear media, Soviet Physics JETP 34 (1972), p.62-69.

Homoclinic Orbits and Bäcklund Transformations for the Doubly Periodic Davey-Stewartson Equation

Y. Li and D.W. McLaughlin

Department of Mathematics, Princeton University, Princeton, NJ 08544, USA

This is a preliminary report of some of the results in a paper under the same title. We want to study the geometric structure of the solution manifold of the doubly periodic Davey-Stewartson equation as an infinite dimensional Hamiltonian system. Specifically we are concerned with the homoclinic structure of the following DS equation

$$\begin{aligned} iq_t - (\partial_x\partial_x + \alpha^2\partial_y\partial_y)q + r_- q = 0, \\ (\partial_x\partial_x - \alpha^2\partial_y\partial_y)r_- = -2\epsilon(\partial_x\partial_x + \alpha^2\partial_y\partial_y)|q|^2. \end{aligned} \tag{1}$$

There are three cases

1. DS1, focusing ($\epsilon = 1$) or defocusing ($\epsilon = -1$), $\alpha^2 = 1$;
2. DS2, focusing: $\alpha^2 = -1$, $\epsilon = 1$;
3. DS2, defocusing: $\alpha^2 = -1$, $\epsilon = -1$.

The corresponding Lax pair ($\hat{\mathbf{L}} - \hat{\mathbf{A}}$) representation of (1) is

$$\begin{aligned} \hat{\mathbf{L}}\psi &= \lambda\psi, \\ \psi_t &= \hat{\mathbf{A}}\psi, \end{aligned} \tag{2}$$

where

$$\hat{\mathbf{L}} = \alpha\partial_y - \sigma_3\partial_x + \begin{pmatrix} 0 & q \\ \epsilon\bar{q} & 0 \end{pmatrix},$$

$$-i\hat{\mathbf{A}} = -2\sigma_3\partial_x\partial_x + 2\begin{pmatrix} 0 & q \\ \epsilon\bar{q} & 0 \end{pmatrix}\partial_x$$

$$+\begin{pmatrix} 0 & q_x + \alpha q_y \\ \epsilon(\bar{q}_x - \alpha\bar{q}_y) & 0 \end{pmatrix} + 1/2\begin{pmatrix} r_+ + r_- & 0 \\ 0 & r_+ - r_- \end{pmatrix},$$

and where r_+ satisfies

$$\left(\partial_x\partial_x - \alpha^2\partial_y\partial_y\right) r_+ = -4\epsilon\alpha\partial_y\partial_x|q|^2.$$

Here $\sigma_3 \equiv diag(1, -1)$. We impose doubly periodic boundary conditions on q

$$q(x + mL_x, y + nL_y, t) = q(x, y, t),$$

for any $m, n \in \mathbf{Z}$. DS2 has a unique solution for the initial value problem under periodic boundary conditions, while DS1 has poseness problem as the following stability calculation indicates. A study of linear stability shows thst the uniform Stokes wavetrain

$$q^s_{(t)} = ce^{-2i\epsilon c^2 t + i\phi} \tag{3}$$

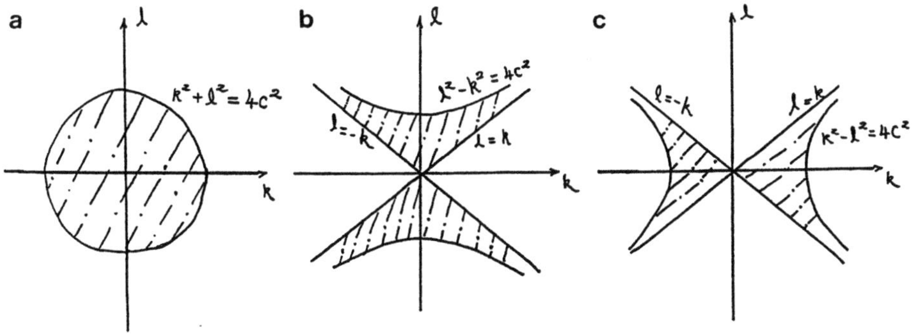

Figure 1: Unstable regions for a) focusing DS2, b) defocusing DS1, c) focusing DS1

where c is a positive constant, ϕ is the initial phase, is stable in DS2 defocusing case, and unstable in other cases. The unstable regions $(---)$ are shown below in Figure 1 (the perturbation: $e^{i(kx+ly+\omega t)}$)

Now we discuss Backlund transformations.

Theorem 1.

The DS2 equation(1) $(\alpha = i)$ and its Lax pair (2) are invariant under the following Backlund/Gauge transformation (in Darboux form) $q \longmapsto Q$:

$$\begin{aligned}
Q - q &= -2b, \\
(\epsilon\bar{Q} - \epsilon\bar{q} &= -2c), \\
R_- - r_- &= 2(\partial_x + i\partial_y)a - 2(\partial_x - i\partial_y)d, \\
R_+ - r_+ &= 2(\partial_x + i\partial_y)a + 2(\partial_x - i\partial_y)d, \\
\hat{\Psi} &= \Gamma\hat{\psi},
\end{aligned}$$

where

$$\begin{aligned}
\Gamma &= \begin{pmatrix} \Lambda + a & b \\ c & \Lambda + d \end{pmatrix}, \\
\Lambda &= i\partial_y - \lambda, \\
a &= \frac{1}{\Delta}\{-\psi_2\Lambda_2\bar{\psi}_2 - \epsilon\bar{\psi}_1\Lambda_1\psi_1\}, \\
b &= \frac{1}{\Delta}\{-\bar{\psi}_2\Lambda_1\psi_1 + \psi_1\Lambda_2\bar{\psi}_2\}, \\
c &= \frac{1}{\Delta}\{\epsilon\psi_2\Lambda_2\bar{\psi}_1 - \epsilon\bar{\psi}_1\Lambda_1\psi_2\}, \\
d &= \frac{1}{\Delta}\{-\bar{\psi}_2\Lambda_1\psi_2 - \epsilon\psi_1\Lambda_2\bar{\psi}_1\}, \\
\Lambda_1 &= i\partial_y - \lambda_0, \\
\Lambda_2 &= i\partial_y + \bar{\lambda}_0, \\
\Delta &= \epsilon|\psi_1|^2 + |\psi_2|^2.
\end{aligned}$$

Remarks:

For DS2 we have the relation $r_- = \bar{r}_-, r_+ = -\bar{r}_+$; for detail discussion of this see [1]. By the substitution $\psi = e^{\frac{\lambda_0}{\alpha}y}\tilde{\psi}$, $\hat{\psi} = e^{\frac{\lambda}{\alpha}y}\check{\psi}$, $\hat{\Psi} = e^{\frac{\lambda}{\alpha}y}\check{\Psi}$, theorem 1 reduces to the case $\lambda_0 = \lambda \equiv 0$, but for the Backlund transformation problem, it is more convenient to keep the λ. By setting $\partial_y = 0$, we arrive at the Backlund transformation for NLS equation [2]

If we apply the above Backlund transformation to the uniform Stokes wave train (3), we find that in the DS2 defocusing case the Backlund transformation on periodic functions reduces to a trivial case [3]; On the other hand, in the DS2 focusing case, there are indeed nontrivial Backlund transformations. Moreover, there is a one-to-one correspondence between linear instabilities and Backlund transformations [2]. More specifically, in the DS2 focusing case, each unstable mode generates a Backlund transform $Q^s_{(x,y,t)}$ of $q^s_{(t)}$ that is a function of x, y, and t which is periodic in both x and y, but not quasiperiodic in t.(It contains "*sech* like" behavior in t.) Moreover, it is homoclinic to $q^s_{(t)}$ up to a torus shift, i.e. as $t \to \pm\infty$

$$\mathbf{Q}^s_{(x,y,t)} \longrightarrow \begin{cases} q^s\{\frac{k_1^+ + ik_2^+}{k_1^+ - ik_2^+}\}, t \to +\infty \\ q^s\{\frac{k_1^- + ik_2^-}{k_1^- - ik_2^-}\}, t \to -\infty \end{cases}$$

where the k's are all real and $k_1^+k_2^+ - k_1^-k_2^- < 0$.(if $k_1^+k_2^+ - k_1^-k_2^- > 0$, the two asymptotic states interchange.) The situation is displayed in Figure 2.

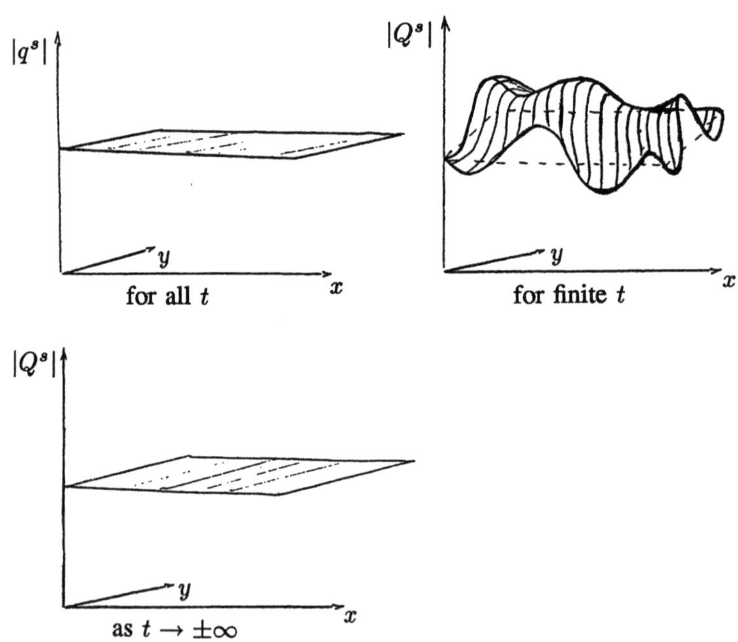

Figure 2: Homoclinic structures connected to constant potentials

Remarks:

For the discussion about Backlund transformations starting from general potential, and the relations between linear instabilities, squared eigenfunctions, special Floquet exponents on the covering surface, Backlund tansformations, and homoclinic structures, see [1].

We close the note by the emphasizing that for focusing DS2 equation, there is homoclinic structure in its solution manifold. Such homoclinic structures should be important when one studies the chaotic behavior of the focusing DS2 equations under perturbations which has not been done in 2+1 dimensions.

REFERENCES

[1] Y. Li and D. W. McLaughlin, Homoclinic orbits and Backlund transformations for the doubly periodic Davey-Stewartson equation,preprint(1991)

[2] Y. Li and D. W. McLaughlin, Morse and Melnikov Functions for NLS Pde's, submitted to Communications in Mathematical Physics (1992)

[3] N. Ercolani, M. G. Forest, and D. W. McLaughlin, Geometry of the Modulational Instability, Part3. Homoclinic orbits for the periodic sine-Gordon equation, PhysicaD43 (1990) 349-384

Volterra Operator Algebra for Zero Curvature Representation. Universality of KP

A. Yu. Orlov

Oceanology Institute, ul. Krasikova 23, 117218 Moscow, Russia

So called integrable equations have a zero measure among other equations, just like a gold section has zero measure among all possible sections. Both cases are marked out because of their beauty, and in the first case beauty show itself via a number of mathematical constructions that have appeared in soliton equations, and also via an infinite set of symmetries which are valuable not only for their own sake but also because they give us a lot of connections. Symmetries can unify different soliton equations into a single integrable hierarchy. Thus we get a large universal space which in future may appear to play a role similar to the role of complex plane in mathematics of nineteenth century. Here I present a few examples of links between different equations, which follow from symmetry approach.

1 Two-dimensional generalisation of Zakharov-Mikhailov and Belinsky–Zakharov systems

Let us recall that to construct a Lax representation for $1+1$ dimensional systems we use operators of the form:

$$\partial\psi(x,\lambda)/\partial x_i = A_i(x,\lambda)\psi(x,\lambda),\ \ A_i \in sl(n,\mathbf{C}),\ \ i=1,2,\ldots, \tag{1}$$

where matrixes A_i are rational functions in λ (for example, see [1] for review). To generalize (1) for $1+1+1$ dimension we take integral-differential operators in the form:

$$S_i = \partial/\partial\beta_i - V_i(x,\vec{b}),\ \ V_i \in gl(\infty)\ i=1,2,\ldots \tag{2}$$

where $\vec{b}=(\beta_1,\beta_2,\ldots)$ and V_i are Volterra operators with a degenerate kernel of rank one:

$$V_i = u_i(x,\vec{b})\int_{x_0}^{x} u_i^*(z,\vec{b})(*)dz, \tag{3}$$

u_i, u_i^* are some matrices. Let us assume that all "scalar products" exist:

$$u_{ik}(x,\vec{b}) = \int_{x_0}^{x} u_i u_k^* dz \tag{4}$$

Thus we obtain an algebra that can be called an algebra of Volterra operators with degenerate kernel of rank one with "fusion rule"(no sum with respect to indices):

$$(u_i\partial^{-1}u_i^*)u_k\partial^{-1}u_k^* = u_iu_{ik}\partial^{-1}u_k^* - u_i\partial^{-1}u_{ik}u_k^* \tag{5}$$

Looking at (5) we see that it is convinient to use operators S_i to construct equations in "zero curvature" form ($\partial_i = \partial/\partial\beta_i$):

$$[\partial_i - u_i\partial^{-1}u_i^*, \partial_k - u_k\partial^{-1}u_k^*] = 0 \tag{6}$$

2 Equations

For matrices u_i, u_i^*, $i = 1, 2, 3$, $u_3 = u_{12}$, $u_3^* = u_{21}$ as a result of (6) we get a system:

$$\partial_1 u_2 = u_1 u_3^*, \;\; \partial_2 u_1 = u_2 u_1, \;\; \partial u_3 = u_2^* u_1, \tag{7}$$

$$\partial_1 u_2^* = u_3 u_1^*, \;\; \partial_2 u_1^* = u_3^* u_2^*, \;\; \partial u_3^* = u_1^* u_2, \tag{8}$$

In contrast to $1+1$ case (1), even in the case of scalar S_i we obtain nonlinear equation: three wave resonant system (for this system see [1] and references therein).

Remark Reductions to $1+1$ dimension corresond to a special choice of x (or of β_i) dependence of matrices $u_i, u_i^*, i = 1, 2, 3$ and of wave function ψ, $(S\psi = 0)$. An example:

$$u_i(x, \vec{b}) = v_i(\vec{b})exp(a_i(\vec{b})x), \;\; u_i^*(x, \vec{b}) = v_i(\vec{b})exp(-a_i(\vec{b})x), \tag{9}$$

$$\psi(\lambda(\vec{b}), x, \vec{b}) = w(\lambda(\vec{b}), \vec{b})exp(\lambda(\vec{b}), x), \tag{10}$$

the choice of $x_0 = \pm\infty$ depends on sign of $Im(a_i - \lambda)$. In such a way we can obtain chiral field models [1] and Zakhaarov–Belinsky representation for gravity equations [2].

3 Operators S_i as symmetry operators

We follow [3]-[4] in this section.

Any $1+1+1$ Zakharov–Shabat integable equatin posseses a double ifinite set of symmetries, which can be parametrized by choice of two functions $u(x)$ and $u^*(x)$. There are arbitrary functions of x (certain reductions may be imposed to keep special type of equation which symmetries we are looking for). $xu(\vec{t})$ and $u^*(\vec{t})$ (where $\vec{t} = \{t_n\} = (x, y, \ldots)$ is a set of higher times) are wave functions, i.e. they are eliminated by L, A (or by L^*, A^* respectively) and all other higher operators form the hierarchy. These symmetries represent $gl(\infty)$ algebra and it can be proved that they form a full set of symmetries. For u and u^* we can take so-called Baker functions:$u = \psi(\lambda, \vec{t}), u^* = \psi^*(\lambda + \epsilon, \vec{t})$, where λ is a special parameter, and calculate a corresponding θ-function dynamics (this correspondence becomes one-to-one if we fix geometrical properties of the Baker functions [4]); then decomposing this special symmetry into double series in λ

and ϵ we obtain symmetries that represent Lie algebra of differential operators on the circle (a central extension appears when we treat the action of these symmetries on the θ function). As the particular cases we get the so-called master symmetries [5] and a full set of contact symmetries [6], [7].

If the higher operators from hierarchy have a form $\partial/\partial t_n - B_n$ then symmetry equations can be written in the Lax representation form:

$$[\partial/\partial t_n - B_n, \partial/\partial\beta - u\partial^{-1}u^*] = 0,\ n = 1, 2, \ldots \tag{11}$$

A double infinite number of symmetry operators S_i provides us with a double ifinite store of symmetries.

In the case of KP equation the generating symmetry equation can be written as a Lagrangian (in variables $u, u^*, \varphi = \int^x q$) system:

$$(\partial/\partial y - \partial^2 - q)u = 0,\ \partial/\partial y + \partial^2 + q)u^* = 0 \tag{12}$$

$$\partial q/\partial\beta = -2\partial(uu^*) \tag{13}$$

which was called the "full KP hierarchy" aand appears to be a degenerate case of Davey–Steuwardson equation [1].

4 Liouville and KP equations

Let us take $u = exp(\varphi/2)exp(\chi/2)$, $u^* = exp(\varphi/2)exp(-\chi/2)$. We rewrite (12, 13) in these variables and get:

$$-q = \varphi_x^2/2 + \varphi_{xx} + Q,\ Q = \chi_x^2/2 - \chi_y \tag{14}$$

$$\partial q/\partial\beta = -2\partial exp(\varphi), \tag{15}$$

$$\varphi_y = \chi_x\varphi_x + \chi_{xx} \tag{16}$$

Let us impose a constraint: $\partial Q/\partial\beta = 0$, then the last equation (16) and variable y will separate. Instead of (14, 15) we obtain:

$$\partial^2\varphi/\partial x\partial\beta = exp(\varphi) + C(\beta)exp(-\varphi) \tag{17}$$

with arbitrary function $C(\beta)$. For $C(\beta) \equiv 0$ we obtain Liouville equation.

Thus $1+1$ dimensional sh-Gordon (and Liouville) equations (17) appear to be the symmetry for KP 2+1 dimensional equation. (17) is a "point in direction y" symmetry.

If we concider symmetries of matrix Zakharov–Shabat systems [8] (so u and u^* are matrices) we can construct an analog of Liouville equation in a following way: $Su = 0, S^*u^* = 0$, or if $\bar\partial \equiv \partial/\partial\beta$ we get

$$\partial(u^{-1}\bar\partial u) = uu^*,\ \partial((\partial u^*)(u^*)^{-1}) = u^*u \tag{18}$$

For small u, u^* (18) may be treated a pair of WZWN equations.

5 An additonal meaning of $gl(\hat{\infty})$ [9] group transformation algebra

In this section for simlicity we treat scalar functions $u_i, u_i^*, i = 1,2,3, u_3 = u_{12}, u_3^* = u_{21}$, which vanish when $x \to -\infty$, and take $x_0 = -\infty$ in the S_i-operators (2,3). We know that KP higher flows correspond to a Cartan subalgebra of $gl(\hat{\infty})$ algebra. In terms of KP τ-function these flows are described via Hirota equations. Let us treat vertex operators [9]:

$$X(\lambda) = exp(\sum \lambda^n t_n) exp(-\sum (n\lambda^n)^{-1} \partial/\partial t_n), \tag{19}$$

$$X^*(\lambda) = exp(-\sum \lambda^n t_n) exp(\sum (n\lambda^n)^{-1} \partial/\partial t_n), \tag{20}$$

where λ is a specral parameter. Let us take linear combinations

$$b_i = \int a_i(\lambda) X(\lambda) d\lambda, \; b_i^* = \int a_i^*(\mu) X^*(\mu) d\mu \tag{21}$$

with the following anticommutators:

$$b_i b_k + b_k b_i = 0, \; b_i^* b_k^* + b_k^* b_i^* = 0, \; b_i b_k^* + b_k^* b_i = b_{ik}, \; b_{ik} \in \mathbf{C}. \tag{22}$$

Let us concider $A_i \equiv b_i b_i^* \in gl(\hat{\infty})$, we introduce a group of times $\bar{x}_i$ that describes the shifts of KP τ-function via A_i:

$$\partial \tau / \partial \bar{x}_i = A_i \tau \tag{23}$$

Let us introduce wave functons (which are linear combinatons of Baker functions):

$$u_i = \tau^{-1} b_i \tau, \; u_i^* = \tau^{-1} b_i^* \tau \tag{24}$$

Theorem 5.1 *If KP τ-function obeys (23) then $\varphi_i = ln(u_i u_i^*)$ obey Liouville equations:*

$$\partial^2 \varphi_i / \partial x \partial \bar{x}_i = exp(\varphi_i) \tag{25}$$

or the same $S_i u_i = 0, \; S_i^* u_i^* = 0$.

Theorem 5.2 *Let us consider two compatible flows (23), for example with $i = 1, j = 2$:*

$$[\partial/\partial \bar{x}_1 - A_1, \partial/\partial \bar{x}_2 - A_2] = 0 \tag{26}$$

Then u_i, u_i^*, $i = 1,2,3$, $u_3 = u_{12}$, $u_3^* = u_{21}$ obeys three-wave system (7, 8) with $\beta_{12} = \bar{x}_{12}$.

The proof is based on straightforward calculation of action of $A_i \in gl(\hat{Inf})$ on the Segal–Wilson [10] τ-function. The main point is an explicit formula of the Green function of $\partial/\partial\bar{\lambda}$ via Baker functions [4].

Let us treat a simple example. Similar to [1] (Appendix,section 5) one can find so-called soluton that depends on functonal parameter on the some background (the latter is described by some KP solution q, some τ-function τ). The corresponding dressing $\hat{K}$ (see [1]) is a dressing with the degenerate kernel which id constructed via a pair of arbitrary wave functions u, u^* decaying at $x = -\infty$:

$$\hat{K}(x,a) = 1 + a(1 + a\int_{-\infty}^{x} u(z)u^*(z)dz)^{-1}u(x)\int_{-\infty}^{x} u^*(z)(*)dz \qquad (27)$$

This dressing may be treated as the Backlund transformation. If we apply this dressing to functions u, u^* themselves,they begin to depend on a:

$$u(x,a) = F^{-1}u(x,0),\ u^*(x,a) = F^{-1}u^*(x,0) \qquad (28)$$

$$F = a^{-1} + \int_{-\infty}^{x} u(z,0)u^*(z,0)dz \qquad (29)$$

Then $\varphi = ln(uu^*)$ obeys Liouville equation,$a = \beta exp(\beta h)$. Then double dressing via u_1, u_1^* and then via u_2, u_2^* with conditions $\int_{-\infty}^{\infty} u_{12}u_{21}dx = 0$, gives us three wave solution in terms of KP waave functions.

6 AKNS system, BKP equation, KdV

Let us treat the following reduction of the full KP hierarchy (12, 13): $\partial/\partial\beta = \partial/\partial x$ [11]. We get AKNS system, KP wave functions u, u^* being AKNS solutions (something in this directon was also made in [12]). The same statement of the BKP [9] equation gives us KdV soluton via BKP wave function. The corresponding Lax representations are $[\partial/\partial y - \partial^2 - uu^*, \partial - u\partial^{-1}u] = 0$ and $[\partial/\partial y - \partial^3 - \partial u, -u\partial^{-1}] = 0$.

Acknowledgments I am grateful to A.Fokas, A.Zabrodin, A.Levin, E.Schulman and especially to P.Grinevich for stimulating talks. I thank V.E.Zakharov and D.Kaup for support and interest, D.Leshiner and V.Rosonoer for help and kind hospitality.

REFERENCES

1. V.E.Zakharov, S.V.Manakov, S.P.Novikov, L.P.Pitaevsky, Soliton Theory,Nauka,Moscow 1980.

2.V.A.Belinsky, V.E.Zakharov, JETP 1(1979),1

3.A.Yu.Orlov, E.I.Schulman, Lett.Math.Phys.12(1986),

A.Yu.Orlov,"Vertex operators,$\bar{\partial}$-problem,variational identities and Hamiltonian formalism in 2+1 integrable systems" in: Proc.Worksh. Plasma ... in Kiev 11987,ed. V.G.Bar'yahtar, World Sci.Pub.,Singapure 1988.

4. P.G.Grinevich, A.Yu.Orlov, "KP theory and Virasoro action on Segal–Wilson τ-function" in: Proc. Work. Quantum Field THeory, Alushta 1989, ed. Belavin, A.Zamolodchikov. Springer 1990;

P.G.Gurevich, A.Yu.Orlov "Flag space in KP theory and Virasoro action on $det\bar{\partial}$ on Riemann surfaces",Cornell Univ. preprint 945/89.

5. W.Devel, B.Fuchssteiner, Phys.Lett. 88A(1982),7

6. D.David, D.Levi, D.Winternitz, Phys.Lett. 118A(1986),890

7. P.Santini, A.Fokas, Comm.Math.Phys,6(1987)

8. V.E.Zakharov, A.B.Shabat, Funk.Analis i ego Pril.3(1974)

9. E.Date, M.Kashiwara, M.Jimbo, T.Miwa, "Group transformation theory of KP", in: Proc.RIMS Symp., ed. M.Jimbo, T.Miwa, World Si Pub., Singapore 1983

10. S.Segal, G.Wilson, Pub.IHES,61(1985),4

11. A.Yu.Orlov, "Symmetries for unifying different soliton systems into a single integrable hierarchy", Preprint BINS/Oce-04/03

12. B.Konopelchenko, V.Strampt, A talk given at Gallipoli, June 1991

Calculation of All Commutation Relations Among Scattering Data Without Using the R-Matrix Approach

G.D. Pang and A.S. Fokas*

Department of Mathematics and Computer Science, Clarkson University, Potsdam, NY 13699-5815, USA
*On leave of absence from Institute of Physics, Chinese Academy of Sciences, Beijing, P.R. China

We have re-derived all the commutation relations among the scattering data for the quantum (1+1)-dimensional nonlinear schrödinger model, without using the R-matrix approach developed by the Leningrad school [1]. We hope that the approach presented here will be useful for (2+1)-dimensional models.

At the early stage in the development of the quantum inverse scattering method (QISM) , there were two approaches. The one of Thacker and Wilkinson [2], and the one of the Leningrad-school (now called the R-matrix approach)[1]. The first approach was more close to the (classical) inverse scattering transform (IST) than the R-matrix approach, but only a part of the commutation relations among scattering data was derived in [2]. On the other hand, the R-matrix approach generated all the desired commutation relations as well as led to a deep insight into the structures of integrable systems.

Recently, there has emerged a considerable interest in the study of quantum (2+1)-dimensional [(2+1)D] integrable systems [3]. Although some progress has been made (see, for example, [4]), many important questions still remain open. One of them is how to generalize the QISM into (2+1)D models. Since the IST for classical (2+1)D equations has been well developed [5], and since almost nothing is known about the R-matrix in (2+1)D, we thought that it may be easier to generalize into (2+1)D the Thacker and Wilkinson's approach instead of the R-matrix. This motivated us to reconsider the Thacker and Wilkinson approach and to extend it appropriately so that all commutator relations for the (1+1)D can be derived within this framework.

We present below the detailed calculations for a concrete model, namely, the nonlinear schrödinger model (NSM).

The related linear equation for the NSM is

$$\partial_x \psi = \frac{1}{2} i\lambda\sigma\psi + :Q\psi: \tag{1}$$

where

$$\sigma = \begin{pmatrix} 1 & 0 \\ 0 & -1 \end{pmatrix}, \qquad Q = \begin{pmatrix} 0 & i\sqrt{c}\phi \\ -i\sqrt{c}\phi^+ & 0 \end{pmatrix},$$

$c > 0$ is the coupling constant, and ϕ and ϕ^+ are field operators which satisfy the following commutation relations

$$[\phi(x), \phi^+(x')] = \delta(x - x') \quad , \qquad [\phi(x), \phi(x')] = 0 \quad . \tag{2}$$

In (1), : : means the normal-ordering, namely, grouping all ϕ^+'s to the left and all ϕ's to the right. Here we also assume that the quantum fields ϕ and ϕ^+ satisfy the asymptotic conditions

$$\Gamma\phi(x) \sim 0 \quad , \qquad \phi^+(x)\Gamma \sim 0 \tag{3}$$

when $x \to \pm\infty$ for an arbitrary operator Γ.

One fundamental solution of (1) is given by

$$u = e_\lambda + G : Qu : \tag{4}$$

where

$$e_\lambda = exp(\frac{1}{2}i\lambda\sigma x) \quad , \qquad Gf(x) \equiv \int_{-\infty}^{x} dx' e^{\frac{1}{2}i\lambda\sigma(x-x')} f(x') \quad . \tag{5}$$

The related scattering data (matrix) is defined by

$$S(\lambda) = I + \int_{-\infty}^{\infty} dx e^{-\frac{1}{2}i\lambda\sigma x} : Qu : \tag{6}$$

where I is the 2×2 unitary matrix. Our purpose is to calculate the commutation relations among the elements of different scattering data $S(\lambda)$ and $S(\lambda')$, based on (2). In order to do that, we first calculate the commutator between $S(\lambda)$ and an arbitrary scalar operator Λ .

Since

$$[: Qu :, \Lambda] =: (1+QG+QGQG+...)Q : \Lambda u - \{: (1+QG+QGQG+...) :\}\Lambda\{: Qu :\} \tag{7}$$

and

$$: Qu := Qu - \frac{1}{2}c\sigma_+\sigma_- u$$

$$: (1 + QG + QGQG + ...)Q := (1 + QG + QGQG + ...)(Q - \frac{1}{2}c\sigma_+\sigma_-), \tag{8}$$

where

$$\sigma_+ = \begin{pmatrix} 0 & 1 \\ 0 & 0 \end{pmatrix}, \qquad \sigma_- = \begin{pmatrix} 0 & 0 \\ 1 & 0 \end{pmatrix},$$

we can easily show that

$$[S, \Lambda] = \int_{-\infty}^{\infty} dx \chi(x)[Q(x), \Lambda]u(x), \tag{9}$$

where

$$\chi(x) = e_{-\lambda} + \int_{x}^{\infty} dx' : \chi(x')Q(x') : e^{\frac{1}{2}i\lambda\sigma(x'-x)}. \tag{10}$$

Eq. (8) is derived using the property of G and setting $\theta(0) = \frac{1}{2}$. Using (9) twice, we can obtain the following expression for the commutator relations among the scattering data

$$[S(\lambda) \overset{\otimes}{,} S(\lambda')]_{il,jm} \equiv [S_{ij}(\lambda), S_{lm}(\lambda')] = c\int_{-\infty}^{\infty} dx[\chi_{i1}u'_{1m}\chi'_{l2}u_{2j} - \chi_{i2}u'_{2m}\chi'_{l1}u_{1j}] \tag{11}$$

where u' and χ' means that the related spectral parameter is λ' instead of λ.

Noting that that $\chi_{ij}(x)$ and $u_{lm}(x)$ commute and that

$$\partial_x(\chi u') = \frac{1}{2}i(\lambda' - \lambda)\chi\sigma u' \quad , \tag{12}$$

we can further write eq.(11) as

$$[S(\lambda) \overset{\otimes}{,} S(\lambda')] = \frac{ic}{\lambda - \lambda'}\{(\chi \otimes \chi')(u' \otimes u)\mathbf{P}\} |_{x=-\infty}^{x=\infty}, \tag{13}$$

where

$$\mathbf{P}_{ij,kl} = \delta_{il}\delta_{kj} \quad , \qquad (A \otimes B)_{ij,kl} = A_{ik}B_{jl} \quad . \tag{14}$$

A crucial step in the quantum theory is to take the proper limits for the right hand side of (13). It can be shown that

$$\partial_x[u' \otimes u] =: \Omega(\lambda', \lambda)[u' \otimes u] : \quad , \qquad \partial_x[\chi \otimes \chi'] = - : [\chi \otimes \chi']\Omega(\lambda, \lambda') : \tag{15}$$

where

$$\Omega(\lambda, \lambda') = (\frac{1}{2}i\lambda\sigma + Q) \otimes I + I \otimes (\frac{1}{2}i\lambda'\sigma + Q) + c\sigma_+ \otimes \sigma_- \quad . \tag{16}$$

From (15) we see that the limit of $u' \otimes u$ is not equal to the tensor product of the limits of $u'(\chi)$ and $u(\chi')$ because of the extra term $c\sigma_+ \otimes \sigma_-$ in (16):

$$u' \otimes u = \begin{cases} e^{x\Omega_0(\lambda',\lambda)}(S(\lambda') \otimes S(\lambda)) & \text{when } x \to \infty \\ e^{x\Omega_0(\lambda',\lambda)} & \text{when } x \to -\infty; \end{cases}$$

similarly

$$\chi \otimes \chi' = \begin{cases} e^{-x\Omega_0(\lambda,\lambda')} & \text{when } x \to \infty \\ (S(\lambda) \otimes S(\lambda'))e^{-x\Omega_0(\lambda,\lambda')} & \text{when } x \to -\infty, \end{cases} \tag{17}$$

where

$$\Omega_0(\lambda, \lambda') = \frac{1}{2}i\lambda\sigma \otimes I + I \otimes \frac{1}{2}i\lambda'\sigma + c\sigma_+ \otimes \sigma_-$$

and

$$e^{x\Omega_0(\lambda',\lambda)} = (e^{\frac{1}{2}i\lambda'\sigma x} \otimes e^{\frac{1}{2}i\lambda\sigma x})(1 + c\sigma_+ \otimes \sigma_- \int_0^x dte^{-i(\lambda'-\lambda)t}) \quad .$$

The expression (17) can be proved by using (3)[6]. Since $\chi_{ij}(x)$ and $u_{kl}(x)$ commute, the limit of the product of $(\chi \otimes \chi')$ and $(u' \otimes u)$ is equal to the product of the limits of $(\chi \otimes \chi')$ and $(u' \otimes u)$, and thus we can calculate the limits in (13) using (17). After some simple calculations, we obtain

$$R_+(\lambda', \lambda)S(\lambda') \otimes S(\lambda) = S(\lambda) \otimes S(\lambda')R_-(\lambda', \lambda) \tag{18}$$

where

$$R_{\pm}(\lambda',\lambda) = \begin{pmatrix} P\frac{1}{\lambda'-\lambda} & 0 & 0 & 0 \\ 0 & \pm i\pi\delta(\lambda'-\lambda) & (\lambda'-\lambda+ic)(\lambda'-\lambda+i0^+)^{-2} & 0 \\ 0 & (\lambda'-\lambda-ic)^{-1} & \mp i\pi\delta(\lambda'-\lambda) & 0 \\ 0 & 0 & 0 & P\frac{1}{\lambda'-\lambda} \end{pmatrix}.$$

Here we have used the well-known formula $\lim_{x\to\infty} P\frac{exp(i\lambda x)}{\lambda} = i\pi\delta(\lambda)$. Eq.(18) is exactly the same as that obtained by using the R-matrix approach, and from it one can construct all the eigenstates for the NSM[1].

In summary, we have re-derived all the commutation relations among the scattering data for the NSM without using the R-matrix approach. We hope that the present approach can be generalized for (2+1)D models.

ACKNOWLEDGEMENTS: One of the author(G.D.Pang) wishes to thank the Department of Mathematics and Computer Science for its hospitality. We appreciate many helpful discussions with A.R. Its. This work was partially supported by the National Science Foundation under Grant Number DMS-8803471, and Air Force Office of Scientific Research under Grant Number 87-0310.

REFERENCES

[1]L.D. Faddeev, Sov. Sci. Rev. Math. Phys. C1(1980)109.

[2] H.B. Thacker and David Wilkinson, Phys. Rev. D19(1979)3660.

[3] C.L. Schultz, M.J. Ablowitz and D. Bar Yaacov, Phys. Rev. Lett. 59(1987)2825; P.P. Kulish and V.D. Lipovsky, Phys. Lett. A127(1988)413.

[4]G.D. Pang, F.C. Pu and B.H. Zhao, Phys. Rev. Lett. 65(1990)3227.

[5]A.S. Fokas and M.J. Ablowitz,in *NonlinearPhenomena*, edited by K.B. Wolf, Lecture Notes in Physics Vol. 189(Springer-Verlag, Berlin, 1983).

[6] We thank Prof. A.R. Its for informing us, after this work was done, about the paper of B. Davies and T.D. Kieu (Inverse Problems 2(1986) 141), where the expression (17) was proved.

Fluctuating Solitons of the KdV Hierarchy

L. Trlifaj

Institute of Physics, Czechoslovak Academy of Sciences,
Na Slovance 2, 180 40 Prague 8, Czechoslovakia

One can associate the AKNS system

$$T_x = 2^{1/2} Z\, T, \qquad T_t = 2^{1/2} A\, T, \tag{1}$$

where the only nonzero elements of the 3 by 3 matrices are

$$\begin{aligned} Z_{12} = 1 = -Z_{23}, \qquad -Z_{21} = \lambda - q = Z_{32}, \\ A_{12} = b = -A_{23}, \qquad -A_{11} = 2^{-1/2} b_x = A_{33}, \\ -A_{21} = \frac{1}{2} b_{xx} + (\lambda - q)\, b = A_{32}, \end{aligned} \tag{2}$$

with every equation of the KdV hierarchy

$$q_t = S(IS)^{m-1} q \qquad (m = 1, 2, 3, \ldots). \tag{3}$$

Here

$$S = \partial_x^3 - 4q\partial_x - 2q_x \tag{4}$$

and $I = \int^x dx$ (general integration constant is set to zero). A general solution of (1a) is

$$\tilde{T} = (w, \ \ 2^{-1/2} w_x, \ \ - \ \frac{1}{2} w_{xx} - (\lambda - q) w), \tag{5}$$

where w is a quadratic form of two independent solutions ψ_1 and ψ_2 to the Schrödinger equation (SE) [1, 2]

$$\psi_{xx} + (\lambda - q)\psi = 0 \tag{6}$$

on the full line. The solution (5) obeys also [3] the equation

$$(S + 4\lambda\partial_x) w = 0. \tag{7}$$

The function b is

$$b = (-4\lambda)^m - 2 \sum_{j=0}^{m-1} (-4\lambda)^j \, (IS)^{m-j-1} q. \tag{8}$$

The first row of (1b) thus reads

$$w_t = -b_x w + b w_x, \tag{9}$$

and determines the time evolution of w. There is the first integral of both (1a) and (1b)

$$2T_1T_3 + T_2^2 = ww_{xx} - \frac{1}{2}w_x^2 + 2(\lambda - q)w^2 = -\frac{1}{2}u^2. \tag{10}$$

The constant $\frac{1}{2}u^2$ is a product of the discriminant of the form w and of the squared Wronskian of ψ_1 and ψ_2. The equation (10) was applied by Milne [4] to the calculation of the bound–state energies.

When the potential q is a periodic (or u.a.p.) function $q(x+c) = q(x)$,

$$u^2 = 4c^2(\lambda_0 - \lambda) \prod_{n=1}^{\infty} (\lambda_n - \lambda)(\mu_n - \lambda)\ell_n^{-2} \qquad (\ell_n = (\pi n/c)^2) \tag{11}$$

for the (only) periodic (with respect to x) solution of (10)

$$w = \prod_{n=1}^{\infty} \Lambda_n(x,t) - \lambda| \; \ell_n^{-1}. \tag{12}$$

$\lambda_0 < \lambda_n \leq \mu_n \quad (n = 1,2,3,...)$ are edges of the allowed (forbidden) bands corresponding to the primary periodic SE (6), simple roots $\lambda_n \leq \Lambda_n(x,t) \leq \mu_n$ oscillate within the SE forbidden bands. Both (11) and (12) are entire functions of λ. This property turns out convenient for solving the inverse problem of SE (6) and the KdV equations as well. In the inverse problem, the potential q is expressed in terms of Λ_n, which obey an infinite system of the first–order differential equations with respect to x (for details see [1, 2]). A similar system for time derivatives of Λ_n follows from (9). This means, at the same time, that the time dependence of q, i. e. the solution of (3) is given. The corresponding formulas are well-known (e. g. see [5]).

By the transformation

$$W = \frac{1}{4}p\;(fw_x - 2gw|^2 - f^2u^2)\;w^{-1} \tag{13}$$

where f, g are some functions of x, t and p is a constant, one introduces a new equation (10) (and correspondingly a new AKNS system) with the transformed potential

$$Q = q + f_{xx} - 2g_x|f^{-1}. \tag{14}$$

The spectral function then is $p^2\;u^2$ (instead of u^2). This transformation can be isospectral or non–isospectral depending on the choice of f, g and p. Choosing

$$f = \sum_{i,j=0}^{r} (\sigma_i - \lambda)^{-1}\;{}_a\mathcal{D}_{ij,x}\;{}_a\mathcal{D}_{ij}^{-1} \qquad (a = \pm 1), \tag{15}$$

$$g = 1 + \sum_{i,j=0}^{r} (\sigma_i - \lambda)^{-1}\;{}_a\mathcal{D}_{ij,x}\;m_a\;{}_a\mathcal{D}_{ij}^{-1}, \tag{16}$$

$$p = \prod_{i=1}^{r} (\sigma_i - \lambda), \tag{17}$$

where

$$
{}_a\mathcal{D}_{ij} = \delta_{i,j} + {}_ah_i\,{}_ah_j a \int_{-a\infty}^{x} dy \exp\{\int_{x_0}^{y} dz m_a(z,t,\sigma_i) + m_a(z,t,\sigma_j)|]\}, \tag{18}
$$

$$
m_a(x,t,\lambda) = \frac{1}{2} w_x(x,t,\lambda) + au(\lambda)w(x,t,\lambda)|w^{-1}(x,t,\lambda)
$$

$$
(a = \pm 1; \quad i,j = 1,2,...,r), \tag{19}
$$

one inserts r isolated eigenenergies $\sigma_i \quad (i = 1,...,r)$ into the forbidden bands, which is the only possibility. In order that

$$
Q(x,t) = q(x,t) - 2\{\ln\det\ {}_a\mathcal{D}(x,t)|]\}_{xx} \tag{20}
$$

corresponding to the choice (15–19) and to (13) may be a solution of (3), it is necessary that

$$
{}_ah_i(t) = \ {}_ah_i(t_0)\exp\{\int_{t_0}^{t} -\frac{1}{2} b_{x_0}(q(x_0,s),\sigma_i) + b(q(x_0,s),\sigma_i)m_a(x_0,s,\sigma_i)|ds\}. \tag{21}
$$

There is an alternative way of solving the preceding non-isospectral problem. One fixes time at $t = t_0$, chooses constant ${}_ah_i$ and solves the inverse problem. In comparison to w there are r additional factors (roots) in the infinite–product expansion of W, which correspond to the inserted eigenvalues $\sigma_i \quad (i = 1,...,r)$. Therefore also r additional terms appear in the sum by means of which Q is reconstructed. The x and t dependence of roots in this sum is given by the infinite systems of the first–order differential equations again.

References

[1] L. Trlifaj, in *Lecture Notes in Physics* **325**, eds. E. Brändas and N. Elander (Springer Verlag, Berlin, 1989), 57.

[2] L. Trlifaj, *Inverse problems* **5** (1989) 1145.

[3] M. Appelle, *C. R. Acad. Sci.* **91** (1880), 211, Paris.

[4] W. E. Milne, *Phys. Rev.* **35** (1930), 863.

[5] B. M. Levitan, *Obratnye zadatchi Sturma–Liouvilla* (Nauka, Moscow, 1984, in Russian)

The Periodic Fixed Points of Bäcklund Transformations

J. Weiss

Aware, Inc., One Memorial Drive, Cambridge, MA 02142, USA

The periodic fixed points of the Bäcklund transformations are finite dimensional invariant manifolds for the flow of the system. The dynamics occur as commuting hamiltonian flows on this finite dimensional manifold. We examine the flow of the KdV periodic fixed points in the neighborhood of steady states and reductions. These are analogous to a flow in the neighborhood of a sequence of heteroclinic points.

The periodic fixed points of Bäcklund transformations also define a natural *factorization* of the two-dimensional Toda lattice as *commuting hamiltonian flows.* These have an interesting connection with caustic surfaces and the Laplace-Darboux transformation [5].

The Korteweg-de Vries equation

$$u_t + u_{xxx} + 3uu_x = 0 \tag{1}$$

has meromorphic singularities of the form

$$u = \phi^{-2} \sum_{j=0}^{\infty} u_j \phi^j \tag{2}$$

about the singularity manifold $\phi(x,t) = 0$. In the above we find that ϕ , u_4 and u_6 are arbitrary functions, and it is required that, for (2) to be well defined, ϕ be *non-characteristic.* That is, $\phi_x \neq 0$ when $\phi = 0$. For locally analytic data we can show that (2) converges in a neighborhood of $\phi = 0$ [2, 3].

Now, we truncate (2) after the constant term to obtain

$$u = u_0/\phi^2 + u_1/\phi + u_2. \tag{3}$$

We find that

$$u = -4\phi_x^2/\phi^2 + 4\phi_{xx}/\phi + u_2 \tag{4}$$

$$u_2 + \lambda = -(\phi_{xxx}/\phi_x) + \frac{1}{2}(\phi_{xx}/\phi_x)^2 \tag{5}$$

$$\phi_t/\phi_x + \{\phi; x\} = \lambda \tag{6}$$

where u and u_2 satisfy the KdV equation. Again,

$$\{\phi; x\} = \phi_{xxx}/\phi_x - \frac{3}{2}(\phi_{xx}/\phi_x)^2 \tag{7}$$

, the Schwarzian derivative [1], is the unique differential invariant [4] of the Moebius group

$$\phi = (a\psi + b)/(c\psi + d). \tag{8}$$

The Moebius group is the unique group of conformal (monodromny preserving) automorphisms of the complex ϕ (Riemann) sphere.

The relations (5) and (6) imply that u_2 satisfy (1) since

$$u_{2,t} = -(\frac{\partial}{\partial x} + V)\frac{\partial}{\partial x}(\frac{\partial}{\partial x} - V)\frac{\phi_t}{\phi_x}$$

where

$$V = \frac{\phi_{xx}}{\phi_x}. \tag{9}$$

This definition of V obtains the modified KdV equation

$$V_t + V_{xxx} - \frac{3}{2}V^2 V_x = \lambda V \tag{10}$$

from (6).

The Bäcklund-Darboux equation (4) may be written in the form

$$u = 4\frac{\partial^2}{\partial x^2}\ln\phi + u_2 \tag{11}$$

where

$$u_2 + \lambda = -\frac{\partial}{\partial x}(\phi_{xx}/\phi_x) - \frac{1}{2}(\phi_{xx}/\phi_x)^2 \tag{12}$$

and ϕ satisfies the Schwarzian-KdV equation (6).

We regard (4) as a *Miura transformation* from (6) to (1). It also has the form of a *Ricati equation* in the variable $V = \phi_{xx}/\phi_x$ and can be linearized by the substitution $V = -2v_x/v$. This obtains the linear equation for v

$$2v_{xx} = (u_2 + \lambda)v \tag{13}$$

and the identification $\phi_x = v^{-2}$. From (6) the additional linear equation

$$v_t = (2\lambda - u_2)v_x + \frac{1}{2}u_{2,x}v \tag{14}$$

is found. By construction (13) and (14) are the Lax pair for the KdV equation and imply u_2 is a KdV solution.

The linearizing substitution for ϕ has the form

$$\phi = v_1/v_2$$

where v_1 and v_2 are solutions of (13).

In terms of the linear equation (13) the Bäcklund transformation is the classical Darboux transformation for adding elements to the spectra [5, 6].

Consider next the Bäcklund-Darboux transformation for the modified KdV equation (10)

$$V = 2\frac{\phi_x}{\phi} + V_1$$

where

$$V_1 = -\frac{\phi_{xx}}{\phi_x},$$

and ϕ is a solution of (6). Comparing (9), (10) to the above obtains the *discrete symmetry* of the modified equation

$$V \Rightarrow -V.$$

Now, consider the Bäcklund transformations for the Schwarzian KdV equation (6). From the invariance of the Schwarzian derivative we have the invariance of (6) under the Moebius group (8). This invariance is also found by examining the singularities of (6), and finding that the truncated expansion

$$\phi = \psi^{-1}\phi_0 + \phi_1 \tag{15}$$

requires constant ϕ_0, ϕ_1 and ψ satisfies (6). In other words, we find the Moebius invariance.

An additional Bäcklund transformation is found from the *discrete symmetries* of the modified equation (10). That is, $V \Rightarrow -V$ implies the invariant transformation

$$\phi_x = 1/\psi_x \tag{16}$$

for (6). The time dependent form of this transformation is found from (6) and (16). That is,

$$\phi_t/\phi_x + \psi_t/\psi_x + (\phi_{xx}/\phi_x)(\psi_{xx}/\psi_x) = 2\lambda \tag{17}$$

and (16) imply, by $\phi_{xt} = \phi_{tx}$ and $\psi_{xt} = \psi_{tx}$, that ϕ and ψ both satisfy (6). Therefore, we have a strong Bäcklund transformation for (6).

We compose $\phi = -1/\psi$ and (16), (17) and get

$$\phi_{j+1,x} = \phi_j^2/\phi_{j,x} \tag{18}$$

$$\frac{\phi_{j+1,t}}{\phi_{j+1,x}} + \frac{\phi_{j,t}}{\phi_{j,x}} = (\frac{\phi_{j,xx}}{\phi_{j,x}})^2 - 4\frac{\partial^2}{\partial x^2}\ln\phi_j + 2\lambda \tag{19}$$

where $j = 1, 2, 3, \cdots, \pmod N$. The periodic fixed points continue to define a strong Bäcklund transformation since

$$\phi_{j+1,xt} = \phi_{j+1,tx}$$

continues to imply that the set $\{\phi_j, j = 1, 2, 3, \cdots, \pmod N\}$ are solutions of (6).

Now, define the variables $\xi_j = \phi_{j,x}/\phi_j$ and find that (16) is $\xi_{j+1}\xi_j = \phi_j/\phi_{j+1}$. By logarithmic differentiation

$$\xi_{j+1,x}/\xi_{j+1} + \xi_{j,x}/\xi_j = \xi_j - \xi_{j+1} \tag{20}$$

where $j = 1, 2, 3, \cdots \pmod N$.

We claim [7] that (20) is a completely integrable hamiltonian system.

First, we introduce the *circulant matrices* [8] A and B where

$$A = \mathbf{circ}[\mathbf{1}, \mathbf{1}, \mathbf{0}, \cdots, \mathbf{0}]$$

$$B = \mathbf{circ}[\mathbf{1}, -\mathbf{1}, \mathbf{0}, \cdots, \mathbf{0}]. \tag{21}$$

We note that $\det B = 0$ for any N and the null vector of B is $(1, 1, \cdots, 1)^t$. We also define the variable $\hat{\beta} = (\beta_1, \beta_2, \cdots, \beta_N)^t$ where $\beta_j = \ln \xi_j$. Then, system (20) is

$$A\hat{\beta}_x = B\hat{\xi}. \tag{22}$$

Applying the null vector of B to (22) we find the *Casimir* integral

$$H_N = \prod_{j=1}^{N} \xi_j. \tag{23}$$

The complete set of independent integrals [7] in involution for the system (22) are

$$H_{N-2m} = L^m \circ H_N \tag{24}$$

,where

$$L = \sum_{j=1}^{N} \frac{\partial^2}{\partial \xi_j \partial \xi_{j+1}}. \tag{25}$$

The higher order flows

$$A\hat{\beta}_x = B diag[\xi_1, \cdots, \xi_N] \nabla H_k \tag{26}$$

,where $k = N, N-2, N-4, \cdots$, $\beta_j = \ln \xi_j$, can be shown to commute [7, 9].

If N is odd then A is invertible and $\Omega = A^{-1}B = [0, -1, 1, -1, 1, \cdots, -1, 1]$ is anti-symmetric. The anti-symmetric matrix

$$M_{\hat{\xi}} = diag[\xi_1, \cdots, \xi_N] \Omega diag[\xi_1, \cdots, \xi_N] \tag{27}$$

defines the bracket

$$\{G, H\} = (\nabla G)^t M_{\hat{\xi}} \nabla H \tag{28}$$

and the systems (26) are

$$\hat{\xi}_x = M_{\hat{\xi}} \nabla_{\hat{\xi}} H_k. \tag{29}$$

For odd N the KdV flow is

$$\hat{\xi}_x = M_{\hat{\xi}} \nabla H_1 \tag{30}$$

$$\hat{\xi}_t = M_{\hat{\xi}} \nabla H_3 \tag{31}$$

where $\{H_1, H_3\} = 0$.

These equations have an interesting type of reduction [3]. Setting two adjacent components to zero

$$\xi_j = \xi_{j+1} = 0$$

reduces a system with N components to a system of the same form with $N-2$ components. Therefore, a system with N components contains, by reduction, all the smaller systems of the same type. The embedded subsystems are analogous to the heteroclinic points in an area preserving map. There are orbits that connect the different subsystems and these and are analogous to the orbits that connect heteroclinic points [3].

For instance, if a specific component is set to zero,

$$\xi_j = 0,$$

then we find that as $t \Rightarrow \infty$, $\xi_{j+1} \Rightarrow 0$, and as $t \Rightarrow -\infty$, $\xi_{j-1} \Rightarrow 0$. Therefore, this orbit connects one reduction at $t = -\infty$ with a different reduction at $t = \infty$.

If ξ_j is small the orbit will visit the neighborhoods of successive reductions (subsystems) in a periodic manner.

To visualize the above orbits we have studied the KdV flows numerically, using the Taylor series method of Y.F. Chang [11]. For ξ_j vanishing or small we solve the commuting KdV flows and plot the field $W(x,t) = \xi_{j-i}\xi_{j+1}$ as a surface parameterized by (x,t). The solutions were obtained for a seven component system (three degrees of freedom) and orbits near five component reductions (two degrees of freedom).

Figure 1 shows ξ_j over (x,t) in the general case. The flow is almost-periodic in t and x.

Figure 2 shows $W(x,t)$ when $\xi_j = 0$.

Figure 3 shows $W(x,t)$ when initially ξ is small.

In general, the periodic fixed points of the Báckkund transformations provide interesting invariant manifolds whose study reveals many qualitative features of the related flow.

The KdV and Boussinesq systems are instances of the general system in component form [9]

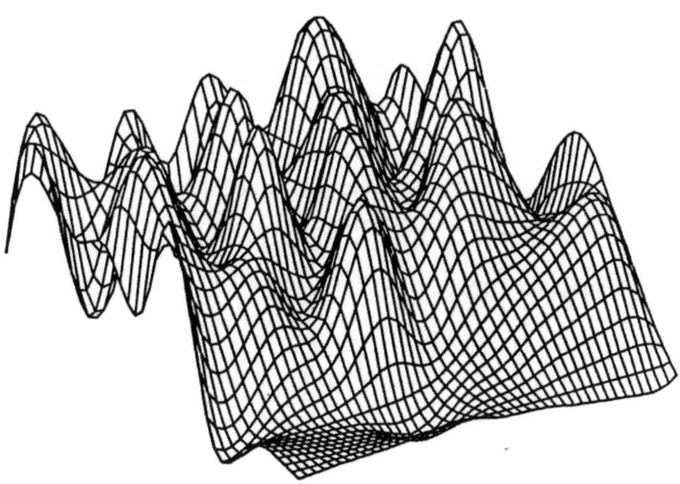

Figure 1: General KdV flow, $\xi_j(x,t)$.

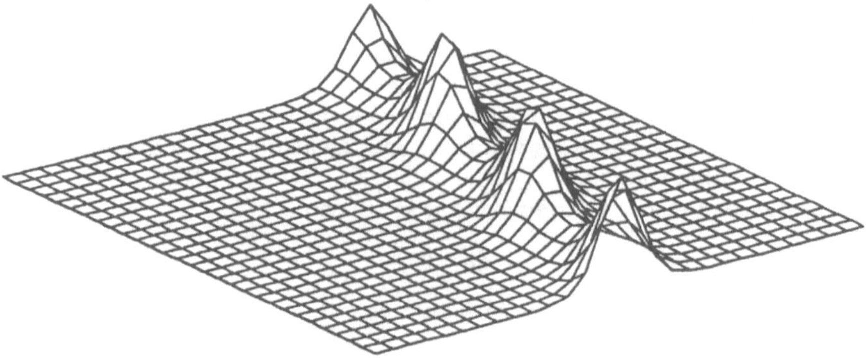

Figure 2: KdV flow, $W(x,t)$ with $\xi_j = 0$.

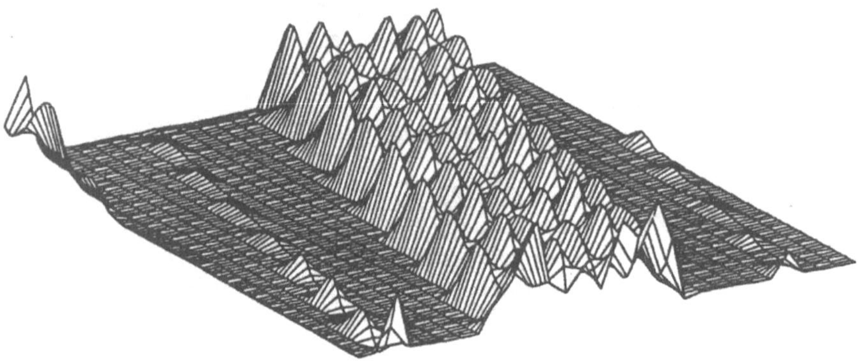

Figure 3: KdV flow, $W(x,t)$ with ξ_j small.

$$\frac{\xi_{j,x}}{\xi_j} + \frac{\xi_{j+1,x}}{\xi_{j+1}} + \ldots + \frac{\xi_{j+p,x}}{\xi_{j+p}} = \xi_j - \xi_{j+p} \tag{32}$$

where $j = 1, 2, \ldots \pmod N$. The KdV systems correspond to $p = 1$ and the Boussinesq to $p = 2$. Let the circulant forward shift matrix [8] be

$$C = \mathbf{circ}[\mathbf{0}, \mathbf{1}, \mathbf{0}, \mathbf{0}, \ldots, \mathbf{0}].$$

In the N-vector form equations (32) are

$$A \begin{pmatrix} \frac{\xi_{1,x}}{\xi_1} \\ \vdots \\ \frac{\xi_{N,x}}{\xi_N} \end{pmatrix} = B\hat{\xi} \tag{33}$$

with

$$\begin{aligned} A &= I + C + \ldots + C^p \\ B &= I - C^p. \end{aligned}$$

The casimir integrals of (32) correspond to the null vectors of B. The null vectors of A produce the constraints.

Associated with the principal Casimir, for any N

$$H_N = \prod_{j=1}^{N} \xi_j$$

we find the principal integrals of (32)

$$H_{N-pm-m} = L^m \circ H_N \tag{34}$$

,where $m = 0, 1, 2, \ldots$ and

$$L = \sum_{j=1}^{N} \partial_{\xi_j} \partial_{\xi_{j+1}} \ldots \partial_{\xi_{j+p}}.$$

The systems (33) have a Hamiltonian structure

$$A \begin{pmatrix} \frac{\xi_{1,x}}{\xi_1} \\ \vdots \\ \frac{\xi_{N,x}}{\xi_N}, \end{pmatrix} = B \begin{pmatrix} \xi_1 & & \\ & \ddots & \\ & & \xi_N \end{pmatrix} \nabla_{\xi} H_1 \tag{35}$$

,where $H_1 = \sum_{j=1}^{N} \xi_j$.

The higher-order equations associated with the integrals (34) are

$$A \begin{pmatrix} \frac{\xi_{1,x}}{\xi_1} \\ \vdots \\ \frac{\xi_{N,x}}{\xi_N} \end{pmatrix} = B \begin{pmatrix} \xi_1 & & \\ & \ddots & \\ & & \xi_N \end{pmatrix} \nabla_{\xi} H_{N-pm-m}. \tag{36}$$

When A is invertible, then

$$\Omega = A^{-1} B$$

is an antisymmetric circulant matrix.

We have the systems

$$\hat{\xi}_{,x} = M_{\hat{\xi}} \nabla_{\hat{\xi}} H_1 \tag{37}$$

and

$$\hat{\xi}_{,x} = M_{\hat{\xi}} \nabla_{\hat{\xi}} H_{N-pm-m} \tag{38}$$

where

$$M_{\hat{\xi}} = \begin{pmatrix} \xi_1 & & \\ & \ddots & \\ & & \xi_N \end{pmatrix} \Omega \begin{pmatrix} \xi_1 & & \\ & \ddots & \\ & & \xi_N \end{pmatrix}$$

is the co-symplectic form.

Darboux [5] studied the periodic fixed points of the Laplace transformation and found that these surfaces are related as a sequence of *focal surfaces.* These are described by the equations for the *invariants* [5]

$$\partial^2 \ln h_j / \partial x \partial y = -h_{j+1} + 2h_j - h_{j-1} \tag{39}$$

, where $j = 1, 2, 3, \cdots \pmod{n}$ and n is the order of the fixed point. The substitution

$$h_j = e^{\theta_{j+1} - \theta_j}$$

obtains the *two dimensional periodic Toda lattice*

$$\theta_{j,xy} = -e^{\theta_{j+1} - \theta_j} + e^{\theta_j - \theta_{j-1}} \tag{40}$$

We now find Bäcklund transformations for the *Darboux equations* (39) and the Toda lattice equations (40). [10] With reference to systems (35) and (36), without loss of generality normalize the casimir, $H_N = 1$, and set

$$\hat{\xi}_{,x} = M_{\hat{\xi}} \nabla_{\hat{\xi}} H_1 \tag{41}$$

$$\hat{\xi}_{,y} = M_{\hat{\xi}} \nabla_{\hat{\xi}} (H_{N-p-1} / H_N) \tag{42}$$

,where $H_{N-p-1} = L \circ H_N$. Then, let $\xi_j = e^{\psi_j - \psi_{j+1}}$ and find that (41), (42) imply

$$\psi_{j,xy} = e^{\psi_{j+p} - \psi_j} - e^{\psi_j - \psi_{j-p}} \tag{43}$$

,where $j = 1, 2, 3, \cdots \pmod{N}$.

When $p = 1$ (43) are the Toda lattice of period N. If N and p are relatively prime (43) is again a Toda lattice of length N. If $N = mp$ (43) is p distinct lattices of length m. When N and p have common factors (43) there is one lattice for each distinct orbit of translation by $p \pmod{N}$. In all cases the set of fields ξ_j are directly related to the set of invariants h_j. When A is not invertible we find for equations (41) and (42) a similar connection with the Toda lattice. In this case one must take into account the *constraints* that apply to these systems to obtain a valid correspondence.

Consideration of the form of (43) and the possible relations between p and N determine that for a lattice of fixed length m there will exist an infinite sequence of distinct Bäcklund transformations. For instance, we have a Bäcklund transformation for a lattice of length m when $N = pm$ for $p = 1, 2, 3, \cdots$. See reference [10] for details.

References

[1] E. Hille, *Ordinary Differential Equations in the Complex Plane,* John Wiley, New York, 1976.

[2] J. Weiss, *J. Math. Phys.* **24** (1983), 1405.

[3] J. Weiss, *The Singular Manifold Method,* in the Proceedings of the conference on **The Painlevé Transcendents, and their physical applications**, St. Adele. 1990, Ed. P. Winternitz, Springer-Verlag (1991).

[4] M. Lavie, *Canadian J. Math.* **21** (1969), 235.

[5] G. Darboux, *Théorie Générale des Surfaces, II,* Chelsea, New York, 1972.

[6] P. Deift and E. Trubowitz, *Comm. Pure Appl. Math.* **32** (1979), 121.

[7] J. Weiss, *J. Math. Phys.* **27** (1986), 2647.

[8] P. Davis, *Circulant Matrices,* Wiley, New York, 1983.

[9] J. Weiss, *J. Math. Phys* **28** (1987), 2025.

[10] J. Weiss, *Phys. Letts.* **137A** (1989), 365.

[11] Y.F. Chang and G. Corliss, *J. Inst. Math. Appl.* **25** (1980), 349.

IST for KPI

Xin Zhou
Yale University, New Haven, CT 06520, USA

The Cauchy problem of the Kadomtsev-Petviashvili I (KPI) equation

$$(u_t + 6uu_x + u_{xxx})_x = 3u_{xx} \tag{1}$$

can be solved by the inverse scattering method with the following Lax pair

$$i\psi_y + \psi_{xx} + u\psi = 0, \tag{2}$$

$$\psi_t + 4\psi_{xxx} + 6u\psi_x + 3(u_x - \frac{i}{2}\partial_x^{-1}u_y)\psi = 0. \tag{3}$$

This problem has been formally solved by Zakharov and Manakov [ZM], [M], Fokas and Ablowitz [FA], Boiti, Leon and Pempinelli [BLP]. The work [ZM]], [M] contains important ideas such as triangular factorizations and the positivity of the scattering operators. The work [FA] includes the lump solutions into the inverse scatering scheme. The work [BLP] derives systematically the relations among different scattering operators. Ablowitz and Fokas also pointed out recently that the operator ∂_x^{-1} in (3) should have the symmetric form

$$\frac{1}{2}\int_{-\infty}^{x} - \frac{1}{2}\int_{x}^{\infty}.$$

Certain rigorous results for the direct problem have been obtained by Segur [S]. More complete rigorous results (in small norm case) for the direct problem and for the inverse problem have been obtained by Zhou [Z]. The basic result of Segur is the following.

Let $\hat{u} = \hat{u}(l; y)$ denote the Fourier transform of u in the x–variable, then under the small norm assumption

$$\|\hat{u}\|_{L^1(\mathbf{R}^2)} < (2\pi)^{1/2}, \tag{4}$$

solutions $\psi_\pm$ of (2) analytic in the upper–lower half plane (introduced by Zakharov and Manakov) exist. The basic results of Zhou are.

1. If one assumes in addition to (4) that,

$$\hat{u} \in L^2(|l|^{-1} dl dy), \tag{5}$$

then the scattering kernels $T^\pm$ are in $L^2(\mathbf{R}^2)$. Note that (5) may be written as $u \in \operatorname{ran}(\partial/\partial x)^{1/2}$. One may compare this condition with the condition

$$\int u dx = 0,$$

which may be written as $u \in \operatorname{ran} \partial/\partial x$.

2. The inverse problem is solved by using the equation

$$\mu = 1 + (C_+ \mathcal{T}^-_{x,y,t} + C_- \mathcal{T}^+_{x,y,t})\mu, \tag{6}$$

where $C_\pm$ are Cauchy integral operators and $\mathcal{T}^\pm_{x,y,t}$ is the integral operator with kernel

$$T^\pm(k,l)e^{i(l-k)x - i(l^2-k^2)y - 4i(l^3-k^3)t}.$$

Since $T^\pm \in L^2$, the integral operator in (6) is compact. A Vanishing Lemma shows that the only homogeneous solution of (6) is zero and hence (6) is solvable. Finally the potential u can be constructed from

$$u(x,y,t) = \frac{1}{\pi}\frac{\partial}{\partial x}\iint dldk(T^+(k,l) + T^-(k,l))e^{i(l-k)x - i(l^2-k^2)y - 4i(l^3-k^3)t}\mu(l,x,y;t).$$

References

[BLP]. M. Boiti, J. JP. Leon and F. Pempinelli, *A new spectral transform for the Kadomtsev–Petviashvili I equation*, Lab. de Phy. Math. (1989).

[FA]. A. S. Fokas and M.J. Ablowitz, *On the inverse scattering of the time–dependent Schrödinger equation and the associated Kadomtsev–Petviashvili (I) equation*, Stud. Appl. Math. 69 (1983), 211–228.

[M]. S. V. Manakov, Physica 3D (1981), 420.

[S]. H. Segur, *Comments on IS for the Kadomtsev–Petviashvili equation, in mathematical metheds in hydrodynamics and integrability in dynamical systems*, AIP Conference Proceedings 88 (1982), 211–228.

[ZM]. V. E. Zakharov and S. V. Manakov, Sov. Sci. Rev. Phys. Rev. 1 (1979), 133.

Part III

Plasmas

A Physical Model for Nonlinear, Supersonic Equatorial Bubbles

W.J. Burke[1] *and T.L. Aggson*[2]

[1]Geophysics Directorate, PL/OL-AA, Hanscom AFB, MA 01731, USA
[2]NASA: Goddard Space Flight Center, Greenbelt, MD 20771, USA

After sunset the bottomside of the equatorial F layer is Rayleigh-Taylor unstable with buoyancy provided by **g x B** drifting ions that accumulate space charge on the walls of depleted flux tubes. The nonlinear equations for bubble motions are those of incompressible fluids[1]. Assuming that buoyancy forces are balanced by drag from shedding vortices into the wake plasma, the model predicts bubbles of radius R have terminal speeds $U \approx 0.5 \sqrt{(Rg)}$. Measurements from DE-2 show a plasma bubble updrafting at ~ 2 km/s, well above model predictions and even accelerating upward at ~ 0.5 **g**. Magnetic perturbations of ~150 nT were detected on the western wall of the deepest plasma depletion. The electromagnetic variations carry downward Poynting flux, too large for shunting of the **g x B** current. The perturbation is Alfvénic, radiating from a generator located near the depleted flux tube's equatorial apex in the plasma at the bubble's leading edge. We model depleted flux tubes of circular cross sections moving upward through the background plasma. The electric field is uniform inside and a 2-D dipole outside the bubble. Without drag, bubbles accelerate at large fractions of **g**. The generator current is driven by inertial forces acting on plasma in thin layer at the bubble's leading edge. Consequent, radiated Alfvén wings[2] act as counter-buoyancy drag forces.

INTRODUCTION

Ionospheric plasma bubbles are responsible for equatorial spread F back-scatter. They originate in the bottomside F layer where plasma-density gradients are linearly unstable to the growth of Rayleigh-Taylor turbulence In the local approximation, the linear growth rate reverses sign at the peak of the F layer. Upward-moving density depletions have been detected below, near and above the peak of the F layer. Computer simulations show that with proper background conditions Rayleigh-Taylor turbulence grows into the nonlinear regime and penetrates through to the topside ionosphere[3].

Plasma bubbles are field-aligned density depletions that percolate through the low-latitude ionosphere during the postsunset hours. As such they are commonly represented as two-dimensional structures whose buoyancy is supplied by eastward electric fields within the depletions. The eastward electric fields result from the divergence of current, carried by **g x B** drifting ions, at the boundaries of depleted flux tubes. In the absence of dissipation, positive (negative) charge continually accumulates on the west (east) wall of depleted flux tubes.

The analytical model for the nonlinear evolution of plasma bubbles, developed by Ott[1] distinguishes between collision-dominated and inertial stages in their upward motion. At low altitudes the buoyancy force is balanced by drag from collisions with neutrals, i.e. Pedersen current leakage. Without drag in the inertial stage the buoyancy force should accelerate bubbles upward without limit. Analogous to bubbles in ordinary fluids, Ott[1] suggested that the bubbles shed vortices, equivalent to elements of space charge, into the background plasma.

The transition to the inertial phase occurs at altitudes where ion-neutral collisions, integrated over the whole flux tube, become negligible. Following Ott, we use 350 km as a working altitude for the transition to the inertial domain. At higher altitudes, the momentum equation for bubbles reduces to that of an incompressible, inviscid fluid. Bubbles accelerate to an upward velocity V_z limited by vortex shedding to

$$V_z = \frac{\sqrt{R g}}{2} \tag{1}$$

where R is the radius of the bubble. Assuming that near the dip equator the radii of bubbles are between 16 and 200 km, Ott's model predicts they should attain terminal upward speeds between 200 and 700 m/s at altitudes > 550 km. The lower number agrees with most reported observations.

Occasionally bubbles do not fit between these limits. This paper describes a single encounter with an equatorial plasma bubble of very large upward speed. It was detected by electric field and plasma instrumentation on the Dynamics Explorer 2 (DE-2) satellite. Electric fields measured in the bubbles indicate updrafting well above Ott's terminal speed. In fact the bubbles were accelerating at a significant fraction of - $\mathbf{g}$. Electromagnetic perturbations not predicted by the model were detected near the wall of the deepest density depletion. We derive equations for bubbles with circular cross sections moving rapidly through a background, magnetized plasma. Intense electric and magnetic fields near the edges of a bubble, predicted by the model, are analogous to the radiation emitted by a satellites moving rapidly across field lines in the equatorial ionosphere[2].

INSTRUMENTATION AND OBSERVATIONS

DE-2 was in a 90° inclination, polar orbit. Its instrumentation payload included a Langmuir probe,[4] an ion drift meter [IDM],[5] a retarding potential analyzer [RPA],[6] a vector electric field detector [VEFI],[7] and a three-axis fluxgate magnetometer.[8] The IDM-RPA combination measured the vector ion drift vector, while the electric field detector measured the two components of the electric field in the orbital plane. When the satellite moved across magnetic meridians, the east-west component of the electric field was estimated from the $\mathbf{E} \cdot \mathbf{B} = 0$ condition.

The supersonic bubble shown in Figure 1 was detected by the DE-2 satellite. DE-2 entered the bubble at 02:50:52 UT on 13 September 1982, at an altitude near 400 km, in the midnight sector. The geographic location of the bubble at 24° S, 55° W, is near the center of the South Atlantic Anomaly where the magnetic field strength is approximately $2.24 * 10^4$ nT and the declination 15° W. Thus, the southward moving DE-2 satellite crossed the depleted flux tube from north-east to south-west in magnetic coordinates. The encounter was at a magnetic latitude of - 13.7°, and the altitude of the flux tube at the dip equator ~ 800 km.

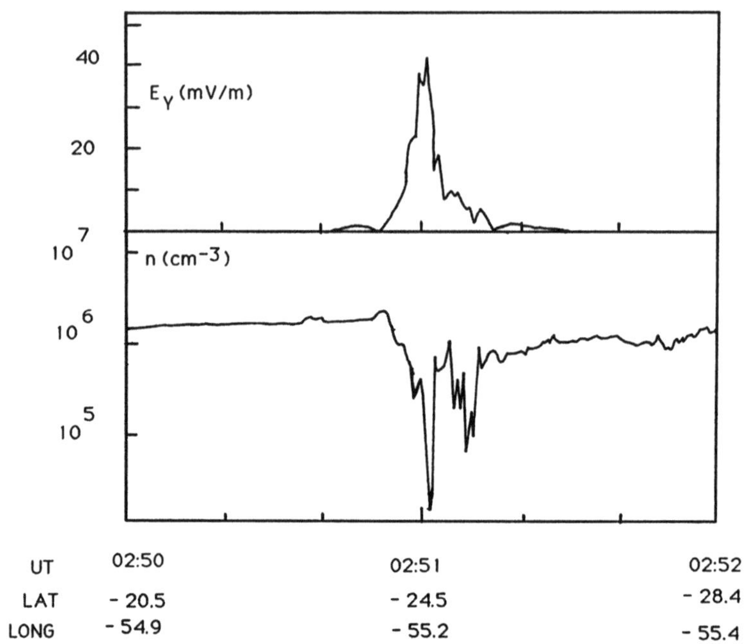

Figure 1. Eastward component of the electric field vector E_y (top panel) and electron density (bottom panel) acquired during a DE-2 satellite crossing of an equatorial plasma bubble.

The density depletion is complex, containing major substructures. In the deepest depletion, centered at 0251:01 UT, the density decreased to $1.5 * 10^4$ cm^{-3}, two orders of magnitude less than the density at the bubble's eastward wall. The electric field in this depletion rose to 43 mV/m. This corresponds to an upward velocity component of 1.9 km/s, significantly greater that the 0.2 to 0.7 km/s terminal speed predicted by Ott's model or the ~ 1 km/s thermal speed of O^+ ions in the topside ionosphere. These measurements agree with those of the IDM and RPA which also detected enhanced ion temperatures and a 0.65 km/s field-aligned plasma flow toward the equator. Combining magnetic declination with the satellite transit time gives the thickness of the deepest depletion to be about 12 km. We also estimate a lower limit for average acceleration needed to achieve the observed updrafting speed. Assuming that the transition from the collision-dominated to the inertial phases occurs at 350 km and the apex of the

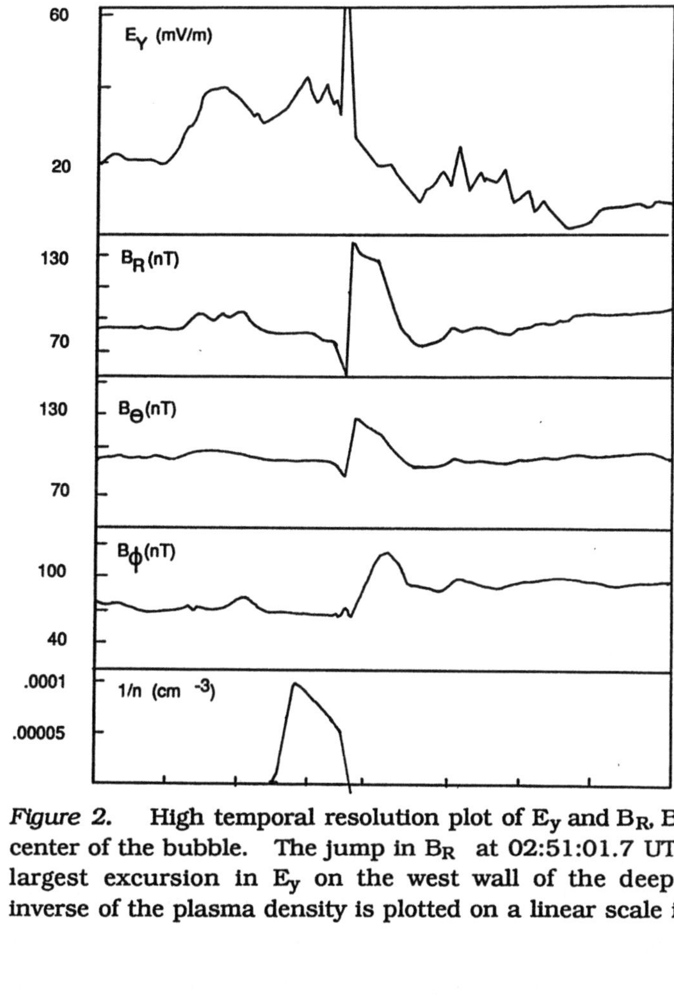

Figure 2. High temporal resolution plot of E_y and B_R, B_Θ and B_Φ near the center of the bubble. The jump in B_R at 02:51:01.7 UT correlates with the largest excursion in E_y on the west wall of the deepest depletion. The inverse of the plasma density is plotted on a linear scale in the bottom panel.

depleted flux tube crosses the dip equator at 800 km gives an acceleration $\dot{V}_z = 4.0\ m/s^2$.

Figure 2 contains plots of the east-west electric field component and the differences between the measured and model magnetic field components for a 68 s interval spanning the deepest plasma depression. From the top to bottom panels the plot contains E_Y, the radial B_R, colatitude B_θ and azimuthal B_ϕ magnetic components and, on a linear scale, the inverse of the plasma density. There are two aspects of these measurements that we emphasize: (1) Within the deepest depletion all three magnetic components are constant. Thus, the depletion's interior is not strongly coupled by field-aligned currents to the lower ionosphere. (2) On the westward wall of the depletion there is a large spike in the electric field and a sharp rise in all three magnetic components. The variation in $|B|$ is < 5 nT, indicating that the responsible currents are along the direction of the main magnetic field vector. The 150 nT excursion is primarily in B_R, the vertical component, and occurs during a 0.2 s interval.

The simplest interpretation is that the magnetic perturbation results from field-aligned currents that close via Pedersen currents in the lower ionosphere. This cannot be the case. From the observed 150 nT increase in B_R we estimate a field-aligned current density of order 80 $\mu A/m^2$, much more current than the local ionosphere can supply by shunting the entire gravity-drift current $j_y = -\rho g/B$, where ρ is the plasma mass density, into the E region along the bubble edges. This extreme assumption only gives $\Delta B = 1 - 10$ nT for $n = 10^5 - 10^6$ cm^{-3}.

Alternatively, the disturbance may be Alfvénic. For plane wave propagation the ratio of the electric to magnetic field perturbations is given by

$$\frac{\Delta E_y}{\Delta B_R} \approx V_A = \frac{B_o}{\sqrt{\mu_o \rho}} \tag{2}$$

where V_A is the Alfvén speed and B_o the magnetic field. For an O^+ dominated plasma with a density just outside the bubble of $\sim 3*10^5$ cm^{-3} we find V_A = 312 km/s. From Fig. 2, with $\Delta B_R \approx 150$ nT and $\Delta E_Y \approx 50$ mV/m we calculate that $V_A \approx 330$ km/s. The Poynting flux $\mathbf{S} = (\Delta \mathbf{E} \times \Delta \mathbf{B})/\mu_o$ can also be estimated from the measured eastward electric and radial magnetic field perturbations. These give a peak Poynting flux of $6 * 10^{-3}$ W/m^2 that is directed southward along magnetic field lines from the equator toward the ionosphere. Since electromagnetic energy flows downward the generator is equatorial rather than thermospheric. In any generator the current and the electric fields must be in opposite directions. Since the electric field is eastward, the driving current in the generator region must be westward.

DISCUSSION

For simplicity we consider the bubble as a depleted flux tube with a circular cross section of radius a, moving upward through the ionospheric plasma as indicated in Figures 3a and 3b. To maintain coherent flux-tube motion the electric field within the bubble must be spatially uniform and directed eastward. We represent the interior field in polar coordinates

(3) $$\mathbf{E} = E_o(t)\,\hat{y} = E_o(t)\,[\mathrm{Cos}\,\chi\,\hat{\mathbf{r}} - \mathrm{Sin}\,\chi\,\hat{\chi}]$$

where $\hat{\mathbf{r}}$ and $\hat{\chi}$ are unit vectors defined in Fig. 3b. Outside, the electric field is a two-dimensional dipole which pushes ambient plasma aside to allow the flux tube's upward motion and refill its wake.

(4) $$\mathbf{E}(r,\chi,t) = -E_o(t)\,\frac{a^2}{r^2}\,[\mathrm{Cos}\,\chi\,\hat{\mathbf{r}} + \mathrm{Sin}\,\chi\,\hat{\chi}]$$

The fluid equation for the conservation of momentum is

(5) $$\rho\,[\frac{\partial \mathbf{V}}{\partial t} + (\mathbf{V}.\nabla)\,\mathbf{V}] - \rho\,\mathbf{g} = \mathbf{j} \times \mathbf{B}$$

where the fluid velocities $\mathbf{V} = (\mathbf{E} \times \mathbf{B})/B^2$ inside and outside the bubble are

(6a) $$\mathbf{V} = V_o(t)\,[\mathrm{Sin}\,\chi\,\hat{\mathbf{r}} + \mathrm{Cos}\,\chi\,\hat{\chi}], \qquad r < a$$

(6b) $$\mathbf{V} = V_o(t)\,\frac{a^2}{r^2}[\mathrm{Sin}\,\chi\,\hat{\mathbf{r}} - \mathrm{Cos}\,\chi\,\hat{\chi}], \qquad r > a$$

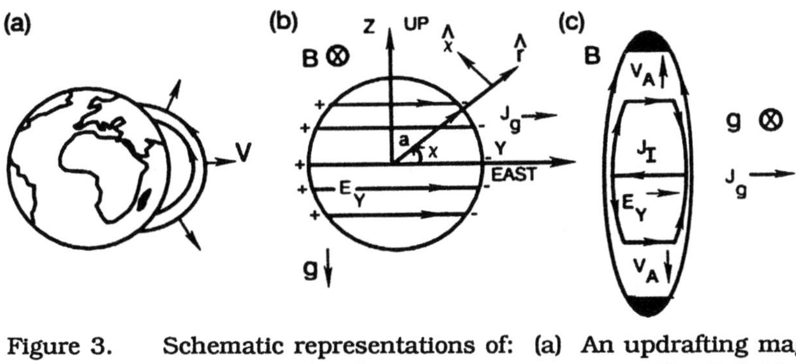

Figure 3. Schematic representations of: (a) An updrafting magnetic flux tube. (b) A plasma bubble of circular cross section. The E field is uniform and eastward inside the bubble and a 2-D dipole outside. (b) The upward moving magnetic flux tube viewed from above. Alfvén wings and associated currents propagate away from a generator in plasma above the bubble's leading edge. The shaded areas are the flux tube's ionospheric foot-prints.

$V_o(t) = E_o(t)/B$ is the updrafting bubble's speed. The cross product of Eq. (5) with **B** shows that the current perpendicular to **B** has three contributors, the gravitational j_g, displacement j_D, and inertial j_I currents

$$j_g = -\frac{\rho\,(\mathbf{g} \times \mathbf{B})}{B^2}$$

$$j_D = \frac{\rho\,(\mathbf{B} \times \frac{\partial \mathbf{V}}{\partial t})}{B^2}$$

$$j_I = \frac{\rho\,(\mathbf{B} \times (\mathbf{V}.\nabla)\mathbf{V})}{B^2}$$

Substituting equations (6a) and (6b) into these expressions and using ρ_o and ρ_1 to represent the plasma density inside and outside the bubble gives

(7a) $$j_g = \frac{\rho_o g}{B}[\mathrm{Cos}\,\chi\,\hat{r} - \mathrm{Sin}\,\chi\,\hat{\chi}] \qquad r < a$$

(7b) $$j_g = \frac{\rho_1 g}{B}[\mathrm{Cos}\,\chi\,\hat{r} - \mathrm{Sin}\,\chi\,\hat{\chi}] \qquad r > a$$

(8a) $$j_D = \frac{\rho_o \frac{\partial V_o}{\partial t}}{B}\,[\mathrm{Cos}\,\chi\,\hat{r} - \mathrm{Sin}\,\chi\,\hat{\chi}] \qquad r < a$$

(8b) $$j_D = -\frac{\rho_1 \frac{\partial V_o}{\partial t}}{B}\frac{a^2}{r^2}[\mathrm{Cos}\,\chi\,\hat{r} + \mathrm{Sin}\,\chi\,\hat{\chi}] \qquad r > a$$

(9a) $$j_I = 0 \qquad r < a$$

(9b) $$j_I = \frac{\rho_1 V_o^2}{B}\frac{a^4}{r^5}[\mathrm{Sin}\,\chi\,\mathrm{Cos}\,\chi\,\hat{r} + (1 + \mathrm{Sin}^2\chi)\,\hat{\chi}] \qquad r > a$$

Since the sum of the gravitational and displacement current densities is continuous across the boundary of the bubble at $r = a$ and $\chi = \pi$, (where $\mathbf{j_I}$ only has a tangential component) we find that

(10) $$\frac{\partial V_o}{\partial t} = -\frac{n_1 - n_o}{n_1 + n_o}\,g$$

For plasma densities inside and on the west wall of the bubble of n_o = 2 *

10^4 cm^{-3} and n_1 = $3^* 10^5$ cm^{-3}, respectively, equation (10) predicts a bubble acceleration of - 0.88 g, about twice the estimated, average value.

This simple two-dimensional treatment for the collisionless updrafting of an equatorial bubble yields a constant upward acceleration. The solutions discussed so far are however, purely electrostatic and does not predict a large magnetic signature with $\Delta E/\Delta B = V_A$. We have shown that the gravitational current is too diffuse to produce the sharp magnetic perturbation observed near the edge of the bubble. The displacement currents, both inside and outside the bubble, are in the same directions as the electric fields. Thus, they cannot power an electromagnetic generator.

The electromagnetic structure detected by DE-2 near the western edge of the supersonic density depletion is reminiscent of the Alfven wings generated by a finite conducting body such as a spacecraft moving upward across field lines in a magnetized plasma.[2] Magnetic perturbations in the vicinity of Io have been interpreted as evidence for the existence of Alfvén wing phenomena in nature (*Southwood et al.*, 1980). In the present case, a depleted flux tube rather than a satellite moves upward through the equatorial ionospheric plasma. As such, it is an insulator rather than as a conductor. Equation (9b) indicates that an inertial current flow in the background plasma near the top edge of the bubble, ($r \approx a$ and $\chi = \pi/2$), and decreases rapidly with distance from the bubble. Since the electric field and the inertial current are oppositely directed, the region just outside the bubble acts as a generator. The rate that mechanical energy of the plasma transforms into electromagnetic energy to radiate Alfvén waves is found by combining equations (4) and (9b)

$$(11) \qquad j_I \cdot E = -2\, n\, m_i\, V_0^3 \frac{a^6}{r^7} \operatorname{Sin} \chi$$

Equation (11) indicates that, in a collisionless plasma with no Joule dissipation, electromagnetic energy is generated along the leading edge of the bubble where Sin $\chi > 0$. The rate at which the bubble's mechanical energy transferred into electromagnetic radiation depends on V^3.

The divergence of the inertial current obtained from Eq. (9b) is

(12)
$$\nabla_{\perp} \cdot j_I = - \frac{2\rho_1 V_o^2}{B} \frac{a^4}{r^6} \operatorname{Sin}\chi \operatorname{Cos}\chi$$

The inertial current deposits negative space charge near the east side ($0 < \chi < \pi/2$) of the bubble's leading edge and positive charge near its west side ($\pi/2 < \chi < \pi$). To preserve quasi-neutrality in the plasma, wherever the perpendicular divergence of j_I is non-zero, field-aligned currents, carried by Alfvén waves, flow away from the generator toward the ionosphere.

The drift motion of a fluid element in an Alfvén wave of amplitude $\Delta\mathbf{B}$ is $\mathbf{V} = \pm \Delta\mathbf{B}/\sqrt{\mu_o \rho}$. Taking the curl of this equation allows us to calculate the current density associated with an Alfvén wave.[10]

(13)
$$j_A = \pm \frac{[(\hat{\mathbf{B}}_o . \nabla)\mathbf{E} - \hat{\mathbf{B}}_o (\nabla . \mathbf{E})]}{\mu_o V_A}$$

where $\hat{\mathbf{B}}_o$ is a unit vector along the magnetic field and $\mathbf{E}$ is the wave electric field. The second term is the field-aligned current carried by the wave. As the wave front moves along the field line space charge, $\nabla . \mathbf{E} \neq 0$, is deposited as required to make the field line a new equipotential. The currents carried by the Alfvén wave, as illustrated in Figure 3c produce radially outward (inward) magnetic field perturbations in the southern (northern) hemisphere. The first term in Eq. (13) is a polarization current that flows in the same sense as electric fields at the leading edge of the Alfvén wave front.

Using equation (11) it is possible to estimate the energy that goes into the Poynting flux S at the satellite altitude. At the time of the observation DE-2 was located at 13° S magnetic latitude or about 1750 km from the equatorial apex of the field line. The maximum energy from the generator should be detected very close to the edge of the bubble $r \approx a$ and $\chi = \pi/4$.

$$S \approx (j_I . \mathbf{E})\, L = \frac{\sqrt{2}\, n_1 m_i V_o^3 L}{a}$$

where L is the half-length of the generator. The maximum value of the Poynting flux estimated from the electric and magnetic perturbations was $6 * 10^{-3}$ W/m^2. With a plasma density of $3 * 10^5$ cm^{-3} in the bubble wall, O^+

as the dominant ion species, a bubble radius and speed of 6 km and 2 km/s, respectively, we only require a generator half-length L of 240 km. Thus, although we do not know the value of χ at which DE exited the bubble, it seems quite likely that the generator region above the satellite is easily capable of producing the observed electric and magnetic radiation fields.

Alfvén waves radiated by plasma at the bubble's leading edge carry off electromagnetic energy at the expense of the bubble's kinetic energy. The Alfvén waves act as brakes on flux tubes. The present emphasis on inertial domination should not obscure the fact that some Joule heating occurs within the flux tube. Figure 3 shows that the ion temperature was elevated by a factor of two higher in the bubble above its outside value. Much of this heating results from ion-neutral collisions in the lower F region and could lead to the development of a local over-pressure which drives the observed, upward field-aligned plasma flow. Alfvén-wing braking and residual Joule dissipation may explain differences between the calculated, average acceleration of the flux tube and the higher values predicted by equation (16).

Drell and coworkers[2] have estimated the magnetic perturbation produced by a moving conductor as

$$\delta B = (V_o/V_A)\, B_o$$

Using a bubble speed of 2 km/s and an Alfvén speed of 330 km/s and B_o = 25,000 nT, the model predicts a magnetic perturbation of 150 nT.

It should be stressed that the supersonic bubble reported here is a very unusual event. In most bubble encounters satellite instruments detect upward propagation speeds under 1 km/s. The detection of the associated electromagnetic pulse may be rarer still. Our model suggests that this signature can only be detected if the satellite crosses the bubble wall on field lines that thread through its upper boundary at the equator. In the present case, the supersonic bubble also accelerated upwards with little dissipation of the polarization charges on its walls. At this time we can only speculate why the ordinary mechanisms for polarization charge dissipation, Pedersen current closure[3] and vortex shedding,[1] did not limit the bubble to a much lower upward speed.

Regarding the apparent lack of charge depletion through field-aligned currents closing in the E region, we note two features of the observations: (1) the bubble was detected with both ionospheres in darkness, and (2) the supersonic portion was embedded in large bubble systems moving upward at subsonic speeds. The first element is necessary for the bubble[3] formation. The second element suggests that the supersonic substructure began late in the life of the larger bubble. Local electric fields transport residual ionization out of the affected nighttime E region. The resulting decrease in height-integrated conductivity inhibits the flow of Pedersen currents, and further destabilizes the flux tube.[11] This suggests a scenario in which large bubbles form and propagate upward. Inhomogeneities in electric fields of the large bubbles produce localized regions of severely depleted E layer conductivity. Within these regions smaller substructures form, develop high levels of polarization charge, and accelerate to great speeds.

Ott's inertial limit of several hundred m/s updrafting velocity runs counter to our observations of supersonic velocities in equatorial bubbles. However, his 1978 study did not have the benefit of the geometrical hindsight of recent measurements or computer simulations that instead of semicircular cross sections, large bubbles often have wedge shaped geometries with the bubble walls lying along magnetic meridians. Elongated bubbles in fluids are called "slugs." Perhaps further interdisciplinary efforts such as Ott's work should be made to carry the "slug" analogy to equatorial wedges to estimate the relative efficiency of dissipation via vortex shedding.

REFERENCES

1. Ott, E., 1978, *J. Geophys. Res.*, 83, 2066.
2. Drell. S.D., H.M. Foley and M.A. Ruderman, 1965, *J. Geophys. Res.*, 70, 3131, 1965.
3. Anderson, D.N. and G. Haerendel, 1979, *J. Geophys. Res.*, 84, 4251.
4. Krehbiel, J. P., L. H. Brace, R. F. Theis, W. H. Pinkus, and R. B. Kaplan, 1981, *Space Sci. Instrum.*, 5, 493.

5. Hanson, W.B., R.A. Heelis, R.A. Power, C.R. Lippincott, D.R. Zuccaro, B.J. Holt, LH. Harmon and S. Sanatani, 1981, *Space Sci. Inst.*, 5, 503.

6. Heelis. R.A., W.B. Hanson, C.R. Lippincott, D.R. Zuccaro, L.C. Harmon, B.J. Holt, J.E. Doherty and R.A. Power, 1981, *Space Sci. Inst.*, 5, 551.

7. Maynard, N. C., E. A. Bielecki, and H. F. Burdick, 1981, *Space Sci. Instrum.*, **5**, 523.

8. Farthing, W. H., M. Sugiura, B. G. Ledley, and L. J. Cahill, Jr., 1981, *Space Sci. Inst.*, 5, 551.

9. Southwood, D.J., M.G. Kivelson R.J. Walker and J.A. Slavin, 1980, *J. Geophys. Res.*, 85, 5959.

10. Kan, J.R. and W. Sun, 1985, *J. Geophys. Res.*, 90, 10,911.

11. Hanson, W.B., S. Sanatani and T.N.L. Patterson, 1983, *J. Geophys. Res.*, 88, 3169.

Nonlinear Dynamics of Electron Cyclotron Heated Plasmas

G.E. Guest

Applied Microwave Plasma Concepts, Inc., Carlsbad, CA 92009, USA

Plasmas that are generated and sustained by electron cyclotron heating (ECH) have been studies experimentally and theoretically for more than thirty years. Even so, many of the fundamental processes that govern the properties of ECH plasmas are strongly nonlinear and are still only poorly understood. In what follows we identify four distinct types of nonlinear dynamical processes that have important consequences for ECH plasmas, briefly describe their effects on the observed properties of the plasmas, and indicate the approaches that have been used to analyze them.

The most rapid of the dynamical processes to be discussed occur in times of the order of a few electron gyroperiods and are associated with spatially localized, gyroresonant wave-particle interactions. The energy transferred to plasma electrons by the RF heating power is usually analyzed using linearized models; and these appear to be satisfactory in moderate-power heating regimes where experimental results are available. [1]But these same experiments indicate that even in the case of moderate incident power densities an associated nonlinear interaction can result in a bunching of electrons in the phase of their gyration. The resulting non-gyrotropic population of heated electrons can drive instabilities in which RF power is emitted at frequencies that are near but distinct from the frequency of the heating power, together with simultaneous bursts of energetic charged particles. This nonlinear phase bunching has been analyzed[2] using elementary methods in the special case of wave propagation exactly parallel to the magnetic field, for which exact constants of the motion can be obtained and used to construct phase-space trajectories. The gyrophase of resonant electrons, for which the Doppler-shifted wave frequency equals the electron gyrofrequency, oscillates about the stable fixed point with a frequency, ω_t, where $\omega_t^2 = kv_\perp \Omega_\perp$. Here k is the propagation constant of the wave, $v_\perp$ is the electron velocity perpendicular to the static magnetic field, and $\Omega_\perp$ is the electron gyrofrequency in the magnetic field of the wave. It appears to be feasible to use iterated mapping techniques to examine the possibility of phase bunching in more general cases; but this approach has not yet been investigated in detail.

On the time scale defined by the electron transit time across the plasma, heated electrons can form a plasma if they are confined in a region that is larger than their mean-free-path for ionizing the ambient gas. If the plasma is to be in equilibrium the loss rates of ions and electrons from this region must be equal, and their common (ambipolar) loss rate must balance the creation rate of ion-electron pairs. Further, the rate at which energy is supplied to the plasma by the ECH power must balance the rate at which it is lost, by ionization, excitation, radiation and transport, for example. The plasma ions are far less mobile than the electrons and act as an inertial source of charge for a positive electrostatic potential that traps the electrons and constrains them to oscillate within the plasma for many "bounce" periods. The equilibrium ambipolar potential is described by Poisson's equation with nonlinear expressions for the negative charge density due to the trapped as well as the untrapped electrons. Water-bag models provide a convenient method of determining self-consistent solutions to this nonlinear dynamical equilibrium problem. The phase-space trajectories of the plasma electrons typically have the form of time-independent vortices, provided specific equilibrium conditions are satisfied.[3]

For times that are longer than electron bounce periods and characteristic of the ionization rate as well as the ion transit time, the plasma equilibrium can be described by nonlinear transport equations that account for the detailed balance of charged particles and power alluded to earlier. Simplified forms of these equations for the plasma electrons are given by

$$dn_e/dt = n_e n_o < \sigma v_e >_i - n_e/\tau_e$$

and

$$d(n_e T_e)/dt = (2/3)(p_{abs} - p_x - p_i - p_{ei}) - n_e T_e/\tau_E.$$

Here n_e and n_o are the densities of electrons and neutral gas particles, respectively; $< \sigma v_e >_i$ is the ionization rate constant averaged over the electron energy distribution; τ_e and τ_E are the electron (particle) and energy confinement times; T_e is the electron temperature; and p_{abs}, p_x, p_i, and p_{ei} are the power densities absorbed in the plasma and lost from the electrons by excitation, ionization, and transfer to ions, respectively. An idealized point-model implementation of these transport equations in terms of spatially averaged plasma parameters can be analyzed by conventional methods and exhibits, for example, a saddle-node bifurcation that is associated with striking physical properties of ECH plasmas;[4] namely, if the ambient gas pressure is reduced below a critical value the time-independent equilibrium disappears and is replaced by relaxation oscillations.

If the preceding equilibrium conditions are satisfied, the spatially varying properties of the plasma may include variations in the static magnetic field strength, the plasma density and temperature, various microscopic kinetic properties, the amplitudes and polarizations of the ECH power, and the density of the ambient neutral gas. These global properties of ECH plasmas can lead to a number of nonlinear phenomena of consequence. For example, the energy of heated electrons can exhibit limit cycles with characteristic times comparable to the electron bounce time.[5] These limit cycles can be destroyed and stochastic heating of electrons restored by increasing the ECH power or by distributing the same power among several closely-spaced frequencies,[6] in which case the efficiency of the heating process is greatly increased. A particularly transparent picture of these limit cycles is provided by a mapping of energy and gyrophase between successive encounters with the resonant interaction zone. The mapping derives from analytic expressions for the gyrophase of electrons trapped in a simple, axisymmetric magnetic mirror. The model magnetic field shape is given by

$$B(z)/B(0) = 0.5[M + 1 - (M - 1)\cos kz],$$

where M is the mirror ratio, z is the distance along the magnetic lines of force, and $k = \pi/L$, where $2L$ is the separation of the magnetic mirrors. The electron bounce motion in this trap is approximated by $z(t) = z_t \sin\omega_b t$, where z_t is the electron turning point and ω_b is the bounce frequency given by

$$\omega_b = (\pi kv/4)[(M-1)/(M_t-1)]^{\frac{1}{2}} K^{-1}[(M_t-1)/(M-1)].$$

Here v is the electron speed, $M_t = B(z_t)/B(0)$ and K is the complete elliptic integral. The mapping can be used to display the superadiabatic limit cycle initially studied by Seidl[7], as well as the null-heating limit cycles of reference 5.

In other respects, the plasma behavior is suggestive of phenomena related to self-organized critical states. In some instances the plasma is able to modify its global equilibrium continuously to remain near the critical state; while in other instances the equilibrium changes abruptly to arrive at a new, sub-critical state.

This work was supported in part by the Air Force Geophysical Laboratory under Contract No. F19628-88-C-0118 (SBIR).

REFERENCES

1. B.H. Quon and R.A. Dandl, "Preferential Electron-Cyclotron Heating of Hot Electrons and Formation of Overdense Plasmas", Phys. Fluids B **1**, 2010 (1989).

2. G.E. Guest, M.E. Fetzer, and R.A. Dandl, "Whistler-Wave Electron Cyclotron Heating in Uniform and Nonuniform Magnetic Fields", Phys. Fluids B **2**, 1210 (1990).

3. R.A. Dandl and G.E. Guest, "Electrostatic Trapping of Plasma Electrons Heated by Whistler-Wave Electron Cyclotron Heating", Phys. Fluids B (in press).

4. R.A. Dandl and G.E. Guest, "On the Low Pressure Mode Transitions in Electron Cyclotron Heated Plasmas", Journal of Vacuum Science and Technology A (in press).

5. see, for example, G.E. Guest and R.A. Dandl, "Prospects for Control of ELMO Rings", Proceedings of the 2nd Workshop on Hot Electron Ring Physics, Dec. 1 - 3, 1981, San Diego, CA, CONF-811203 Vol. 1, p. 393.

6. B.H. Quon, R.A. Dandl, W. DiVergilio, G.E. Guest, L.L. Lao, N.H. Lazar, T.K. Samec, and R.F. Wuerker, "Impact of Multiple-Frequency Heating on the Formation and Control of Diamagnetic Electron Rings in an Axisymmetric Mirror", Phys. Fluids **28**, 1503 (1985).

7. M. Seidl, "High-Frequency Heating of Electrons in a Mirror Machine" Plasma Physics (Journal of Nuclear Energy Part C) 6, 597 (1964).

Nonlinear Evolution of Alfvén Waves in Space Plasmas

T. Hada

Department of Physics, College of General Education,
Kyushu University, Fukuoka, Japan

Alfvén waves play essential roles in various space plasma environment. With emphasis on introducing relevant articles to interested readers outside of the field, several important nonlinear aspects of Alfvén waves are briefly surveyed. In particular, mathematical and physical properties of the derivative nonlinear Schrödinger equation (DNLS) that models the evolution of quasi-parallel Alfvén waves are discussed.

1. Introduction

The nonlinear properties of finite amplitude Alfvén waves are of great interest not only to basic plasma physics but also for their applications to space plasma physics and astrophysics[1]. In particular, the solar wind, the collisionless neutral plasma embedded in the vast interplanetary space, may be regarded as an ideal natural laboratory for testing theories of nonlinear Alfvén waves, since the scalelengths of plasma inhomogeneities in the solar wind are far greater than a typical Alfvén wavelength, and the solar wind is free from boundary effects that could modify wave evolution in a complicated manner. In the solar wind, a variety of electromagnetic instabilities[2] are thought to amplify Alfvén waves to an amplitude large enough (the wave magnetic field is of the order of the background field), that nonlinearity plays an essential role in the evolution of the waveform.

Large amplitude Alfvén waves have been observed by spacecraft experiments in the solar wind[3], near planetary bow shocks[4], near interplanetary shocks[5], and near comets[6]. The observed Alfvén waves show a variety of waveforms, from almost monochromatic circularly polarized waves, to steepened, compressional waves, whose complex envelopes suggest that nonlinear wave evolution is in progress. Some of the more prominent features of the observed nonlinear Alfvén waves include wave steepening, discrete wave packets, mixed polarization, discrete phase changes, and solitary wave packets. Since obliquely propagating hydromagnetic waves suffer from heavy Landau and ion-cyclotron damping in a Maxwellian plasma[7], the quasi-parallel Alfvén waves are the most likely candidates for long-time evolution, although there are some exceptions[8].

2. Alfvén waves in the MHD system

Parallel, circularly polarized, finite amplitude Alfvén waves are the exact solutions of the magnetohydrodynamic (MHD) system. Such Alfvén waves, however, can be unstable to several nonlinear instabilities. In the MHD system, the parallel Alfvén waves may decay via induced Brillouin scattering, or the decay instability[9]. In this process, the parent Alfvén wave decays into two daughter Alfvén waves (Stokes and anti-Stokes waves) and the sound wave. Both right- and left-hand polarized (RHP/LHP) Alfvén waves are subject to this instability, although high plasma β ratio (ratio of the plasma pressure to the magnetic pressure) strongly reduces the growth rate. Long time evolution of the instability is studied by numerical simulations using the "hybrid" (particle ions + fluid electrons) code[10]. Extension of the analysis to a geometry where oblique daughter waves are allowed shows that the decay process is most efficient in the 1-dim. geometry[11].

The stability of parallel, circularly polarized Alfvén waves are also studied with respect to the filamentation instability[12]. In this process, the parent Alfvén wave couples to transverse non-oscillatory compressional daughter waves. The relevance of this instability with solar and astrophysical problems has been discussed[12].

3. The Alfvén system - the DNLS

Although the MHD system includes all the essential elements of nonlinear Alfvén wave evolution except for kinetic effects, equations in the MHD system are highly nonlinear that it is difficult to study long time evolution of Alfvén waves analytically. An effort has been made to construct a reduced system of Alfvén waves, which is simple enough that rigorous mathematical treatment can be applied, but is still rich in physics. By using the so-called quasi-static approximation, which implies that hydrodynamic nonlinearities and steepening are weak, Rogister[13] derived an equation describing long time evolution of the Alfvén waves, using kinetic theory. Essentially the same equation was subsequently obtained by several authors, using fluid theory[14]. The equation, the derivative nonlinear Schrödinger equation (DNLS), reads,

$$b_t + \alpha(|b|^2 b)_x + i\gamma b_{xx} = 0 \tag{1}$$

where x is the main axis aligned with both the wave vector and the background magnetic field (=1), $b=b_y+ib_z$ is the complex wave magnetic field, and α and γ are real coefficients determined by statistical properties of the plasma. In the DNLS, only Alfvén waves propagating in the same direction are described, and further, the couplings to backward propagating Alfvén waves and to the sound waves are neglected (A Zakharov-like equation for Alfvén waves that contains these effects are discussed[15].) Thus, the DNLS is not valid when the Alfvén speed $C_A \sim$ sound speed C_s, as it is also clear from the fact that, in the fluid formulation, α is singular at $C_A=C_s$. Hereafter we assume C_A and C_s are well separated, and let $\alpha=\gamma=1$, without losing generality.

The DNLS describes weakly nonlinear[16], weakly dispersive, quasi-parallel Alfvén waves. In this limit, RHP and LHP Alfvén waves are nearly degenerated. Thus, the DNLS can deal with the coupling between these two modes, or in other words, the "shear Alfvén" waves and the "magnetosonic" waves. The DNLS system contains rich variety of physics. It has oblique limits which match the Korteweg-de Vries (KdV) description of the magnetosonic wave and the modified Korteweg-de Vries (mKdV) description of the Alfvén wave[17]. For large wave numbers, it has a limit which matches the nonlinear Schrödinger description of the left- and right-hand cyclotron modes[17].

The DNLS is so far known to be completely integrable by the inverse scattering transform (IST) under four different boundary conditions(b.c.): the "vanishing" b.c.[18] ($b\to 0$ as $x\to\pm\infty$), the "finite" (or "oblique") b.c.[19] ($b\to b_\pm$ as $x\to\pm\infty$, where $b_\pm$ is a constant), the "circularly polarized" b.c.[20] ($b\to b_0 e^{ikx}$ as $x\to\pm\infty$, where b_0, k(=real) are constants), and the periodic b.c.[21] (In ref. 21, the initial profile is assumed to be a so-called N-band potential with the solutions expressed in terms of hyper-elliptic functions which have the standard Jacobi elliptic functions as a special case). In applying the IST, for any of the four cases above, one first solves the "Kaup-Newell" scattering problem for given initial condition, b(x,0),

$$\Psi_x = D(\lambda,x,t)\Psi, \qquad D = i\lambda\begin{bmatrix} -\lambda & -ib \\ -ib^* & \lambda \end{bmatrix} \tag{2}$$

where the asterisk denotes complex conjugate. Time dependency of the scattering data is given by

$$\rho(\lambda,t)=\rho(\lambda,0)e^{-4i\lambda^2 t} \;;\; \lambda_n(t)=\lambda_n(0) \;;\; c_n(t)=c_n(0)e^{-4i\lambda_n^2 t} \tag{3}$$

where ρ is the reflection coefficient associated with the continuous spectrum λ, λ_n's are the discrete eigenvalues, and c_n is the normalizing factor of the reflection coefficient associated with λ_n(c.f. ref. 18). Finally, one solves the Gel'fand-Levitan integral equations to reconstruct b(x,t) from the scattering data. Only in this last step the equations used are different for different boundary conditions[18-21].

The scattering problem (2) allows only complex discrete eigenvalues for "vanishing" b.c., and both complex and real discrete eigenvalues for "oblique" and "circularly polarized" b.c. Structure of solitons associated with the real eigenvalues (bright and dark "one-parameter solitons")[14,22] and the complex eigenvalues ("two-parameter solitons")[19,23] have been discussed. Riemann problem of the DNLS has been studied for various initial waveforms[24]. Further, spacecraft data of magnetic field and plasma density fluctuations are analyzed with respect to the theory of the DNLS and the Alfvén solitons[25].

In the physical context, the DNLS has been used to study modulational instabilities of the Alfvén waves. When $C_A>C_s$, the parallel, circularly polarized Alfvén waves are modulationally unstable when $k>b_0^2/2$, where k is the wave number with positive (negative) sign corresponding to the LHP(RHP) mode, and b_0 is the wave amplitude[26]. This criterion is reversed when $C_A<C_s$, although the criterion is significantly modified when kinetic effects are taken into account[27]. Among a family of obliquely propagating, elliptically polarized travelling Alfvén waves[17,28], those with "mixed polarization" are shown to be modulationally unstable[23].

4. Variants of the DNLS

An extension of the DNLS to multi-dimensions has been discussed[27,29-30]. In 2-dim., it reads,

$$\begin{aligned} b_t + ((|b|^2+2p)b)_x - (|b|^2+2p)_y + ib_{xx} &= 0 \\ p_x + \frac{1}{2}(b+b^*)_y &= 0 \end{aligned} \tag{4}$$

where y is the transverse axis, $b=b_y+ib_z$, and p is the x-component of the magnetic field. In (4), the dependence on the transverse coordinate y is assumed to be weak compared to the x-dependence. In the opposite limit, the evolution of parallel Alfvén waves is described by the nonlinear Schrödinger equation with y being the main variable, and based on this equation the self-focusing and subsequent evolution of Alfvén wave polarization have been discussed[31].

Analysis of the transverse stability of the one-parameter solitons based on (4) shows that the dark solitons are unstable[30], in a similar way as the KP solitons in the case of positive dispersion are[32]. The stability analysis of the circularly polarized parallel Alfvén waves with respect to oblique modulations indicates that modulational instabilities exist for nearly all (b_0, k), both LHP and RHP waves[33].

Other extensions include various kinds of damping and driving terms. Already the first derivation of the DNLS equation started from a Vlasov description, and the resulting equation included a nonlinear Landau damping term[13]. The same term for (4) was obtained[29]. The influence of nonlinear Landau damping on a single soliton is discussed[34], and a numerical study was made of the nonlinear Landau damping of an initially incompressible (i.e., circularly polarized) wave train[35]. In the incompressible state, the Landau damping vanishes, while if this state is perturbed, and it is modulationally unstable, the result is that the Landau damping increases along with the developing modulations. Other kinetic modifications to the fluid version of the DNLS include generation of a field-aligned flow for parallel-propagating wave packets[36], which may play a major role in the wave nonlinearity. The DNLS in a CGL-fluid[37] with finite Larmor radius effects is discussed[38], and the DNLS with Braginskii's pressure tensors[39] is obtained[40].

A model equation based on the DNLS including the source (S) and the damping (D) terms,

$$b_t + (|b|^2 b)_x + i b_{xx} = S + D\,, \tag{5}$$

has been used to study Alfvén waves and shocks in realistic space plasmas. A perturbation method based on the perturbed scattering data approach is discussed, and is applied to a study of the influence of ohmic resistance ($D=\nu b_{xx}$) on the Alfvén solitons[41].

It is important to look at shock structures in (5) by letting S=0 and $D=\nu b_{xx}$. The equation now has three stationary points, and thus it allows six types of transitions which connect two out of the three stationary points. Further requiring that the entropy increases as one passes the transition from upstream to downstream, three shock solutions remain physical. When $C_A>C_s$ ($C_A<C_s$), one of the shock solutions corresponds to the MHD fast (slow) shock, and the other two to the intermediate shocks[42]. In the limit $\nu \to 0$, the intermediate shocks are non-evolutionary, i.e., the intermediate shocks in this limit are unstable when Alfvénic perturbations with infinitesimal amplitude are superposed[43]. However, it was recently shown that nonzero ν allows structurely stable intermediate shocks[42]. This surprising finding is connected to the fact that the DNLS is not strictly hyperbolic, due to the confluence of two velocities of propagation[44]. The structure and the stability of the intermediate shocks in the kinetic formulation is still under debate[45].

Equation (5) with the ohmic dissipation and a source term modeling a micro-instability of the plasma has been used to study nonlinear Alfvén waves and turbulence in a space plasma containing free energy source[46]. Ths source term, or the driver, is assumed to have the form

$$S = \int \Gamma(x-x')b(x')dx' \tag{6}$$

where Γ is the Fourier transform of the linear growth rate of the model micro-instability, which is frequently assumed to have a single δ-function type peak at wave number k_0.

One particular application of the driven DNLS is the problem of "shocklets" -- nonlinear, periodic, quasi-parallel Alfvénic shock trains frequently found in the region upstream of earth's bowshock and near comets[4]. Shocklets are believed to be generated by steepening of periodic Alfvén waves, and the steepening could be either due to quadratic nonlinearity of oblique Alfvén waves[47], or due to cubic nonlinearity of parallel Alfvén waves in a presence of elliptic Γ[48]. When the driver is circular, various qualitatively different types of wave evolution can result depending on the values of ν, k_0, and the magnitude of the driver; these include monotonically growing circular Alfvén waves(unstable system), travelling periodic waves[49](reduction of the number of degrees of freedom), and a system with interacting Alfvén solitons[50](Alfvénic turbulence).

Acknowledgment

The author wishes to thank useful discussions with B. Buti, F. V. Coroniti, R. L. Hamilton, C. F. Kennel, E. Mjølhus, M. Nambu, R. Pellat, and C. C. Wu.

References

[1]See C. F. Kennel, in *From Particles to Plasmas, Lectures Honoring Marshall N. Rosenbluth*, ed. by J. W. Van Dam, pp. 203-226, Addison-Wesley, New York, 1989, for a discussion on roles of Alfvén waves in astrophysical plasmas.

[2]A. Barnes, *Cosmic Electrodyn. 1*, 90 (1970); D. Winske and M. Leroy, *J. Geophys. Res. 89*, 2673 (1984); D. Sentman *et al., J. Geophys. Res. 86*, 7487 (1981).

[3]L. F. Burlaga, *Rev. Geophys. 21*, 363 (1983).

[4]M. M. Hoppe *et al., J. Geophys. Res. 84*, 4471 (1981); D. H. Fairfield and K. W. Behannon, *J. Geophys. Res. 81*, 3897 (1976); B. T. Tsurutani, Geophysical Monograph 34, pp. 1-36, American Geophysical Union, Washington DC, 1985.

[5]C. F. Kennel *et al., J. Geophys. Res. 89*, 5419 (1984); *ibid, J. Geophys. Res. 89*, 5436 (1984).

[6]F. L. Scarf *et al., Science 232*, 382 (1986); B. T. Tsurutani and E. J. Smith, *Geophys. Res. Lett. 13*, 259 (1986); *ibid, Geophys. Res. Lett. 13*, 263 (1986).

[7]A. Barnes, *Phys. Fluids 9*, 1483 (1966); T. Hada and C. F. Kennel, *J. Geophys. Res. 90*, 531 (1985).

[8]B. T. Tsurutani *et al., Geophys. Res. Lett. 6*, 757 (1990); *ibid, Geophys. Res. Lett. 11*, 1817 (1990).

[9]R. Z. Sagdeev and A. A. Galeev, *Nonlinear Plasma Theory*, ed. by T. O'Neil and D. Book, pp. 8, W. A. Benjamin, New York (1969); C. N. Lashmore-Davies, *Phys. Fluids 19*, 587 (1976); M. L. Goldstein, *Astrophys. J. 219*, 700 (1978); N. F. Derby, Jr., *Astrophys. J. 224*, 1013 (1978); J. Sakai and B. U. Ö. Sonnerup, *J. Geophys. Res. 80*, 9069 (1983); H. K. Wong and M. L. Goldstein, *J. Geophys. Res. 91*, 5617 (1986); M. Longtin and B. U. Ö. Sonnerup, *J. Geophys. Res.* 91, 6816 (1986).

[10]T. Terasawa *et al., J. Geophys. Res. 91*, 4171 (1986).

[11]T. Hada and M. Longtin, submitted to *Phys. Fluids*, 1991.

[12]S. P. Kuo *et al., J. Geophys. Res. 93*, 9621 (1988); L. Stenflo and P. K. Shukla, *J. Geophys. Res. 93*, 4115 (1988).

[13]A. Rogister, *Phys. Fluids 14*, 2733 (1971).

[14]E. Mjølhus, Rep. 48, Dept. of Appl. Math., Univ. of Bergen, Norway, 1974; K. Mio *et al., J. Phys. Soc. Jpn. 41*, 265 (1976); M. Wadati *et al., Rocky Mn. J. Math. 8*, 323 (1978).

[15]C. R. Ovenden *et al., J. Geophys. Res. 88*, 6095 (1983).

[16]R. H. Cohen and R. M. Kulsrud, *Phys. Fluids 17*, 2215 (1975).

[17]C. F. Kennel *et al., Phys. Fluids 31*, 1949 (1988).

[18]D. J. Kaup and A. C. Newell, *J. Math. Phys. 19*, 798 (1978).

[19]T. Kawata and H. Inoue, *J. Phys. Soc. Jpn. 44*, 1968 (1978).

[20]T. Kawata et al., *J. Phys. Soc. Jpn. 48*, 1371 (1980).

[21]A. K. Prikarpatskii, *Theor. Math. Phys. 47*, 487 (1981).

[22]Y. H. Ichikawa *et al., J. Phys. Soc. Jpn. 48*, 279 (1980).

[23]E. Mjølhus, *Phys. Scr. 40*, 227 (1989).

[24]E. Mjølhus, *J. Plasma Phys. 19*, 437 (1978); Y. H. Ichikawa and Y. Abe, *Prog. Theor Phys., Suppl. 94*, 128 (1988); R. L. Hamilton, Ph.D. thesis, U.C.L.A., 1991.

[25]V. I. Patel and B. Dasgupta, *Physica 18D*, 151 (1986); *ibid, Physica 27D*, 387 (1987). S. Spangler and S. Fuselier, *J. Geophys. Res. 93*, 845 (1988).

[26]E. Mjølhus, *J. Plasma Phys. 16*, 321 (1976); Mio *et al., J. Phys. Soc. Jpn. 41*, 2667 (1976).

[27]E. Mjølhus and J. Wyller, *Phys. Scr. 33*, 442 (1986).

[28]T. Hada et al., *J. Geophys. Res. 94*, 65 (1989).

[29]E. Mjølhus and J. Wyller, *J. Plasma Phys. 40*, 299 (1988).

[30]M. S. Ruderman, *Fluid Dyn. 22*, 299 (1987) (*Isv. Akad. Nauk SSSR, Mekh. Zhid. i Gaza 2*, 159).

[31]H. Washimi, *J. Phys. Soc. Jpn. 34*, 1373 (1973); M. Hoshino, *Phys. Fluids 31*, 3271 (1988).

[32]B. B. Kadomtsev and V. I.Petviashvili, *Dokl. Akad. Nauk SSSR 192*, 753 (1970).

[33]E. Mjølhus and T. Hada, *J. Plasma Phys. 43*, 257 (1990).

[34]J. Wyller *et al., Physica D39*, 405 (1989).

[35]T. Flå *et al., Phys. Scr. 40*, 219 (1989).

[36]S. R. Spangler, *Phys. Fluids B1*, 1738 (1989); *ibid, Phys. Fluids B2*, 407 (1990).

[37]G. F. Chew *et al, Proc. Roy. Soc. A236*, 112 (1956).

[38]M. Khanna and R. Rajaram, *J. Plasma Phys. 28*, 459 (1982).

[39]S. I. Braginskii, in *Reviews of Plasma Physics*, ed. by M. A. Leontovich, Vol. 1, pp. 218, Consultants Bureau, New York, 1965.

[40]B. K. Som *et al., Phys. Fluids B1*, 2340 (1989).

[41]J. Wyller and E. Mjølhus, *Physica 13D*, 234 (1984); V. I. Karpman, *Phys. Scr. 20*, 462 (1979).

[42]C. F. Kennel *et al., J. Plasma Phys. 42*, 299 (1989); *ibid, Phys. Fluids 2*, 253 (1990); C. C. Wu, *J. Geophys. Res. 95*, 8149 (1990).

[43]A. R. Kantrowitz and H. E. Petschek, in *Plasma Physics in Theory and Application*, ed. by W. B. Kunkel, pp. 148-206, McGraw-Hill, New York (1966).

[44]C. C. Wu, *Geophys. Res. Lett. 14*, 668 (1987); *ibid, J. Geophys. Res. 93*, 987 (1988).

[45]C. C. Wu and T. Hada, *J. Geophys. Res. 96*, 3755 (1991); *ibid, J. Geophys. Res. 96*, 3769 (1991); P. Richter and M. Scholer, *Geophys. Res. Lett. 16*, 1257 (1989); L. C. Lee *et al., J. Geophys. Res. 94*, 8813 (1989).

[46]S. R. Spangler, *Phys. Fluids 29*, 2535 (1986); S. Gosh and K. Papadopoulos, *Phys. Fluids 30*, 1371 (1987).

[47]T. Hada *et al., J. Geophys. Res. 92*, 4423 (1986).

[48]C. F. Kennel *et al., JETP Lett. 48*, 79 (1988); M. A. Malkov *et al., Phys. Fluids 3*, 1407 (1991).

[49]T. Hada *et al., Phys. Fluids 11*, 2581 (1990).

[50]T. Hada *et al.*, in *Nonlinear Processes in Space Plasma Physics*, ed. by V. Stefan, La Jolla, California, 1991.

Alfvén Solitons and the DNLS Equation

R.L. Hamilton[1,*], *C.F. Kennel*[1,*], *and E. Mjølhus*[2]

[1]Institute for Theoretical Physics, University of California, Santa Barbara, CA 93106, USA
[2]Institute for Mathematical and Physical Sciences, University of Tromsø, Tromsø, Norway

*Permanent address: Institute for Geophysics and Planetary Physics, Department of Physics, University of California, Los Angeles, USA

There is a significant body of literature related to the analytic modeling of Alfven waves in the solar wind which takes dispersive magnetohydrodynamics as an idealized basis. In this context, the derivative nonlinear Schrodinger (DNLS) equation has been found by several authors [1-5] to describe the evolution of small amplitude Alfven waves. It may be scaled to the form

$$\frac{\partial b}{\partial t} + \frac{\partial}{\partial x}(|b|^2 b) + i\frac{\partial^2 b}{\partial x^2} = 0 \tag{1}$$

where b(x,t) is the complex representation of the magnetic field perpendicular to the direction of propagation, x. Although the DNLS neglects a rich variety of mechanisms which affect the propagation of Alfven waves in the solar wind, it does provide a powerful tool for studying their underlying nonlinear behavior.

The value of the DNLS as a theoretical basis for studying Alfven waves is in large part due to its integrability through the inverse scattering transformation (IST). Kaup and Newell [6] found the appropriate Lax pair and developed the IST for so-called vanishing boundary conditions with $b(x) \to 0$ for $x \to \pm\infty$. These boundary conditions are appropriate for a localized perturbation travelling parallel to a static magnetic field. Kawata and Inoue [7] developed the IST with non-vanishing boundary conditions appropriate for oblique Alfven waves and Kawata et al. [8] went on to treat the case of a localized perturbation travelling on a circularly polarized carrier wave. Lastly, Prikarpatskii [9] has dealt with the DNLS under periodic boundary conditions.

While for applications to space physics the IST is a particularly promising approach to the DNLS, the equation itself is amenable to other forms of integrability tests. For example, Mikhailov et al. [10] list it amongst a class of equations which pass a test for integrability based on a symmetry approach. By writing two Hamiltonian decompositions of the DNLS, we show here that the DNLS also satisfies a beautiful formalism for integrable systems set forth by Magri [11]:

$$b_t = M\, Q_1(b) = L\, Q_2(b)$$

$$M = i\,\partial_x^2 + \partial_x[\, b\, \partial_x^{-1} (\,\bar{b}\, \partial_x + b\, \bar{\partial_x}\,)] \qquad Q_1(b) = b$$

where

$$L = \partial_x \qquad Q_2(b) = b^2 \bar{b} + i\, b_x$$

$$\partial_x^{-1} = \frac{1}{2}\left(\int_{-\infty}^{x} + \int_{+\infty}^{x}\right) dx$$

and M and L are symplectic operators with respect to the bilinear form:

$$\langle f, g \rangle = \int_{+\infty}^{-\infty} [\bar{f}\, g + f\, \bar{g}]\, dx$$

This immediately leads to the recursion operator, $L^{-1}M$, for the infinite sequence of conservation laws of the DNLS.

An advantage of the IST, over other techniques for the analytic study of the DNLS, is that it allows one to consider the relation of initial conditions to the formation of solitons. This has been done by Ichikawa and Abe [12] , Mjolhus [13] and Dawson and Fontan [14] for the case of parallel Alfven waves. Numerical studies of soliton formation have also been carried out for this case by a number of authors [15-19]. In these works, the modulational instability, discussed in an early paper by Mjolhus [20], appears as a useful criterion for the formation of Alfven solitons.

The importance of non-vanishing boundary conditions for space physics applications has been stressed [21,22] due to the possible existence of oblique Alfven waves or the possible refraction of initially parallel Alfven waves [23]. As noted by Kawata and Inoue [7], under non-vanishing boundary conditions the DNLS has a two-parameter family of solitons corresponding to a set of discrete complex eigenvalues of the scattering problem and a one-parameter family whose solitons are either bright or dark and correspond to a set of discrete real eigenvalues. Of these soliton families, only the two-parameter family continues to exist for parallel propagation.

The existence of the one-parameter family is connected to the unique physical setting of oblique Alfven wave propagation [5,24,25]. To sketch this setting, we note that the DNLS describes the mode coupling between the Alfven and magnetosonic waves. For strictly parallel propagation, the underlying MHD phase speeds of these two waves coincide. As the angle of propagation is increased, however, the underlying wave speeds will separate and thus gradually diminish the role of mode coupling in the nonlinear development of these waves. Even so, the DNLS provides an accurate description of the Alfven and magnetosonic waves in this oblique regime. In fact, depending on the choice of waveframe, the DNLS has been shown [5,24] to reduce to either the KdV description of the magnetosonic wave [26] or the MKdV description of the Alfven wave [27]. Moreover, the velocity of a one-parameter DNLS soliton is not only bounded by the underlying MHD magnetosonic and Alfven wave speeds, but as its speed approaches either the magnetosonic or the Alfven speed, it will deform into a KdV or MKdV soliton respectively [5].

The formation of Alfven solitons under non-vanishing boundary conditions has been considered by Hamilton, Kennel and Mjolhus [25] through an analytic study of the scattering data for a set of initial field profiles. It is found that one-parameter solitons are formed in trains which can unambiguously be identified as either Alfvenic (with the slowest soliton speed approaching the Alfven speed) or magnetosonic (with the fastest soliton speed approaching the magnetosonic speed). The solitons in a given train are either all dark (rarefactive) or all bright (compressive) and, if both an Alfvenic and magnetosonic train are formed from a given initial profile, then the fastest soliton in the Alfvenic train must be slower than any soliton of the magnetosonic train. It was found, though, that the eigenvalues of the fastest Alfvenic soliton and the slowest magnetosonic soliton may coalesce if the wavenumber of the initial profile is large enough make a section of the profile modulationally unstable. The coalescence of two real eigenvalues forms a degenerate and structurally unstable soliton which will bifurcate into a two-parameter soliton as the wavenumber of the initial profile is increased. More detailed observations of the formation of Alfven solitons are contained in reference [25]. The Gelfand-Levitan equations for the degenerate one-parameter soliton have been solved in reference [28].

Acknowledgments

This research was supported by NASA under grant No. NASAW-1624. One of the authors, RLH, would like to acknowledge and thank the Institute of Mathematical and Physical Sciences of the University of Tromso for support and hospitality during his stay there. While at the Institute for Theoretical Physics, RLH was supported in part by the National Science Foundation under grant No. PHY 89-04035. RLH would like to thank Dr. Y. H. Ichikawa and Dr. T. Hada for their comments.

References

1. Rogister, A., Phys. Fluids **14**, 2733 (1971).
2. Mjolhus, E., Report no. 48, Department of Applied Mathematics, University of Bergen, Norway (1974).
3. Mjolhus, E., Wyller, J., J. Plasma Physics **40**, 299 (1988).
4. Ruderman, M. S., Fluid Dyn. 22, 299 (1987), (Izv. Akad. Nauk SSSR, Mekh. Zhid. i Gasa, 2, 159 (1987)).
5. Kennel, C. F., Buti, B., Hada, T., Pellat, R., Phys. Fluids **31**, 1949 (1988).
6. Kaup, D. J., Newell, A. C., J. Math. Phys. **19**, 798 (1978).
7. Kawata, T., Inoue, H., J. Phys. Soc. Japan **44**, 1968 (1978).
8. Kawata, T., Kobayashi, N., Inoue, H., J. Phys. Soc. Japan 46, 1008 (1979).
9. Prikarpatskii, A. K., Theor. Mat. Phys. **47**, 487 (1981).
10. Mikhailov, A. V., Shabat, A. B, Sokolov, V. V., in "What is Integrability?", V. E. Zacharov (Ed.), Springer-Verlag (1991).
11. Magri, F., J. Math. Phys. **19**, 1156 (1977).
12. Ichikawa, Y. H., Abe, Y., Prog. Theor. Phys. Supp., No. 94, 128 (1988).
13. Mjolhus, E., J. Plasma Physics **19**, 437 (1978).
14. Dawson, S. P., Fontan, C. F., Phys. Rev. A, 39, 5289 (1989).
15. Spangler, S. R., Sheerin, J. P., Payne, G. L., Phys. Fluids **28**, 104 (1985).
16. Spangler, S. R., Astrophys. J. **299**, 122 (1985).
17. Ghosh, S., Papadopoulos, K.,Phys. Fluids **30**, 1371 (1987).
18. Dawson, S. P., Fontan, C. F., Phys. Fluids **31**, 83 (1988).
19. Dawson, S. P., "Solitons and Radiation from the Derivative nonlinear Schrodinger Equation", preprint 1990.
20. Mjolhus, E., J. Plasma Phys. **16**, 321 (1976).
21. Kennel, C. F., Buti, B., Hada, T., Pellat, R., Phys. Fluids **31**, 1949 (1988).
22. Mjolhus, E., Wyller, J., Physica Scripta **33**, 442 (1985).
23. Hada, T., Kennel, C. F., Terasawa, T., J. Geophys. Res. **88**, 4423 (1987).
24. Mjolhus, E., Physica Scripta **40**, 227 (1989).

25. Hamilton, R. L., Kennel, C. F., Mjolhus, E., Institute for Theoretical Physics Preprint No. NSF-ITP-91-131.
26. Kakutani, T., Ono, H. , Taniuti, T., Wei, C.-C., J. Phys. Soc. Japan **24**, 1159 (1968).
27. Kakutani, T., Ono, H. , J. Phys. Soc. Japan **26**, 1305 (1969).
28. Hamilton, R. L., Kennel, C. F., Mjolhus, E., Institute for Theoretical Physics Preprint No. NSF-ITP-91-130.

On the Analytical Theory for Self-Focusing of Radiation

V.M. Malkin

Institute of Nuclear Physics, 630090 Novosibirsk, Russia

Abstract

The investigation of dynamics of laser beam in a medium, dielectric susceptibility of which depends on the beam power, represents one of the classical problems of nonlinear physics. The solution of this problem is difficult to obtain even at the most favorable conditions. In the present paper the problem is solved analytically in the framework of the two dimensional Schroedinger equation with focusing cubic nonlinearity and small defocusing nonlinearity of fifth order. It is shown that the model remains valid during the whole process of self-focusing, which goes in the region of applicability of macroscopic description and consists in decreasing oscillations of the waveguide width. The shape of these oscillation and their envelope are studied in detail. Parameters of the final state, which is a homogeneous waveguide, are determined.

Ionospheric Langmuir Turbulence Driven by an Electromagnetic Pump Below the Upper-Hybrid Frequency

D.L. Newman and M.V. Goldman

APAS Department, University of Colorado, Boulder, CO 80309, USA

In ionospheric modification experiments, an intense ground-launched radio wave of frequency ω_0 reflects from the critical altitude in the ionosphere where $\omega_p = \omega_0$ setting up a standing Airy pattern with the maximum electric field intensity occuring below the critical layer ($\omega_p < \omega_0$) and with the electric field polarized approximately in the direction of the background magnetic field vector $\mathbf{B}_0$ near the critical altitude. The driving of the ionosphere can be approximated by considering a spatially uniform ($\mathbf{k} = 0$) electrostatic pump $\mathbf{E}_0$ parallel to $\mathbf{B}_0$ and at frequency ω_0. A driver of this form has been used as a source in numerical simulations of ionospheric turbulence using the Zakharov equations.[1–3] The simulations of Refs. 1 and 2 treated a parameter regime different from that considered in Ref. 3. Specifically, Refs 1 and 2 modeled the pump at the altitude of the first Airy maximum whereas Ref. 3 modeled the pump at a lower altitude corresponding to the matching height of the incoherent scatter radar diagnostic used in the Aracebo experiments. As a result, the pump amplitude was lower and the frequency mismatch ($\omega_0 - \omega_p$) larger under the assumptions of Ref. 3. In addition, a weaker ion-acoustic damping ratio was assumed in Ref. 3. All of these difference favor the existence of a weak-turbulence backscatter cascade under the conditions modeled in Ref. 3 and, in fact, such a cascade was observed in Ref. 3 but not in Refs 1 and 2. However, whereas Ref. 3 used a 2D simulation, only 1D simulations were used in Refs 1 and 2. It is well know that certain important properties of the Langmuir turbulence cannot be properly modeled in 1D: (1) The influence of the background magnetic field is absent. (2) Wave collapse in strong Langmuir turbulence occurs only in two or more dimensions. (3) The turbulence can assume a more complex geometrical and topological structure in higher dimensions. We have therefore undertaken two- and three-dimensional simulations of ionospheric turbulence in a spatially homogeneous plasma for parameters similar to those considered in Refs 1 and 2 with the further restriction that the pump frequency satisfies the condition $\omega_p < \omega_0 < \omega_{UH}$ where $\omega_{UH} = (\omega_p^2 + \Omega_e^2)^{1/2}$ is the upper-hybrid frequency and Ω_e is the electron cyclotron frequency. To demonstrate the importance of the role of the background magnetic field $\mathbf{B}_0$, we contrast the results of these simulations with those in which there is no magnetic field (i.e., $\Omega_e = 0$).

The simulations described here are based on the standard electrostatic Zakharov equations[4] with corrections to the Langmuir dispersion from the background magnetic field for $\Omega_e^2 \ll \omega_p^2$:

$$\nabla \cdot (i\partial_t + i\hat{\gamma} + \frac{3}{2}\omega_p \lambda_D^2 \nabla^2 + \frac{1}{2}\frac{\Omega_e^2}{\omega_p} \sin^2\theta)\nabla\phi = \frac{\omega_p}{2n_0}\nabla \cdot (\delta n \nabla\phi)$$

$$(\partial_t^2 + 2\hat{\nu}_i \partial_t - c_s^2 \nabla^2)\delta n = \frac{1}{16\pi m_i}\nabla^2 |\nabla\phi|^2$$

where ϕ is the low-frequency envelope of the Langmuir electrostatic potential and δn is the quasineutral density perturbation. These simulations were performed on a 128^2 grid in 2D and a 64^3 grid in 3D with a mass ratio $m_i/m_e = 4m_H/m_e = 7344$, $c_s^2 = T_e/m_i$ frequency mismatch $(\omega_0 - \omega_p)/\omega_p = 1.7 \times 10^{-3}$, heavy ion-acoustic damping $[\nu_i(k) = kc_s]$, and pump intensity $|E_0|^2/(8\pi n_0 T_e) = 0.001$ (2D) or 0.002 (3D). The Langmuir damping consisted of a wavelength-independent collisional term $\gamma_c/\omega_p = 2\times10^{-5}$ plus a power-law component to model transit-time damping of the wave packet at small scales. Because there is a large dynamical range separating the injection scale (governed by the pump frequency ω_0) and the dissipation scale of 10–20 λ_D as determined by particle-in-cell simulatons of wave collapse,[5–7] we have chosen artificially large power law damping coefficients ($\gamma/\omega_p = A(k\lambda_D)^6$ where $A = 10,000$ and 50,000 in 2D and 3D, respectively) in order to adequately model the large-scale structures of the turbulence while arresting collapse before the wave packet reaches the resolution limit of the grid.

It is well know from linear theory that for $\omega_0^2 > \omega_p^2 + \Omega_e^2 \sin^2\theta$, the plasma can be subject to both the parametric-decay instability (PDI) and the modulational instability (MI), the former for $k < k_0$ and the later for $k > k_0$ where $(k_0\lambda_D)^2 = (2/3)[(\omega_0 - (\omega_p^2 + \Omega_e^2 \sin^2\theta)^{1/2}]/\omega_p$. Thus, for $\theta > \theta_c$ where $\sin^2\theta_c = (\omega_0^2 - \omega_p^2)/\Omega_e^2$, the PDI is forbidden. Figure 1 show contour plots of the linear instability growth rates $\Gamma(\mathbf{k})$ for our 2D simulation parameters with $\Omega_e = 0$ and $\Omega_e = 0.3\omega_p$. Whereas in the absence of an external magnetic field the instability extends to large angles from the direction of polarization of the pump wave, the instability is confined to a very narrow angular range for $\Omega_e = 0.3\omega_p$.

The driven/damped system described above was evolved in two and three dimensions from initial conditions consisting of random-phase noise. A statistical steady state was eventually reached, which will now be described. Only the results of the 2D simulations are presented here; however in those aspects where the 3D results are qualitatively different, those differences will be described. At any time after the turbulence has reached a statistical steady state, there are many regions in the plasma that are either undergoing or have just undergone wave collapse. Although at any instant these collapse centers appear to be randomly located within the simulation volume, it becomes evident upon averaging $|\mathbf{E}(\mathbf{x})|^2$ over a period long compared with the collapse and burnout of a single wave packet that the collapses occur preferentially in well-delineated regions. Figure 2 shows $|\mathbf{E}|^2$ averaged over a period containing many wave collapses for $\Omega_e = 0$ and $\Omega_e = 0.3$. Although not shown here, the corresponding time-averaged plots of $-\delta n$ are practically identical. For both values of Ω_e, the Langmuir field intensity and the underlying density depressions lie along approximately one-dimensional filaments or channels (i.e., regions of below-average density), the orientation of which depend on the magnetic field strength. For $\Omega_e = 0$, these channels are approximately parallel to the polarization direction of the pump, whereas for $\Omega_e = 0.3$ they form an intersecting network oblique to the direction of the pump (and, hence, $\mathbf{B}_0$). The perpendicular scale is greatly compressed in Fig. 2b; thus, these oblique density channels are in fact nearly perpendicular to the direction of $\mathbf{B}_0$ and the individual collapsing wave packets are highly elongated in the perpendicular direction as well. Density filaments are observed also in 3D simulations. However, the topology of the filaments is more complex in 3D than in 2D because the orientation of a filament can have an arbitrary azimuthal angle relative to the direction of $\mathbf{B}_0$.

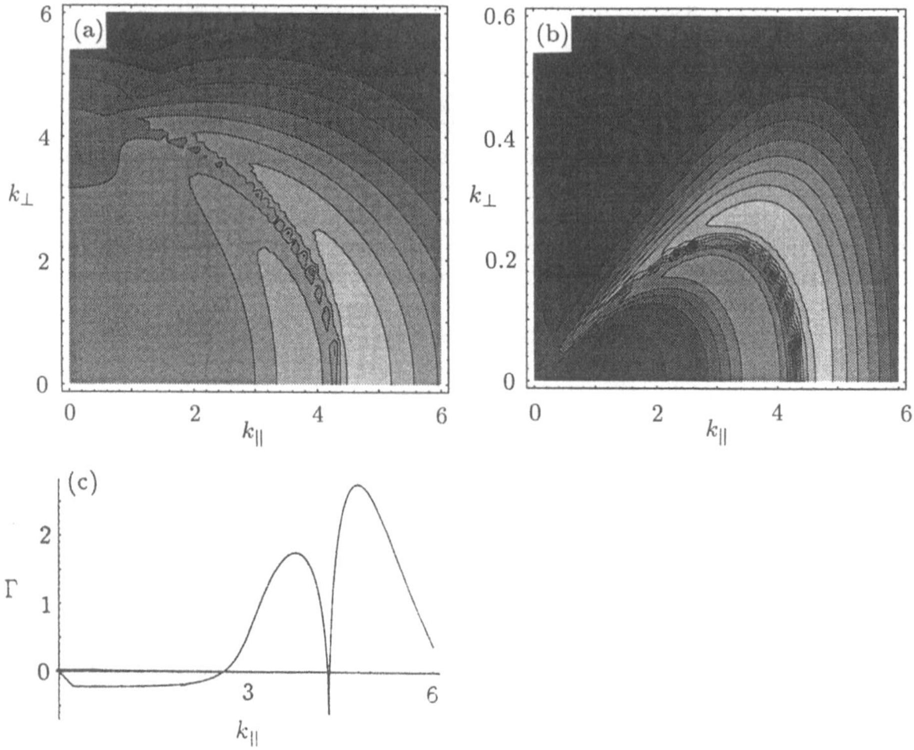

Fig. 1 Contours of constant linear growth rate for (a) $\Omega_e = 0$ and (b) $\Omega_e = 0.3$, and (c) a cut along $k_\perp = 0$. Lighter shading corresponds to stronger instability. k and Γ are in units of $(4m_e/9m_i)^{1/2}k_D$ and $(2m_e/3m_i)w_p$, respectively where $m_e/m_i = 1/7344$. Thus the maximum growth rate is $\Gamma_{max} = 2.5 \times 10^{-4}\omega_p$ at $k_\parallel = 3.7 \times 10^{-2}k_D$. Note that the $k_\perp$ axis has been magnified by a factor of 10 in (b).

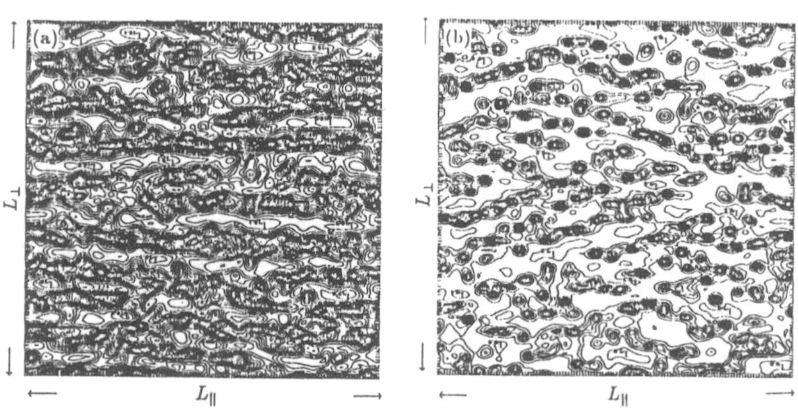

Fig. 2 Contour plots of $|\mathbf{E}|^2$ (in arbitrary units) for (a) $\Omega_e = 0$ in a region of dimension $L_\parallel = 1485\lambda_D$ by $L_\perp = 2970\lambda_D$ and (b) $\Omega_e = 0.3\omega_p$ with $L_\parallel = 1485\lambda_D$ by $L_\perp = 11,891\lambda_D$.

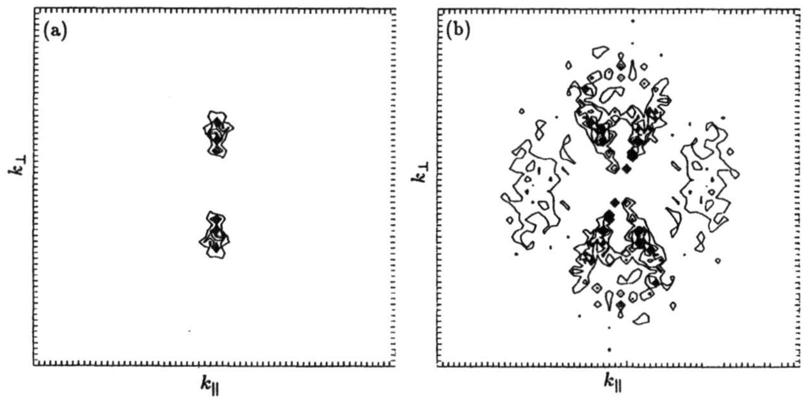

Fig. 3 Contour plots of $|n_{\mathbf{k}}|$ (in arbitrary units) for (a) $\Omega_e = 0$ in a region defined by $-.135 \leq k_\parallel \lambda_D \leq 0.135$; $-0.067 \leq k_\perp \lambda_D \leq 0.067$, and (b) $-0.135 \leq k_\parallel \lambda_D \leq 0.135$; $-0.0011 \leq k_\perp \lambda_D \leq 0.0011$.

In order to understand the origin of the filamentary structure in Fig. 2, it is necessary to consider the wave-number density spectra $|n_{\mathbf{k}}|$ shown in Fig. 3 for a region greatly magnified in $k_\perp$. These are averages over the same time period as for the spatial structure of Fig. 2. By taking a time average, the contribution to the spectra from individual wave packets is suppressed relative to the contribution from the long-lived channels. As expected, the maxima of the spectra lie at angles complementary to the orientation of the corresponding channels. Furthermore, in the case of $\Omega_e = 0.3$, the spectral maxima are also at approximately the critical angle θ_c discussed in conjunction with Fig. 1b. Thus, in the magnetized case, the density channels appear to be associated with the purely growing density perturbations of the linear modulational instability in the vicinity of $\theta = \theta_c$ in k space.

The interpretation of the density channels in the unmagnetized case is not so evident since there is no linear pump-driven instability along the $k_\perp$ axis. Instead, these density channels are a consequence of a secondary process whereby the Langmuir waves with $k_\perp \ll k_\parallel$ driven unstable by the pump resonantly beat to produce density modulations with $k_\parallel \approx 0$ in a process closely related to the perpendicular modulational intability driven by a propagating Langmuir wave in a multidimensional plasma. In both the magnetized and unmagnetized cases, the formation or nucleation[8] of wave packets, which then undergo collapse, is favored in the low-density channels relative to the higher than average density regions between channels. The ponderomotive force associated with the Langmuir waves serves to maintain the low density in the channels, thus contributing to their longevity.

This research is supported by grants from NSF; computations were performed at the National Center for Atmospheric Research.

[1] D. R. Nicholson, G. L. Payne, R. M. Downie, and J. P. Sheerin, *Phys. Rev. Lett.*, 1984, 52, 2152.

[2] G. L. Payne, D. R. Nicholson, R. M. Downie, and J. P. Sheerin, *J. Geophys. Res.*, 1984, 89, 10921.

[3] D. F. DuBois, H. A. Rose, and D. Russell, *Phys. Rev. Lett.*, 1991, **66**, 1970.
[4] V. E. Zakharov, *Sov. Phys. JETP*, 1972, **35**, 908.
[5] A. I, D'yanchenko, V. E. Zakharov, A. M. Rubenchik, R. Z. Sagdeev, and V. F. Shvets, *Sov. Phys. JETP*, 1988, **67**, 513.
[6] V. E. Zakharov, A. N. Pushkarev, A. M. Rubenchik, R. Z. Sagdeev, and V. F. Shvets, *Sov. Phys. JETP*, 1989, **69**, 334.
[7] D. L. Newman, R. M. Winglee, P. A. Robinson, J. Glanz, and M. V. Goldman, *Phys. Fluids B*, 1990, **2**, 2600.
[8] D. Russell, D. F. DuBois, and H. A. Rose, *Phys. Rev. Lett.*, 1988, **60**, 581.

Mode Conversions in Ionospheric Modification Experiments

N.N. Rao[1;*] *and D.J. Kaup*[1]

[1]Department of Mathematics, Clarkson University, Potsdam, NY 13699, USA
*Permanent address: Theoretical Physics Division, Physical Research Laboratory, Navrangpura, Ahmedabad 380009, India

ABSTRACT: We investigate resonance excitation of electrostatic and electromagnetic electron cyclotron harmonic waves by mode conversion during the recent ionospheric F–region modification experiments at Tromsø (Norway) and Arecibo (Puerto Rico). Frequency band width for quenching of DM feature in the electromagnetic emission spectra from ionosphere is found to be about a few kHz which is in good agreement with the experimental results.

I INTRODUCTION

Since the experimental discovery of "Stimulated Electromagnetic Emission" (SEE) from the auroral ionosphere above Tromsø [1], the SEE phenomenon has been extensively studied under different geophysical conditions and experimental parameters. Recent experiments at Tromsø (Norway) and Arecibo (Puerto Rico) have underscored for the first time the fundamental role played by the upper–hybrid (UH) resonance layer in significantly modifying the SEE spectra from the F–region. For example, the generally dominant "Downshifted Maximum" (DM) feature is found to be totally quenched for certain pump frequencies.[2] New spectral features called "Broad Symmetrical Structures" (BSS) have also been reported.[3] On the theoretical side, it has been suggested that many of the spectral features could arise due to three–wave as well as four–wave interactions involving the UH waves. For example, the DM feature has been shown[4] to originate due to the parametric decay instability[5]

$$\text{Upper–Hybrid (UH)} \rightarrow \text{Electromagnetic} + \text{Lower–Hybrid (LH)},$$

which has a finite amplitude threshold. The UH wave itself is generated in the interaction region from the pump O–mode by either a four–wave interaction[6] or a linear conversion process[7]. In either case the UH wave generation is facilitated by the presence of the geomagnetic field aligned, small–scale density striations[8] that are excited during the modification experiments.

Recent results on the SEE spectral features are obtained by varying pump frequency (f_0) in small steps (typically, about 10 kHz) around the harmonics (third, fourth and fifth) of the local electron cyclotron frequency (f_{ce}) which is about a MHz. Consequently, the UH wave frequency will also be near the cyclotron harmonic (CH) frequencies. It is therefore expected that both electrostatic as well as electromagnetic CH harmonic waves could be excited by the mode conversion of either the incident pump wave or the UH wave. We present below some theoretical results and compare them with the experiments.

II ELECTROSTATIC CH WAVES

The electrostatic CH waves, also called the electrostatic Bernstein (EB) modes, are resonantly excited by mode conversion of the UH wave when $f_0 \sim mf_{ce}$. The transmission coefficient for the UH wave can be calculated[9] by applying the theory of Cairns and Lashmore–Davies[10] to the kinetic dispersion relation,

$$1 - \frac{\omega_{pe}^2}{k^2 v_{te}^2} e^{-\sigma} \sum_{m=-\infty}^{+\infty} I_m(\sigma) \left[\frac{m\Omega_e}{\omega - m\Omega_e} + \frac{\omega Z'(\zeta_m)}{2(\omega - m\Omega_e)} \right] = 0 , \tag{1}$$

where $\sigma = k_\perp^2 \rho_{te}^2$, $\zeta_m = (\omega - m\Omega_e)/\sqrt{2}\, k_\parallel v_{te}$, $\omega_{pe} \equiv 2\pi f_{pe} = (4\pi n_e e^2/m_e)^{1/2}$, $k = (k_\perp^2 + k_\parallel^2)^{1/2}$, $v_{te} = (k_B T_e/m_e)^{1/2}$, $\Omega_e \equiv 2\pi f_{ce} = eB/m_e c$, $\rho_{te} = v_{te}/\Omega_e$, $I_m(\sigma)$ is the Bessel function with imaginary argument, $Z'(\zeta_m)$ is the derivative with respect to the argument of the plasma dispersion function $Z(\zeta_m)$ and all the remaining symbols have their usual meanings[9,11]. Neglecting the damping term, (1) yields for $\sigma \ll 1$ the coupled dispersion relation

$$(\omega^2 - \omega_{UH}^2)(\omega^2 - \omega_{CH}^2) = \frac{m^2(m^2-1)}{2^{m-1} m!} \omega_{pe}^2 \Omega_e^2 \sigma^{m-1} , \tag{2}$$

where $m \geq 3$, $\omega_{UH}^2 = \omega_H^2 + 3\omega_{pe}^2 \Omega_e^2 \sigma (\omega_{pe}^2 - 3\Omega_e^2)^{-1}$, $\omega_H^2 = \omega_{pe}^2 + \Omega_e^2$, $\omega_{CH} = m\Omega_e$, and $\omega_{pe}^2 \neq 3\Omega_e^2$. The transmission coefficient (T) for the UH mode conversion to EB modes is[9] $T = \exp(-\mu L_B)$ where

$$\mu = \sqrt{3}\,\pi \frac{(m^2-1)}{6^m m!} \frac{1}{\rho_{te}} (\alpha^2 - 3)^{m-1/2} \left(\frac{m^2-1}{\alpha^2} - 1 \right)^{m-3/2} , \qquad L_B = \left| \frac{1}{\Omega_e} \frac{\partial \Omega_e}{\partial x} \right|^{-1} , \tag{3}$$

and $\alpha = \omega_{pe}/\Omega_e$. For Tromsø and Arecibo parameters, μ is typically of the order of about 1 cm^{-1} except when $\alpha^2 \approx 3$ or, for a given m, $\alpha^2 \approx m^2 - 1$ whereas even the most conservative value for the effective scale length (L_B) is about a few meters or more. Thus, the mode conversion process, when it occurs, is almost complete with the UH wave energy being transferred to the EB mode which can either decay or be damped depending on the UH wave frequency. On the other hand, the consequent damping of the UH wave seems to explain, as discussed below, the quenching of the DM feature.

The conversion process UH → EB takes place only when the pump frequency and hence the UH frequency is within a narrow band width around the CH frequency. For $m \geq 3$ the band width is given by[11]

$$\Delta f_m(\sigma) \equiv f_0 - m f_{ce} = \frac{3}{2m} \frac{m^2-1}{m^2-4} f_{ce} \sigma . \tag{4}$$

To estimate σ, we assume that the wavelength $\lambda_\perp \equiv 2\pi/k_\perp$ is given by the transverse scale length of the density striations.[8] This yields[11],

$$\Delta f_m = 10^5 \times \frac{m^2-1}{m^5(m^2-4)} f_{ce}\, a b^2 \exp\left(-\frac{2b}{m}\right) \quad \text{kHz}, \tag{5}$$

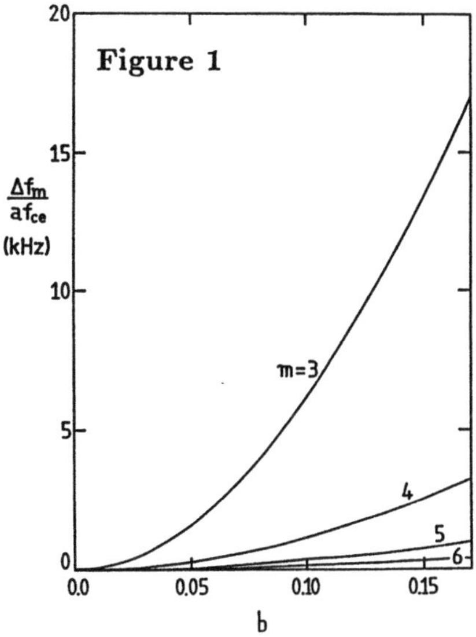

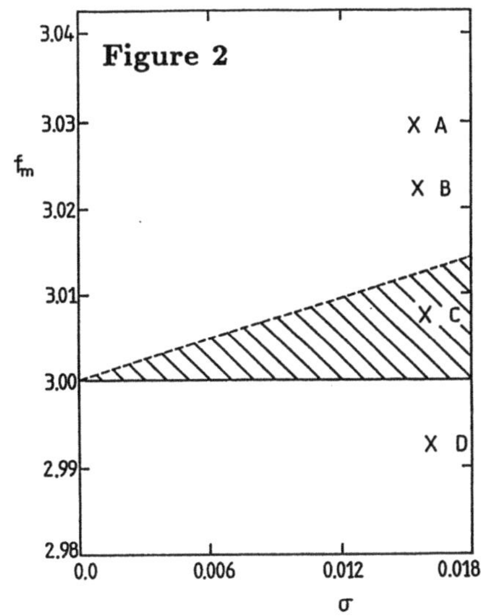

Figure 1: Plot of frequency band width $\Delta f_m (\equiv f_0 - m f_{ce})$ for the occurrence of electrostatic mode conversion as a function of b for different harmonic mode numbers (m).

Figure 2: Plot of normalized pump frequency $f_m (\equiv f_0/f_{ce})$ as a function of $\sigma(\equiv k_\perp^2 \rho_{te}^2)$ for the third harmonic $(m = 3)$. Theory predicts Bernstein mode excitation in the shaded region. For points A, B and D, the SEE spectrum in the Tromsø experiments is dominated by the DM feature whereas for C the latter is quenched.

where f_{ce} is in units of MHz. The parameters a and b are defined in Ref. [11]. Typical values are $a = 1.936$ and $b = 0.104$ for the Tromsø parameters, and $a = 0.331$ and $b = 0.137$ for Arecibo.

Figure 1 shows a plot of (5). In Figure (2), we compare the predictions of the mode conversion theory with the experimental results for the case of $m = 3$. For the points A, B and D, experiments show that DM is the dominant feature in the lower side band of the SEE spectrum; for point C the DM feature is either extremely weak or totally quenched. Theory predicts mode conversion to take place only in the shaded region which is in accordance with the experimental observations. Similar results follow for the other harmonics $m = 4$ and $m = 5$ as well as for the Arecibo parameters. On the other hand, for the same values of the various parameters, the width due to the cyclotron damping is too small to be in accordance with the experiments. Thus, these results seem to establish the important role played by the mode conversion process in the phenomenon of DM quenching. Also, knowing the frequency for the onset of DM quenching, it is possible to calculate using (5) the geomagnetic field in the interaction region. When the pump frequency is near the second harmonic, (2) gets modified but leads to enhanced mode conversion over a much wider range of frequencies around the CH frequency. However, no experimental data are currently available for these parameters.

III ELECTROMAGNETIC CH WAVES

In the case of electromagnetic CH waves, we have shown earlier[9] that the conversion efficiency for the O–mode is negligible for the Tromsø parameters. However, X–mode can be converted into CH waves as well as UH waves depending on certain parameter values. The latter may be relevant for the E–region heating observed in some of the Tromsø experiments.

(1) X–Mode Cyclotron Waves:

The kinetic dispersion relation for the X–mode wave is $D_1 D_2 = D_3^2$ where

$$D_1 = 1 - 2\omega_{pe}^2 \frac{e^{-\sigma}}{\sigma} \sum_{m=1}^{+\infty} \frac{m^2 I_m}{\omega^2 - m^2\Omega_e^2}\ , \quad D_3 = 2\omega_{pe}^2 \frac{\Omega_e}{\omega} e^{-\sigma} \sum_{m=1}^{+\infty} \frac{m^2(I_{m+1} - I_m + mI_m/\sigma)}{\omega^2 - m^2\Omega_e^2}\ , \tag{6}$$

$$D_2 = D_1 - \frac{2\omega_{pe}^2}{\omega^2} \sigma e^{-\sigma} (I_0 - I_1) - \frac{k^2c^2}{\omega^2} - 4\omega_{pe}^2 e^{-\sigma} \sum_{m=1}^{+\infty} \frac{(\sigma - m)I_m - \sigma I_{m+1}}{\omega^2 - m^2\Omega_e^2}\ , \tag{7}$$

In the limit $\sigma \to 0$, one obtains the usual cold fluid dispersion relation for the X–mode,

$$\frac{k^2c^2}{\omega^2} = 1 - \frac{\omega_{pe}^2}{\omega^2} \frac{\omega^2 - \omega_{pe}^2}{\omega^2 - \omega_H^2}\ . \tag{8}$$

For $m \geq 2$, the coupled dispersion relation is[11],

$$(\omega - \omega_\pm)(\omega - \omega_{CH}) = \frac{\alpha^2}{[(1-\sigma\beta)^2 + 4\alpha^2]^{1/2}} \frac{(\pm\Theta_M)}{2^{m+1}m!} \sigma^{m-1}\ , \tag{9}$$

where

$$\Theta_M = (m-1)^2 \frac{[m(m+1) - \sigma^2]^2}{(m^2-1) - \sigma^2} \Omega_e^2\ , \tag{10}$$

and the subscripts '+' and '–' in (9) refer to the fast and the slow branches of X–mode valid, respectively, when $0 < \alpha^2 < m(m-1)$ and $(m^2-1) < \alpha^2 < m(m+1)$. The transmission coefficient is given, as earlier, by $T = \exp(-\mu L_B)$ with

$$\mu = \frac{\pi(m-1)^2}{2^m m!} \frac{\alpha^2}{\beta^{(m-1/2)}\rho_{te}} \left[\frac{m(m+1) - \alpha^2}{\pm(m^2-1) \mp \alpha^2}\right]^{m+1/2} \left[\pm m(m-1) \mp \alpha^2\right]^{m-3/2}\ , \tag{11}$$

where upper (lower) sign should be used for $\omega_+(\omega_-)$ in (9).

For the typical ionospheric parameters, $\beta >> 1$ and the maximum absorption occurs at $m = 2$. From (11) it follows that the mode conversion efficiency is appreciable only when $\alpha^2 \sim m^2-1$. Defining ϵ by $\alpha^2 = (m^2-1)+\epsilon$, the wavenumber k_M at the mode conversion point is obtained from $k_M^2\rho_{te}^2 = (m^2-1)/\epsilon\beta$ which requires $\epsilon > 0$. Thus, only the slow branch of the X–mode can be converted into a CH wave. For the Tromsø parameters, $\mu \sim 10^{-2}$ cm^{-1} indicating appreciable conversion even for the most conservative estimate for L_B. It should however be noted that the slow X–mode reflection region cannot be directly accessed from the ground because of the reflection from lower regions. However, since the slow

X–mode connects to the O–mode for propagation at an angle to the magnetic field, it may be possible to excite the former by suitably launching the latter from ground.

(2) Upper–Hybrid Waves:

When $\alpha^2 < m^2 - 1$, the fast branch of the X–mode can be converted into UH waves. The coupled dispersion relation for this process is

$$(\omega - \omega_+)(\omega - \omega_{UH}) = \Omega_e^2 F(\alpha, \sigma_M) , \tag{12}$$

where,

$$F(\alpha, \sigma_M) = \frac{18\alpha^4(\alpha^2 - 3)}{[(\alpha^2 - 3) + 3\sigma_M]\,[(\alpha^2 - 3)^2 + 3\alpha^2\sigma_M]} \times$$
$$\times \frac{\sigma_M^2}{\left\{4\alpha^2 + (1 - \sigma_M\beta)^2 + (1 + 2\alpha^2 + \sigma_M\beta)\,[4\alpha^2 + (1 - \sigma_M\beta)^2]^{1/2}\right\}} , \tag{13}$$

$$\sigma_M = \frac{1}{2}\frac{3 - \alpha^2}{[3\alpha^2 + \beta(3 - \alpha^2)]}\left\{1 \pm \left[1 + 4\alpha^2 + \frac{4\beta}{3}(3 - \alpha^2)\right]^{1/2}\right\} . \tag{14}$$

The transmission coefficient for (12) can be calculated[11] as earlier and is given by $T = \exp(-2\pi\eta)$ where, for $\beta \gg 1$, η is approximated by

$$\eta = \frac{6}{\rho_{te}}\left(\frac{3}{\beta^7}\right)^{1/4}\frac{\alpha^2(1 + \alpha^2)}{(3 - \alpha^2)^{5/4}}\frac{1}{\left|2\alpha^2 L_n^{-1} + 3L_B^{-1}\right|} , \tag{15}$$

where L_n and L_B are the effective scale lengths for density and magnetic field inhomogeneities, respectively. For appreciable wave absorption, we need $\eta \sim 1$ which requires $(3 - \alpha^2)^{5/4} \sim (8\,L_0/\rho_{te})\,(3/\beta^7)^{1/4}$, where we have assumed $L_n \sim L_B \equiv L_0$. Using the Tromsø parameters, one gets, $\sqrt{3}f_{ce} - f_{pe} \sim 0.1$ kHz. For typical daytime ionospheric parameters, the latter is more likely to be satisfied in the E–region, thereby suggesting the possible excitation of UH waves in this region which is consistent with the heater induced 1–meter irregularities observed in some of the Tromsø experiments using X–mode pump wave.[12]

IV SUMMARY

We have investigated the possible excitation of electrostatic as well as electromagnetic electron cyclotron harmonic waves during the recent ionospheric modification experiments at Tromsø and Arecibo. Theoretical predictions about the frequency band width for the occurrence of upper–hybrid mode conversion to the Bernstein waves are in good agreement with the experimental results on the onset/quenching of the DM feature in the stimulated emission spectra from the ionosphere. Excitation of the upper–hybrid waves in the E–region is shown to be possible for X–mode pump wave.

ACKNOWLEDGMENTS

This work is supported by the University Research Initiative Grant AFOSR–89–0510 and ONR Grant No. N000 14–88–K–0153.

REFERENCES

[1] B. Thidé et al., Phys. Rev. Letts., 1982, 49, 1561.

[2] T.B. Leyser et al., Phys. Rev. Letts., 1989, 63, 1145.

[3] P. Stubbe and H. Kopka, Phys. Re. Letts., 1990, 65, 183.

[4] T.B. Leyser, Geophys. Res. Letts., 1991, 18, 408.

[5] G. Murtaza and P.K. Shukla, J. Plasma Phys., 1984, 31, 423.

[6] M.C. Lee and S.P. Kuo, J. Plasma Phys., 1983, 30, 463.

[7] S.N. Antani, N.N. Rao and D.J. Kaup, Geophys. Res. Letts., 1991, (in Press).

[8] A.C. Das and J.A. Fejer, J. Geophys. Res., 1979, 84, 6701.

[9] N.N. Rao and D.J. Kaup, J. Geophys. Res., 1990, 95, 17245.

[10] R.A. Cairns and C.N. Lashmore-Davies, Phys. Fluids, 1983, 26, 1268.

[11] N.N. Rao and D.J. Kaup, J. Geophys. Res., 1991, (in Press).

[12] P. Høeg et al., J. Geophys. Res., 1986, 91, 11309.

On the Superstrong Wave Collapse

V.F. Shvets

Department of Physics, Auburn University, Auburn, AL 36849-5311, USA

1. The development of wave collapse (WC) results in many cases in the formation of a long-living hot spot that draws continuously waves from the surrounding space. This important and widespread mechanism of nonlinear dissipation of wave energy in strong-turbulent processes has been lately studied intensively[1-8] in the model of nonlinear Schroedinger equation (NLS)

$$i\psi_t + \Delta\psi + |\psi|^s\psi = 0 \tag{1}$$

that describes the various types of WC in the d-dimensional space for sufficiently intensive initial conditions (see e.g.[6,8] and references therein). For $4 \le sd < 2s+2$ the short-living collapses (strong collapse with fixed energy trapped into singularity at $sd=4$ and weak collapse with continuously decreasing energy at $sd > 4$) take place. These conclusions can be illustrated through the self-similar solution of Eq. (1) $\psi(r,t)=(t_0-t)^{-1/s-i\kappa}g[r(t_0-t)^{-1/2}]$ (t_0 is the moment of collapse) with the wave energy captured into the singularity given by the expression

$$N_{sing} \propto \int|\psi|^2 dr \propto (t_0-t)^{d/2-2/s}, \qquad t \to t_0, \tag{2}$$

and they also have been confirmed by many computer experiments[8-11]. However, for $sd \ge 2s+2$, a small-size dissipative zone (hot spot or black hole), which absorbs quasisteadily energy, is formed as a result of a weak collapse. This effect of long collapse, named as a "superstrong wave collapse" (SWC) in[3,6] is critically important for understanding of many macroscopic displays of strong wave turbulence (optical, plasma and others). The SWC can be developed in the four different regimes[3,5]. Corresponding stationary singular solutions (SSS) are described in the symmetrical case ($\Delta = \Delta_r \equiv d^2/dr^2 + [(d-1)/r]\, d/dr$) by the following expressions as $r \to 0$:

$$|\psi| \to (2/s^2)^{1/s} r^{-2/s} |\ln r|^{-1/s}, \qquad sd = 2s+2 \tag{I}$$

$$|\psi| \to [2(sd-2s-2)/s^2]^{1/s} r^{-2/s}, \qquad 2s+2 < sd < 2s+4 \tag{II}$$

$$|\psi| = cr^{-2/s}, \quad c^4[c^s - (2/s)^2] = P^2, \qquad sd = 2s+4 \tag{III}$$

$$|\psi| \to [P^2 r^{-2(d-1)}]^{1/(s+4)}, \qquad 2s+4 < sd, \tag{IV}$$

where the exact stationary solution (III) and quasi-qlassical asymptotics (IV) are determined by the constant energy flux into singularity $P = |\psi|^2 r^{d-1}(\arg\psi)_r < 0$ while for regimes (I) and (II) the flux appears in the additional terms of the asymptotics. These SSS are valid for physically small distances $r > r_0$ where $r_0 << 1$ is the characteristic size of the dissipative region. The transition (for $sd \ge 2s+2$) of stable self-similar weak collapse to the quasistationary state, after the dissipation begins, has been demonstrated[3] by direct computer simulations of the supercritical WC dynamics in the NLS with strong nonlinear damping near the singularity

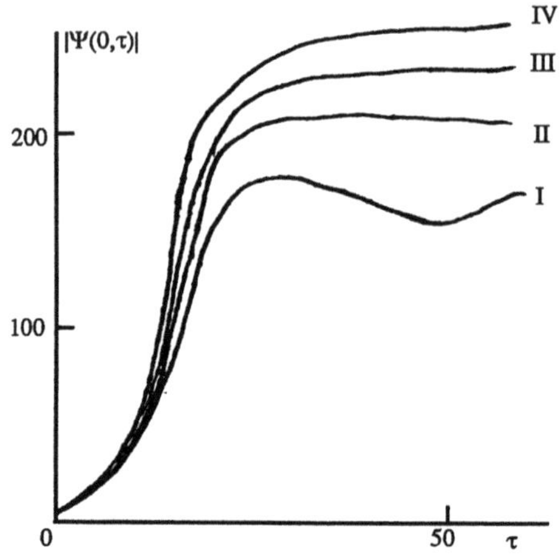

Fig. 1[3].

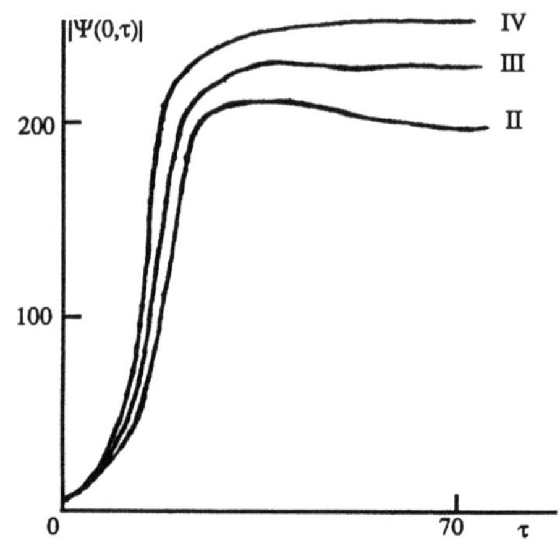

Fig. 2[7].

$$i\psi_t + \Delta_r\psi + |\psi|^s\psi + i\beta|\psi|^m\psi = 0, \qquad \beta <<< 1, \quad m > s. \tag{3}$$

These computations were performed using adaptive technique[9,8] and were reproduced recently[7] by another adaptive numerical method[10,11]. The results of both simulations are shown for comparison for the same runs ($s=2$, $m=6$, $\beta=10^{-9}$, $3 \leq d \leq 5$) in Figs. 1 and 2 (the numbers of curves indicate regimes of the SWC, τ is the "straightened" time in adaptive variables). One can see not only the direct numerical proof of the of weak collapse going out to quasistationary regime but also a convincing agreement between the results of the different computations presented.

Note that the SWC phenomena may exist in equation (1) independently of s at $d>2$. This means that the convenient and widely used model of forced damped NLS for studies of strong (collapsing) wave turbulence in nonlinear media[12-15] can not take into account this significant effect for $d \leq 2$. Since long 3D computations of turbulence with multiple collapses are extremely expensive, there is an important problem of creating of the SWC models in lower dimensions[6].

2. One of the possible models is associated with effective increase in dimensionality due to anisotropy of nonlinear oscillations. It was shown[16] that for 3D equation for the envelope of the HF potential of upper-hybrid plasma oscillations in a weak magnetic field[17,18]

$$(i\psi_t + \psi_{zz})_{zz} - \Delta_\perp\psi + (|\psi_z|^2\psi_z)_z = 0 \tag{4}$$

(magnetic field is supposed to be z-directed) the dimension which corresponds to that of the NLS (1) with analogous collapsing properties as in (4) is $d_{eff}=5$. Moreover, the existence of the quasi-classical SWC in Eq. (4) with a flux into anisotropic singularity has been discovered[16]. Since we are interested in low-dimensional SWC, consider the following d-dimensional power s generalization of the equation (4) ($\psi \to \psi_z$):

$$(i\psi_t + \psi_{zz} + |\psi|^s\psi)_{zz} - \Delta_\perp\psi = 0, \qquad d \geq 2, \; s \geq 1 \tag{5}$$

This equation also permits the self-similar substitution

$$\psi(z, r_\perp, t) = (t_0-t)^{-1/s-i\kappa} g[z(t_0-t)^{-1/2}, r_\perp(t_0-t)^{-1}] \tag{6}$$

which in particular means that wave energy trapped into singularity changes according to the law

$$N_{sing} \propto \int|\psi|^2 dz dr_\perp \propto (t_0-t)^{d-2/s-1/2}. \tag{7}$$

Thus the WC condition is $2sd \geq s+4$ and strong collapse at $2sd=s+4$ is expected. It follows from (7) and (2) that the equation (5) corresponds to the NLS with the effective dimension

$$d_{eff} = 2d-1 \tag{8}$$

that gives for 3D case the same as in[16] value $d_{eff}=5$. The substantional increase in the effective dimension due to anisotropy leads to the existence of the 2D SWC in the Eq. (5). Indeed, using (2), (7), (8) and taking account of result[16] we may conclude that the hierarchy of wave collapses in 2D equation (5) should be similar to that in the 3D NLS: strong collapse at $s=4/3$, weak collapse for $4/3 < s < 2$, and superstrong collapse for $s \geq 2$ with transition to the quasi-classical regime for $s > 4$. Even though all known quasi-classical solutions are unstable[16,19], one can expect realization of the SWC in the 2D computer simulations of the equation (5) for at least $2 \leq s \leq 4$.

3. Another possibility of the existence of the low-dimensional SWC follows from the NLS model with stronger nonlinearity than the power one. It was shown recently[6] that in the generalized NLS

$$i\psi_t + \Delta\psi + f(|\psi|^2)\psi = 0 \tag{9}$$

the SWC does not occur for any function of nonlinearity in dimensions $d < 2$. In 2D case it is possible, however, to constract the symmetrical SSS with the flux that is related to the nonlinearity function $f(x)$ by the equation

$$A^4 f(A^2) = e^{2y}(P^2 - A^3 A_{yy}), \quad A = |\psi|, \quad y = \ln(1/r), \tag{10}$$

where $A(y)$ increases not faster than linearly. This equation generates various types of rapidly growing functions of nonlinearity in the 2D NLS (9), each of those allows different regimes of the SWC. The least growing function $f(|\psi|^2)$ corresponds to the two-dimensional SWC model

$$i\psi_t + \Delta\psi + \exp(\lambda|\psi|^q)\,|\psi|^{-s}\psi = 0, \quad \lambda > 0, \quad s > 0, \tag{11}$$

that has the exact fluxless singular solution $\psi_0(r)=\{c\ln[c(\alpha^2-\alpha)^{1/2}/r]\}^\alpha$ for $0<q<1$, $s=2q$ $(\alpha=1/q,\ c=2/\lambda)$ and several families SSS[6] with the constant energy flux, that correspond to the parameters q and s in the following ranges: $q=1$, $s \geq 4$; $1 < q < 2$, $s=2q$; $q \geq 2$, $s=4$. There are two exact solutions among them: $|\psi|=c\ln(P/r)$ for $q=1$, $s=4$ and $|\psi|=\{c\ln[(P^2+c^2/4)^{1/2}/r]\}^{1/2}$ for $q=2$, $s=4$; for $q \geq 2$, $s=4$ the SWC passes to the quasi-classical stage. Thus, the behaviour of the SSS of the equation (11) and the regimes of the SWC are qualitatively similar to those of the equation (1), except for logarithmical space dependence of the solutions of equation (11).

4. Thus, the results of recent computations confirm the existence and variety of regimes of the superstrong wave collapse in the NLS (1) for $d > 2$. The two 2D nonlinear wave evolution models which describe the SWC phenomena are proposed: one-parameter (s) model (5) at $d=2$ and two-parameter (q,s) model (11). Both models include the ranges of parameters corresponding to the quasi-classical SWC regimes. The parametric flexibility of proposed models is convenient in particular for their computer simulation which seems to be an important and interesting problem.

I thank B.J. LeMesurier for collaboration in computations[7].

References

[1]S.N. Vlasov, L.V. Piskunova, and V.I. Talanov, *Nonlinear and Turbulent Processes in Physics*, (Naukova Dumka, Kiev, 1988), Vol. 2, p. 210.

[2]V.M. Malkin, JETP Lett. **48**, 653 (1988).

[3]V.E. Zakharov, N.E. Kosmatov, and V.F. Shvets, JETP Lett. **49**, 432 (1989).

[4]S.N.Vlasov, L.V.Piskunova, and V.I.Talanov, Sov. Phys. JETP **68**, 1125 (1989).

[5]V.M. Malkin and E.G. Shapiro, Sov. Phys. JETP **70**, 102 (1990).

[6]V.F. Shvets, Phys. Lett. A **150**, 74 (1990).

[7]B.J. LeMesurier and V.F. Shvets, 1990 (unpublished).

[8]N.E. Kosmatov, V.F. Shvets, and V.E. Zakharov, Physica D **52**, 16 (1991).

[9]N.E.Kosmatov, I.V.Petrov, V.F. Shvets, and V.E. Zakharov, Preprint 1365, Space Research Institute, Moscow, 1988.

[10]D.W. McLaughlin, G.C. Papanicolaou, C. Sulem, and P.L. Sulem, Phys. Rev. A **34**, 1200 (1986).

[11]B.J.LeMesurier, G.C. Papanicolaou, C.Sulem, and P.L.Sulem, Physica D **32**, 210 (1988).

[12]A.C. Newell, D.A. Rand, and D. Russel, Physica D **33**, 281 (1988).

[13]M. Bartucelli, P.Constantin, C.R. Doering, J.D. Gibbon, and M. Gisselfalt, Phys. Lett. A **142**, 349 (1989).

[14]V.F. Shvets and V.E. Zakharov, *Nonlinear World*, (World Scientific, Singapore, 1990), Vol. 1, p. 671.

[15]H. Pietch, R. Blaha, E.W. Laedke and A. Kumar, Europhys. Lett. **15**, 173 (1991).

[16]E.A. Kuznetsov and S.K. Turitsyn, Sov. J. Plasma Phys. **16**, 524 (1990).

[17]V.V. Krasnosel'skikh and V.I. Sotnikov, Sov. J. Plasma Phys. **3**, 491 (1977).

[18]E.A. Kuznetsov and M.M. Scoric, Phys. Rev. A **38**, 1422 (1988).

[19]V.E. Zakharov and E.A. Kuznetsov, Sov. Phys. JETP **64**, 773 (1986).

A Moving Boundary Problem in Plasma Physics

N. Sternberg[1,*] *and V.A. Godyak*[2]

[1]Department of Mathematics and Computer Science, Clark University, Worcester, MA 01610, USA

[2]GTE Laboratories, Inc., Waltham, MA 02254, USA

Introduction. To a large degree, the electromagnetic properties of capacitive rf discharges are determined by the space-charge sheaths along the electrode boundaries of the rf discharge plasma. Understanding the behavior of the electrode sheaths and predicting their properties and parameters are crucial in the analysis of capacitive rf discharges. Different models have been developed to describe the characteristics of the rf sheaths. A review of these models can be found in Ref.[1]. A self-consistent hydrodynamic model for the sheaths of a symmetric rf discharge was formulated in Ref.[2]. In Ref.[1], this model has been generalized and fully worked out. In the model, under the action of the rf field, the position of the plasma-sheath interface at any given time θ can be described by a periodic function $\lambda(\theta)$ that achieves its maximum λ_1 at some time $\theta = \theta_1$ and its minimum λ_2 at $\theta = \theta_2$. In other words, the plasma-sheath interface oscillates between its two extreme positions λ_1, the maximal sheath width, and λ_2, the minimal sheath width. To the right of the moving interface (in the plasma), the electron density is equal to the ion density. To the left of the moving interface (in the sheath), the electron density is equal to zero. In Ref.[1], the generalized model from Ref.[2] was worked out under the assumption that the function λ is symmetric with respect to its minimum λ_2. A comparison of the analytical and numerical solutions of the mathematical model with the available experimental results for mercury vapor have demonstrated a good agreement in corresponding values. Ref.[1], however, did not include the mathematical analysis, which underlies the numerical results that were obtained. The purpose of the present paper is to provide more of the details.

Formulation of the mathematical model. The dynamic model of the electrode sheath in symmetrically driven rf discharges can be described by the following moving boundary problem: the Poisson equation

$$\frac{d^2u(\xi,\theta)}{d\xi^2} = \begin{cases} y(\xi)\,, & \text{if } \xi \le \lambda(\theta) \\ 0\,, & \text{if } \xi > \lambda(\theta) \end{cases} \quad , \tag{1}$$

the equation for the ion distribution $y(\xi)$

$$\frac{d\langle u\rangle}{d\xi} + \frac{1}{(1+\alpha)y^3}\frac{dy}{d\xi} + \frac{\alpha}{(1+\alpha)y^2} = 0 \quad , \tag{2}$$

and the equation for the sheath current

* Supported by the U. S. Air Force Office of Scientific Research under contract No. AFOSR 88–0232

$$\sigma\Big(\gamma y(\lambda(\theta))\exp(u(\lambda(\theta),\theta)-u(0,\theta))\ -\ (1+\alpha)^{-1/2}\Big)$$
$$+\ \frac{y(\lambda(\theta))}{\rho}\frac{d\lambda(\theta)}{d\theta} = \cos\theta\ . \qquad (3)$$

In the equations listed above, $0 \leq \xi \leq \lambda_1$ denotes the space coordinate and $\theta \in \mathbb{R}$ denotes the time coordinate, u is the normalized instantaneous potential, y is the normalized ion distribution, $\langle\ \rangle$ denotes the time average over an rf period, γ is a constant that depends on the gas (for instance, $\gamma = 242$ for mercury vapor), $0 \leq \alpha < \infty$ is the collision parameter, $0 < \sigma < 1$ is the sheath conductivity parameter, and $\rho > 0$ describes the discharge current. Since the rf discharge is symmetric, the dc component of the current in each sheath must be equal to zero, *i.e.*,

$$\Big\langle y(\lambda(\theta))\exp(u(\lambda(\theta),\theta)-u(0,\theta))\Big\rangle = \gamma^{-1}(1+\alpha)^{-1/2}\ . \qquad (4)$$

The conditions at the plasma-sheath boundary are

$$u(\lambda_1,\theta) = 0\,,\qquad \frac{d}{d\xi}u(\lambda_1,\theta) = -1\,,\qquad y(\lambda_1) = 1\,. \qquad (5)$$

It is given that λ is periodic and symmetric with respect to λ_2.

Analysis. We make the following observations.

1. The function λ is 2π-periodic. Let T be the period of λ. Then, since y does not depend on θ, the function u is T-periodic in θ. Setting $d\lambda(\theta+T)/d\theta = d\lambda(\theta)/d\theta$, we obtain from equation (3) that $\cos\theta = \cos(\theta+T)$ and thus $T = 2\pi$.

2. Equation (3) implies that the point $\theta_1 \in [0,\pi]$ such that $\lambda_1 = \lambda(\theta_1)$ is obtained from the equation

$$\theta_1 = \arccos\Big(\sigma\gamma\exp(-u(0,\theta_1)) - \sigma(1+\alpha)^{-1/2}\Big)\ . \qquad (6)$$

3. The first two observations together with the symmetry of λ imply that the time average $\langle\ \rangle$ can be taken over half of an rf period, $\theta_2 = \theta_1 - \pi$, and equation (4) is equivalent to

$$\frac{1}{\pi}\int_{\theta_1-\pi}^{\theta_1} y(\lambda(\theta))\exp(u(\lambda(\theta),\theta)-u(0,\theta))d\theta = \gamma^{-1}(1+\alpha)^{-1/2}\ . \qquad (7)$$

4. Consider equation (1). Integrating this equation for $\xi > \lambda(\theta)$ with the boundary condition (5), we find that $du/d\xi = -1$ and $u(\xi,\theta) = -\xi+\lambda_1$ for all $\xi > \lambda(\theta)$ and $\theta \in [\theta_1-\pi,\theta_1]$. Continuity of u and of $du/d\xi$ imply that $u(\lambda(\theta),\theta) = -\lambda(\theta)+\lambda_1$ and $du(\lambda(\theta),\theta)/d\xi = -1$. From now on, we will let $v(\xi,\theta) = du(\xi,\theta)/d\xi$.

5. Integrating equation (1) for $0 \leq \xi \leq \lambda(\theta)$, we obtain for all $\theta \in [\theta_1-\pi,\theta_1]$:

$$v(\xi,\theta) = -1 - \int_\xi^{\lambda(\theta)} y(\psi)d\psi = -1 - \int_\xi^{\lambda_1} y(\psi)d\psi + \int_{\lambda(\theta)}^{\lambda_1} y(\psi)d\psi$$
$$= v(\xi,\theta_1) - v(\lambda(\theta),\theta_1) - 1\ , \qquad (8)$$

and integrating that same equation once again, using (8),

$$
\begin{aligned}
u(\xi,\theta) &= -\lambda(\theta)+\lambda_1 - \int_{\xi}^{\lambda(\theta)} v(\psi,\theta)\,d\psi \\
&= -\lambda(\theta)+\lambda_1 - \int_{\xi}^{\lambda(\theta)} v(\psi,\theta_1)\,d\psi + \int_{\xi}^{\lambda(\theta)} \Big(v(\lambda(\theta),\theta_1)+1\Big)d\psi \\
&= -\lambda(\theta)+\lambda_1 - \int_{\xi}^{\lambda_1} v(\psi,\theta_1)\,d\psi + \int_{\lambda(\theta)}^{\lambda_1} v(\psi,\theta_1)\,d\psi \\
&\quad + \Big(v(\lambda(\theta),\theta_1)+1\Big)\Big(\lambda(\theta)-\xi\Big) \\
&= \lambda_1 - \xi + v(\lambda(\theta),\theta_1)\Big(\lambda(\theta)-\xi\Big) + u(\xi,\theta_1) - u(\lambda(\theta),\theta_1)\ . \qquad (9)
\end{aligned}
$$

Thus it is sufficient to solve the given system for $\theta = \theta_1$ and then use formulas (8) and (9) in order to obtain the solutions for all θ.

6. In order to find $\langle u\rangle$, one has to know the behavior of u as a function of θ. Let ξ, with $\lambda_2 \le \xi \le \lambda_1$, be arbitrary but fixed. Then, since λ is strictly monotone increasing on the interval $[\theta_1-\pi,\theta_1]$, there exists a unique $\theta_\xi = \lambda^{-1}(\xi)$ such that

$$
\frac{d^2u(\xi,\theta)}{d\xi^2} = \begin{cases} y(\xi)\,, & \text{if } \theta_\xi \le \theta \\ 0\,, & \text{if } \theta_\xi > \theta \end{cases}\quad, \qquad (10)
$$

where $\theta \in [\theta_1-\pi,\theta_1]$. Thus we obtain for $\lambda_2 \le \xi \le \lambda_1$,

$$
\frac{d^2\langle u\rangle}{d\xi^2} = \frac{1}{\pi}\int_{\theta_\xi}^{\theta_1} y(\xi)\,d\theta = \frac{1}{\pi}(\theta_1-\theta_\xi)\,y(\xi)\ , \qquad (11)
$$

and, for $0 \le \xi \le \lambda_2$, we obtain

$$
\frac{d^2\langle u\rangle}{d\xi^2} = y(\xi)\ . \qquad (12)
$$

One could now find y and $\langle u\rangle$ by solving equations (2), (11) and (12) if, for each $\xi \in [\lambda_2,\lambda_1]$, the corresponding θ_ξ could be found. This we do next.

7. Consider $\lambda_2 \le \xi \le \lambda_1$. Introduce a new space variable $\varphi \in [\theta_1-\pi,\theta_1]$ and describe $\xi = \xi(\varphi)$ by the equation

$$
\frac{d\xi}{d\varphi} = \frac{\rho\cos\varphi}{y(\xi(\varphi)+\lambda_1)} + \frac{\rho\sigma(1+\alpha)^{-1/2}}{y(\xi(\varphi)+\lambda_1)} - \rho\sigma\gamma\exp(-\xi(\varphi)-u(0,\varphi)) \qquad (13)
$$

with the initial condition $\xi(\theta_1)=0$. It follows from equation (3) and observation 4 that $\xi(\varphi)=\lambda(\theta)-\lambda_1$ if and only if $\varphi=\theta$ and, since λ is strictly monotone increasing on the interval $[\theta_1-\pi,\theta_1]$, $\xi(\varphi)<\lambda(\theta)-\lambda_1$ if and only if $\varphi<\theta$. Using parameterization (13), we introduce new variables $\tilde{y}(\varphi)=y(\xi)$ and $\tilde{u}(\varphi,\theta)=u(\xi,\theta)$. In these new variables, equation (1) has the following form (with the tildas omitted)

$$
\frac{d^2u(\varphi,\theta)}{d\varphi^2} = \begin{cases} y(\varphi)\,(d\xi/d\varphi), & \text{if } \varphi \le \theta \\ 0\,, & \text{if } \varphi > \theta \end{cases}\quad, \qquad (14)
$$

equation (2) changed according to the chain rule will be called (2a), equation (11) becomes

$$\frac{d^2\langle u\rangle}{d\varphi^2} = \frac{1}{\pi}(\theta_1 - \varphi)\, y(\varphi)\frac{d\xi}{d\varphi}\,, \tag{15}$$

and equation (7) has the form

$$\frac{1}{\pi}\int_{\theta_1-\pi}^{\theta_1} y(\varphi)\exp(-\xi(\varphi) - u(0,\varphi))\,d\varphi = \gamma^{-1}(1+\alpha)^{-1/2}\,. \tag{16}$$

In solving the original system (1–3) for various parameters ρ, α, and σ, we consider two regions between the plasma and the electrode. In the first, $\lambda_2 \leq \xi \leq \lambda_1$, and in the second, $0 \leq \xi \leq \lambda_2$. We proceed by solving the system given by equations (13–15, 2a) with the initial condition (5) and obtain the values of the solution at $\xi = \lambda_2$, $(\varphi = \theta_1 - \pi)$. These values are then used as initial conditions in order to solve the original system in the second region given by the equations (1,2,12). The values for λ_2 and $u(0,\theta)$ are obtained through an iteration method using equation (16).

References

1. V.A. Godyak and N. Sternberg, Physical Review A, 1990, 42, 2299.
2. V.A. Godyak, *Soviet Radio Frequency Discharge Research* (Delphic, Falls Church, VA, 1986).

Proton-Whistler Interactions in the Radiation Belts

E. Villalón[1] *and W.J. Burke*[2]

[1]Center for Electromagnetics Research, Northeastern University, Boston, MA 02115, USA

[2]Geophysics Directorate, PL/OL-AA, Hanscom AFB, MA 01731

The interactions of whistlers with radiation belt protons is investigated. In the inhomogeneous geomagnetic field, near the equator, the spacing between cyclotron resonances is very small. After crossing multiple harmonic resonances, a significant change of particle energy takes place, and the protons pitch-angle scatter toward the atmospheric loss cone. A test-particle hamiltonian formalism is investigated for first and second order resonant protons. Quasilinear theory is applied for first-order resonant particles to obtain bounce-averaged, diffusion coefficients. The Fokker Planck equation, containing pitch-angle, energy and the cross energy/ pitch-angle diffusion terms, is investigated to calculate diffusion life times.

I. INTRODUCTION

We consider the interaction of plasmaspheric electrons and protons with whistler waves. The particles are trapped within the earth's radiation belts moving back and forth along field lines between magnetic mirror points. We call τ_B the bounce period, the time required for a particle to go from one mirror point to the other and return. In the region of interest, the geomagnetic field, $\mathbf{B_o}$, is described as a dipole. The interaction region is limited to the plasmasphere, $L < 4$, where L is the equatorial distance of the field line measures in Earth radii (R_E). The plasmashere is made up of cold particles of ionospheric origin whose distribution is isotropic and Maxwellian. During magnetic storms the radiation belts fill with energetic, trapped particles whose density is much smaller than that of the cold plasma. Whistlers are right-hand polarized electromagnetic waves whose magnetic field, $|\mathbf{B_k}| \ll \mathbf{B_o}$. Often they propagate in field-aligned ducts due to density depletions in local flux tubes. They can either be launched from ground sources or be generated in the plasmasphere. The dielectric properties for wave propagation are determined by the magnetized cold plasma distribution. These waves interact with the energetic particles, if the Doppler-shifted frequency of the waves is some harmonic of the gyrofrequency. For electron-whistler interactions the waves and particles travel in opposite directions. For protons they travel in the same direction and the wave phase velocity is very close to the proton parallel velocity. The situation is depicted in the Figure 1.

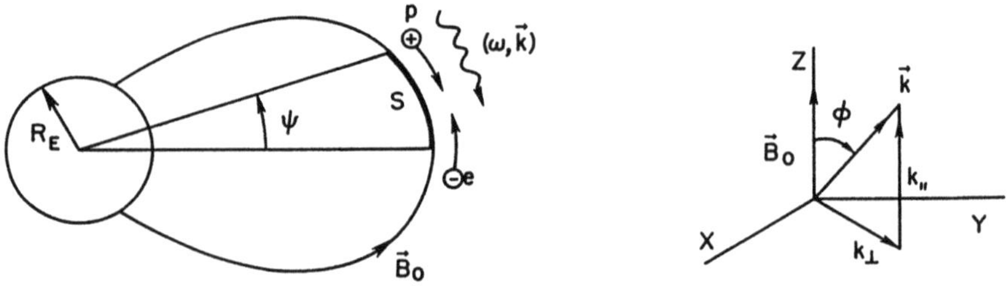

Figure 1. Schematic representation of whistler $(\omega, \mathbf{k})$, interacting with electrons and protons near the equator. The coordinate system used in this paper is depicted here.

Whistler-electron interactions has been extensively study over the years [1,2,3]. The electrons typically have energies between 10 to 50 keV. The interaction occurs mainly at the first gyroharmonic of the electron gyrofrequency, although higher gyroharmonics may also be important[4]. The electron energies change very little during these interactions. The electron pitch angle is θ, where $\tan\theta = v_{\perp}/v_{\parallel}$, the ratio between the parallel and perpendicular components of the particle velocity. The pitch angle can be significantly changed and, as a result, the particle is scattered into the loss cone and precipitate into the ionosphere. Because large numbers of electrons interact with the waves, they grow in amplitude to values whose limits depend on the degree of anisotropy of the electron distribution function[2]. Detailed analyses are given in the papers by Villalón and coworkers[4,5]. These investigations where based on relativistic, quasilinear theory that simultaneously considers wave growth and particle depletion from the radiation belts.

Proton-whistler interactions have not received as much attention. Recents experiments have shown[6,7] that protons whose energies are in the hundreds of keV range, can be scattered from the radiation belts by analogous interactions. The frequency of the wave must be close to the equatorial electron gyrofrequency. The particle energy changes significantly during the interactions[8]. Thus, the changes in pitch angle is due to both direct pitch angle and energy diffusion. Because of the small population of high-energy protons we neglect their effects on the amplitudes of the waves. We present a study of proton whistler interactions by using a test particle formalism and a statistical approach based on the Fokker-Planck equation. In Sec. II, we present the main dielectric properties of whistler waves; because the whistler protons interactions require large refractive indices, we limit ourselves to the pararesonance mode[9]. Sec. III presents the resonance condition for multiple harmonics of the gyrofrequency. The geomagnetic latitudes of high harmonic resonances are obtained based in a parabolic approximation for the near equatorial geomagnetic field. We show that the distance between subsequence resonances is very small. The crossing of multiple resonances near the equator makes the interactions very effective[10]. Sec IV contains the equations for the test particle in a varying geomagnetic field using hamiltonian formalism. Sec V studies the evolution of the action (I) angle (ξ_ℓ) variables as function of the distance (s) along the flux tube using Taylor expansions around isolated resonances. Let us expand ξ_ℓ around the equator: $\xi_\ell(s) = \xi_\ell(0) + \xi_\ell^{(1)} s + \xi_\ell^{(2)} s^2$. First-order resonant particles are such that $\xi_\ell^{(1)} = 0$ (i. e., at the equator $d\xi_\ell/ds = 0$). This is the resonance condition as given in Eq. (5). The second-order term $\xi_\ell^{(2)} \sim dB_o/ds + O(B_k)$. For large wave amplitudes $O(B_k)$ is larger than the contribution of the inhomogeneous geomagnetic field dB_o/ds. In this case, we say that protons which are in gyroresonance (i. e. $\xi_\ell^{(1)} = 0$), satisfy the second-order resonance condition. This is because to zero order in the electric field amplitudes $d\xi_\ell/ds = d^2\xi_\ell/ds^2 \simeq 0$. For first-order resonant particles, the change in action is proportional to the electric field amplitude. For second-order resonant protons the change in action is proportional to the square root of the electric field amplitude. The second-order resonance condition is met when the field amplitude is large[11,12], the threshold is calculated in this paper. Sec. VI contains a quasilinear formulation for the distribution function of first order resonant protons. We assume that the protons are unmagnetized in time scales of the order of $2\pi/\omega$, where ω is the frequency of the whistler wave. They are magnetized in times comparable to the bounce period. Because diffusion occurs over many bounce periods, we average the diffusion equation along the flux tube. The bounce averaged, Fokker-Planck equation contains the diffusion coefficients for the pitch angle, energy, and the cross energy/ pitch angle terms. These coefficients are shown to have the same orders of magnitude. We reduce the equation to a one-dimensional diffusion equation to be solved for the energy part of the distribution function. This eigenvalue equation gives the diffusion life-times of protons in the radiation belts.

II. QUASI-ELECTROSTATIC WHISTLER WAVES

We consider a whistler wave of frequency ω and wave vector $\mathbf{k}$, propagating in a field aligned duct. The geomagnetic field $\mathbf{B_o}$ is along the z direction and ϕ is the angle between $\mathbf{k}$ and $\mathbf{B_o}$. The dispersion relation for the refractve index $\eta = ck/\omega$ is

$$\eta^2 = 1 + \frac{(\omega_p/\omega)^2}{(\Omega_e/\omega)\,|\cos\phi| - 1} \tag{1}$$

where ω_p and Ω_e are the electron plasma and gyro frequencies, respectively.

The electric fields components are denoted by $\mathcal{E}_x = \mathcal{E}_1$, $\mathcal{E}_y = i\mathcal{E}_2$, and $\mathcal{E}_z = -\mathcal{E}_3$, where

$$\frac{\mathcal{E}_2}{\mathcal{E}_1} = \frac{1}{\eta^2 - 1}\;\frac{(\omega_p/\omega)^2}{(\Omega_e/\omega) - |\cos\phi|} \tag{2}$$

$$\frac{\mathcal{E}_1}{\mathcal{E}_3} = \frac{1 - (\omega_p/\omega)^2 - (\eta\,\sin\phi)^2}{\eta^2\,\sin\phi\,\cos\phi} \tag{3}$$

For the case where $\omega \sim \Omega_e(L)|\cos\phi|$, the equatorial refractive index $\eta^2(L) \gg 1$, then $\mathcal{E}_2/\mathcal{E}_1 \ll 1$, and $\mathcal{E}_1/\mathcal{E}_3 \sim -\sin\phi/\cos\phi$. The wave becomes quasi-electrostatic, i.e. $\mathbf{E}$ is in the direction of **k**, and the group velocities $v_g \sim 1/\eta$ are very small. These waves can interact with protons which energies are in the hundreds of keV.

Near the equator, the Earth's magnetic field approximates a parabolic profile

$$\frac{\Omega}{\Omega(L)} = 1 + \left(\frac{s}{r_L}\right)^2 \tag{4}$$

where $s \simeq R_E L \psi$, R_E is the Earth's radius, L is the magnetic shell and ψ is the geomagnetic latitude (see the figure), and $r_L = (\sqrt{2}/3) R_E L$. The equatorial gyrofrequency is $\Omega(L)$; Ω stands for the gyrofrequencies either for electrons or protons, along the field line.

III. RESONANT PROTON-WHISTLER INTERACTIONS

For whistler waves to interact strongly with protons near equatorial regions, they must satisfy the resonance condition

$$\omega - k_{\|} v_{\|} - \ell \Omega_p = 0 \tag{5}$$

where, $\ell = 0, 1, 2...$; Ω_p is the proton gyrofrequency, and $k_{\|}$ and $v_{\|}$ are the parallel components of the wave vector and particle's velocity, respectively. We call $\mu = \sin^2 \theta_L$, where θ_L is the equatorial pitch angle. Here $\theta_L > \theta_c(L)$, where $\theta_c(L)$ is the pitch angle at the boundary of the loss cone, and μ_c the corresponding value of μ. As function of the L shell, the mirror ratio is $\sigma = 1/\mu_c = L^3(4 - 3/L)^{1/2}$. To zero order in electric field amplitudes, the first adiabatic invariant is conserved. Then we may write for the parallel and perpendicular components of the particle velocity v: $v_{\|} = v[1 - \mu\Omega/\Omega(L)]^{1/2}$, $v_{\perp} = v[\mu\Omega/\Omega(L)]^{1/2}$.

If we assume that at the equator the protons interact with the harmonic $\ell = 1$, the energy of resonant particles is found solving for the equation: $\omega - k_{\|}\, v_{\|} - \Omega_p(L) = 0$. We show

$$\frac{v}{c} = \frac{1}{\eta(L)\, \cos\phi}\, \frac{1}{(1-\mu)^{1/2}}\, \left(1 - \frac{m_e}{m_p} f_e\right) \tag{6}$$

where L denotes equatorial values, $m_{e,p}$ are the electron, proton masses, and $f_e = \Omega_e(L)/\omega$.

By solving for Eq. (5), using the parabolic profile in Eq. (4) , we find the geomagnetic latitude ψ_ℓ of higher order resonances (i.e., $\ell \geq 1$),

$$\psi_\ell^2 = \frac{4}{9}\, \frac{m_e}{m_p}\, (\ell - 1)\, (f_e\, |\cos\phi| - 1)\, \frac{1}{g(\mu)} \tag{7}$$

where

$$g(\mu) = \frac{\mu}{1-\mu}\, \left(|\cos\phi| - \frac{1}{f_e}\right) + |\cos\phi|. \tag{8}$$

The distance along the flux tube where resonant interactions take place is given by, $s_\ell = R_E L \psi_\ell$. The distance between sequential resonances is $\Delta s_\ell = R_E L(\psi_{\ell+1} - \psi_\ell)$.

For example, we take $L = 3.5$, $\omega_p/\Omega_e(L) = 7.9$, $\omega/\Omega_e(L) = 0.75$, and $\theta_L = 10^o$. For $\phi = 37^o$, we show that $\eta(L) = 41.4$ and the energy of the resonant protons is 437 keV.

The location along the geomagnetic field of the gyroresonances are: $\psi_2 = 0.25^o, \psi_3 = 0.35^o, ..., \psi_{17} = 1.^o$. As another example we take $\phi = 40^o$, then $\eta(L) = 72$ and the proton energy is 158.6 keV. The location of the gyroresonances are: $\psi_2 = 0.15^o, \psi_3 = 0.21^o, ..., \psi_{47} = 1.^o$. Thus there are multiple resonances crossings (17 for the first and 47 for the second examples) within one degree of the magnetic equator, which makes the proton whistler interactions very efficient.

IV. THE HAMILTONIAN EQUATIONS

We normalize time t to $\Omega_p(L)$, velocity v to c^{-1}, and length s to r_L^{-1}, and from now on we always refer to these normalized variables. Let us define

$$\xi_\ell = \ell\lambda + \int_0^s ds'\, r_L\, k_\parallel(s') - \frac{\omega}{\Omega_p(L)}\, t \tag{9}$$

where $\tan\lambda = v_y/v_x$, and $v_{x,y}$ are the components of the particle velocity in the x and y directions, respectively. The dimensionless electric field amplitudes are

$$\varepsilon_i = \frac{q\mathcal{E}_i}{m_p c\omega} \tag{10}$$

for $i = 1, 2, 3$, and where q is the proton charge. The action-angle variables are (I, λ), where

$$I = \frac{v_\perp^2}{2}\frac{\Omega_p(L)}{\Omega_p} \tag{11}$$

To first order in the electric field amplitudes ε_i, the normalize, time-dependent hamiltonian, as function of the canonical pairs, $(v_\parallel, s)$, and action-angle variables, is

$$\mathcal{H} = \frac{v_\parallel^2}{2} + I\,\frac{\Omega}{\Omega(L)} + \sum_{\ell=-\infty}^{\infty} \sin\xi_\ell \left\{\varepsilon_3\, v_\parallel\, J_\ell(k_\perp\rho) - [\frac{I\Omega}{2\Omega(L)}]^{1/2}\,\Gamma_\ell\right\} \tag{12}$$

Here

$$\Gamma_\ell = (\varepsilon_1 - \varepsilon_2)\, J_{\ell+1}(k_\perp\rho) + (\varepsilon_1 + \varepsilon_2)\, J_{\ell-1}(k_\perp\rho) \tag{13}$$

where J_ℓ are Bessel functions whose arguments are $k_\perp\,\rho = (ck_\perp/\Omega_p)\,[2I\Omega/\Omega(L)]^{1/2}$. If, in addition to the electromagnetic wave, there is an electrostatic potential ϕ_o, then we replace in Eq. (12), ε_3 by $\varepsilon_3 + \varepsilon_o/v_\parallel$, where $\varepsilon_o = q\phi_o/m_pc^2$.

For particles crossing a single isolated cyclotron resonance, we consider only one term ℓ in the summation in Eq. (12). In this case, we find the following constant of motion

$$C_\ell = \ell\mathcal{H} - \frac{\omega}{\Omega_p(L)}\, I \tag{14}$$

By calling $\chi = (\omega/\Omega_p)\,\sin^2\theta(s)$, where $\theta(s)$ is the local pitch angle, we find

$$\chi = \frac{\ell\omega}{\Omega_p(L)}\;\frac{I}{C_\ell + [\omega/\Omega_p(L)]\, I} \tag{15}$$

This defines the evolution of the pitch angle as a function of the action I.

By defining v_o so that $\mathcal{H} = v_o^2/2 + I\Omega/\Omega(L)$, we obtain

$$v_o = \left\{\frac{2}{\ell}\,[C_\ell + I\,(\frac{\omega}{\Omega_p(L)} - \ell\frac{\Omega}{\Omega(L)})]\right\}^{1/2} \tag{16}$$

We can now reduce the problem to one-dimension, in which case we find

$$\begin{aligned} v_\parallel &= v_o + \sin\xi_\ell \left\{-\varepsilon_3\, J_\ell(k_\perp\rho) + \frac{1}{v_o}\,[\frac{I\Omega}{2\Omega(L)}]^{1/2}\,\Gamma_\ell\right\} \\ \frac{ds}{dt} &= v_o + \frac{1}{v_o}\,[\frac{I\Omega}{2\Omega(L)}]^{1/2}\,\Gamma_\ell\,\sin\xi_\ell \end{aligned} \tag{17}$$

To zero order in ε_i, the dimensionless length $s = t\,v_o$. The equation of motion for I as a function of s is

$$\frac{dI}{ds} = \ell\,\cos\xi_\ell\,\Upsilon_\ell(I, v_o) \tag{18}$$

$$\Upsilon_\ell(I, v_o) = -\varepsilon_3\,J_\ell(k_\perp\rho) + \frac{1}{v_o}\,[\frac{I\Omega}{2\Omega(L)}]^{1/2}\,\Gamma_\ell \tag{19}$$

As $\varepsilon_i \to 0$, then

$$\frac{d\xi_\ell}{ds} \to k_\parallel r_L + \frac{\ell\Omega_p - \omega}{\Omega_p(L)v_o} \tag{20}$$

The gyroresonance condition is obtained by setting Eq. (20) equal to zero. When this is satisfied $s = s_\ell$ (the resonance length) which is defined as $s_\ell = 3/\sqrt{2}\,\psi_\ell$ and ψ_ℓ is given in Eq. (7).

By assuming that the protons are in gyroresonance, we show that ξ_ℓ satisfies the second order differential equation

$$\frac{d^2\xi_\ell}{ds^2} = \alpha_\ell + \frac{(k_\parallel r_L)^2}{v_o}\,\frac{1}{\ell}\,\frac{dI}{ds} \tag{21}$$

Here

$$\alpha_\ell = \frac{\beta_\ell}{\Omega(L)}\,\frac{d\Omega}{ds} \tag{22}$$

$$\beta_\ell = \Omega_p(L)\,r_L\,\frac{m_p}{m_e}\,\frac{dk_\parallel}{d\Omega_e} + \frac{1}{v_o}\,[\frac{\ell}{2} + \frac{C_\ell}{v_o^2}] \tag{23}$$

where $d\Omega/ds = 2s\,\Omega(L)$.

V. SECOND-ORDER RESONANCE

We next solve the pair of coupled Eqs. (18) and (21) under the assumption that s is very close to the resonance length s_ℓ. The parallel velocity v_o is given by setting Eq. (20) equal to zero, i.e.

$$v_o(R) = \frac{\omega}{ck_\parallel}\,\frac{c}{\Omega_p(L)\,r_L} \tag{24}$$

In this case we may use a Taylor expansion around s_ℓ, then

$$I \simeq I_\ell(R) + (\frac{dI}{ds})_{(R)} \; (s - s_\ell) \tag{25}$$

$$\xi_\ell \simeq \xi_\ell(R) + \xi_\ell^{(1)} \; (s - s_\ell) + \frac{\xi_\ell^{(2)}}{2} \; (s - s_\ell)^2 \tag{26}$$

where $I_\ell(R), \xi_\ell(R)$ are constants, and R denotes values at the resonance ($s = s_\ell$). Here $(dI/ds)_{(R)}$ is given by Eqs. (18) and (19), with $\xi_\ell = \xi_\ell(R), I = I(R)$, and $v_o = v_o(R)$, evaluated for resonant values. For protons satisfying the resonance condition, $\xi_\ell^{(1)} = 0$. For convenience we choose $\cos[\xi_\ell(R)] = 1$.

Using Eq. (14) and setting $v_o = v_o(R)$, we find

$$I(R) = \frac{f_p}{[1 - \ell f_p \, \Omega(R)/\Omega(L)]} \; \left\{ -C_\ell + \frac{\ell}{2} \, v_o^2(R) \right\} \tag{27}$$

where $f_p = \Omega_p(L)/\omega \ll 1$, and $\Omega(R)/\Omega(L) = 1 + s_\ell^2$. By substituting Eq. (26) into Eq. (21) we show

$$\xi_\ell^{(2)} = \alpha_\ell(R) + \frac{(k_\| r_L)^2}{v_o(R)} \, \frac{1}{\ell} \, (\frac{dI}{ds})_{(R)} \tag{28}$$

where $\alpha_\ell(R)$ is evaluated at the resonance.

The change of the action I after crossing the ℓ'th resonance, ΔI, obtained by integrating Eq. (18), is approximately

$$\Delta I = (\frac{dI}{ds})_{(R)} \; \delta s_\ell \tag{29}$$

The change in χ after crossing a resonance is

$$\Delta\chi = \chi(R) \left[\frac{1}{I(R)} - \frac{1/f_p}{C_\ell + I(R)/f_p} \right] (\frac{dI}{ds})_{(R)} \; \delta s_\ell \tag{30}$$

where $\chi(R)$ is given by Eq. (15) setting $I = I(R)$.

The resonance length δs_ℓ is defined as

$$\delta s_\ell = \int_{-\infty}^{+\infty} ds \cos \xi_\ell \tag{31}$$

By combining Eqs. (26), (28), and integrating along s we show

$$\delta s_\ell = \Gamma(1/2) \; \cos(\pi/4) \; \left[\frac{2}{|\xi_\ell^{(2)}|} \right]^{1/2} \tag{32}$$

The condition of isolated resonances is $\delta s_\ell < \Delta s_\ell$, where $\Delta s_\ell = 3/\sqrt{2} \, (\psi_{\ell+1} - \psi_\ell)$ and ψ_ℓ is given in Eq. (7).

In the case where the inhomogeneity of the magnetic field is larger than the contribution of the resonance, we may neglect the term proportional to $(dI/ds)_{(R)}$ in Eq. (28), we get

$$\Delta I = (\frac{dI}{ds})_{(R)} \; \Gamma(1/2) \; \cos(\pi/4) \; \left[\left| \frac{1}{\beta_\ell(R) \, s_\ell} \right| \right]^{1/2} \tag{33}$$

where $\beta_\ell(R)$ is given by Eq. (23) and must be evaluated at resonance. From the definition of Γ_ℓ in Eq. (13), the change in the action is proportional to the electric field amplitudes.

For interactions such that the contribution of $\alpha_\ell(R)$ in Eq. (28) is smaller than the contribution of $(dI/ds)_R$, we get

$$\Delta I = \pm \left[|\ell \, (\frac{dI}{ds})_{(R)}| \right]^{1/2} \Gamma(1/2) \, \frac{[2|v_o(R)|]^{1/2}}{k_\| r_L} \, \cos(\pi/4) \tag{34}$$

where the $\pm$ sign depends on the sign of $(dI/ds)_R$. We see that the change in particle momentum I is now proportional to the square root of the electric field amplitudes, i.e. $\sqrt{\varepsilon_i}$. We call this the second order resonance condition because to zero order in the electric field amplitudes $d^2\xi_\ell/ds^2 \simeq 0$. For the case of equatorial interactions ($s_\ell = 0$), the condition for the validity of this approximation is

$$\left[\frac{k_\| r_L}{\sqrt{2v_o}} \left(\frac{1}{\ell} \, |(\frac{dI}{ds})_{(R)}| \right)^{1/2} \right]^3 \gg \beta_\ell(R) \, \Gamma(1/2) \, \cos \pi/4 \tag{35}$$

Note that for a fix value of ω the second order resonance condition is most likely satisfied for equatorial interactions, because then the inhomogeneity of the magnetic field is small. Thus the first harmonic will dominate the second-order interactions. If we allow ω to be a function of s, then

$$\alpha_\ell = \beta_\ell \, \frac{1}{\Omega(L)} \, \frac{d\Omega}{ds} + r_L \, \frac{dk_\|}{d\omega} \, \frac{d\omega}{ds} \tag{36}$$

By changing ω so that $\alpha_\ell(R) = 0$ for $s_\ell > 0$, the second-order resonance condition is satisfied for other harmonics, and the change in the particle velocity is proportional to $\sqrt{\varepsilon_i}$. This should be contrasted with the result in Eq. (33) where the change in action is linear with the electric fields and thus smaller than when the condition for second order resonance is satisfied.

VI. QUASILINEAR THEORY

The distribution function of protons which satisfy the first order resonance condition is given by solving for the quasilinear equation Lyons and Williams (1984):

$$(\frac{1}{\tau_{atm}} + \frac{\partial}{\partial t}) \, f = \pi q^2 \sum_{\ell=-\infty}^{+\infty} \int \frac{d^3\mathbf{k}}{(2\pi)^3} \, [\hat{G} + \frac{\omega - k_\| v_\|}{\omega p_\perp}] \, \delta(k_\| v_\| - \ell\Omega_p - \omega) \, \Theta_\ell(\mathbf{k}) \, \hat{G} \, f \tag{37}$$

where p is momentum and τ_{atm}, the atmospheric loss time is defined in[1]. By assuming that $\omega/\Omega p \ll \sin^2\theta_c$ (where θ_c is the local pitch angle at the loss cone boundary) we may approximate

$$\hat{G} = -\frac{2}{p} \, \frac{\Omega_e(L)}{\Omega_e} \, (\frac{p_\perp}{p})^3 \, \frac{\partial}{\partial \mu} + \frac{p_\perp}{p} \, \frac{\partial}{\partial p} \tag{38}$$

$$\hat{G} + \frac{\omega - k_\| v_\|}{\omega p_\perp} = \frac{1}{p^2} \, \frac{\partial}{\partial p} \, p^2 \sin\theta - \frac{2}{p} \, \frac{\Omega_e(L)}{\Omega_e} \, \frac{p_\|}{p} \, \frac{\partial}{\partial \mu} \, (\frac{p_\perp}{p})^3 \, \frac{p}{p_\|} \tag{39}$$

$$\sum_{\ell=-\infty}^{+\infty} \delta(k_\parallel v_\parallel + \ell\Omega_p - \omega)\, \Theta_\ell(\mathbf{k}) \simeq (2\pi)^3\, \delta(k_\parallel v_\parallel - \omega)\, \frac{\omega\Omega_e}{\omega_p^2}\, \frac{W_k(\phi,t)}{|\cos\phi|}\, b(\phi) \tag{40}$$

where

$$b(\phi) = 1 + \cos^2\phi + \frac{1}{2}\left[\frac{p_\parallel}{p_\perp}\, \frac{\omega}{\Omega_e}\, \sin\phi\right]^2 \tag{41}$$

If B_k is the wave magnetic field ($B_k \ll B_o$, the geomagnetic field), then the energy density of waves is

$$W_k(\phi,t) = \frac{1}{8\pi}\left(\frac{B_k}{2\pi}\right)^2 \tag{42}$$

We assume that diffusion occurs on time scales such $t \gg \tau_B$, where τ_B is the proton bounce time between ionospheric conjugates. We integrate the diffusion equation along the flux tube by applying the operator $1/\tau_B \int dz/v_\parallel$ to both sides of Eq. (37). The bounce-averaged diffusion equation, in terms of equatorial pitch-angles θ_L and particle momentum, is

$$\begin{aligned}\left(\frac{1}{\tau_{atm}} + \frac{\partial}{\partial t}\right) f &= \frac{1}{p\sin\theta_L\,\cos\theta_L}\,\frac{\partial}{\partial\theta_L}\,\sin\theta_L\,\cos\theta_L\left[D_{\theta,\theta}\,\frac{1}{p}\,\frac{\partial f}{\partial\theta_L} + D_{\theta,p}\,\frac{\partial f}{\partial p}\right] + \\ &\quad \frac{1}{p^2}\,\frac{\partial}{\partial p}\left\{p\left[p\, D_{p,p}\,\frac{\partial f}{\partial p} + D_{p,\theta}\,\frac{\partial f}{\partial\theta_L}\right]\right\}\end{aligned} \tag{43}$$

The bounce-averaged diffusion coefficients are

$$D_{\theta,\theta} = \tan^2\theta_L\; D_{p,p} \tag{44}$$

$$D_{\theta,p} = D_{p,\theta} = -\tan\theta_L\; D_{p,p} \tag{45}$$

The energy-diffusion coefficient is

$$D_{p,p} = \frac{\pi q^2}{v\tau_B}\int_0^\infty k^2\, dk \int_{-\pi/2}^{+\pi/2} \sin\phi\; \Lambda(k,\phi)\, d\phi \tag{46}$$

where

$$\Lambda(k,\phi) = \frac{4\pi\Omega_e(L)^3}{\omega_p^2}\,\frac{W_k(\phi,t)}{|\cos\phi|}\,\left(\frac{ds}{d\Omega}\right)_{(R)}\left[\frac{\Omega(R)}{\Omega(L)}\right]^2 \frac{p_\parallel}{p}\, b(\phi) \tag{47}$$

Here R denotes values at the resonance where $v_\parallel \sim v$, and $\omega - k_\parallel\, v \simeq 0$. Note that for small values of ϕ, we can neglect the contribution of the parallel component of the wave field in $b(\phi)$ (see Eq. (41)), then $D_{p,p}$ is approximately independent of μ, the equatorial pitch angle, and we write

$$f = F(t)\,\mu^\sigma\, K(p) \tag{48}$$

where $\sigma > 0$ is a free parameter. We define the precipitation lifetime as

$$\tau_p = -\left[\frac{1}{F}\,\frac{dF}{dt}\right]^{-1} \tag{49}$$

By combining Eqs. (43) through (45) and Eq. (48), we show

$$[\frac{2\kappa_c}{\tau_B} - \frac{1}{\tau_p}]K(p) = \frac{4\sigma(\sigma+1)}{p^2} \mathcal{D}_{p,p} K + \frac{d}{dp}[\mathcal{D}_{p,p} \frac{dK}{dp}] - \frac{4\sigma}{p} \mathcal{D}_{p,p} \frac{dK}{dp} - \frac{2\sigma}{p^2} K \frac{d}{dp}[p\, \mathcal{D}_{p,p}] \tag{50}$$

where $\kappa_c = \mu_c^{(\sigma+1)}$. This is an eigenvalue equation for τ_p as a function of the free parameter σ. The eigenfunction $K(p)$ is such that must be regular as $p \to 0$, and well behaved for large p, i. e. as $p \to \infty$ then $K \ll p^{-2}$.

VII. SUMMARY AND CONCLUSIONS

We have presented a theoretical analysis of proton-whistler interactions near the equator in the plasmasphere. Whistler waves which are near the pararesonance mode[9], can interact with protons whose energies are in the hundreds of keV. In an inhomogeneous geomagnetic field, we show that the spacing between subsequence cyclotron resonances is very small. Because of that, protons are scattered into the atmospheric loss cone after crossing multiple resonances. A test-particle hamiltonian formalism is given in terms of the action (I), angle (ξ_ℓ), variables as function of the distance (s) along the flux tube. We show that for second-order resonant protons, $d\xi_\ell/ds = d^2\xi_\ell/ds^2 = 0$, and the change in the particle's momentum is proportional to the square root of the electric field amplitudes. The thresholds in electric fields for second-order resonance conditions are calculated. A quasi-linear formulation for the distribution function of first-order resonant protons is presented. The bounce-averaged diffusion equation contains diffusion coefficients for the pitch angle, energy, and cross energy/ pitch angle terms. They are shown to be of the same orders of magnitude. We reduce the diffussion equation to a one-dimensional energy dependent equation to be solved for the precipitation life times of protons in the Radiation Belts.

Acknowledgements. This work has been supported in part by the U. S. Air Force contract F19628-89-K-0014 with Northeastern University.

REFERENCES

1. L. R. Lyons and D. J. Williams, Quantitative Aspects of Magnetospheric Physics, D. Reidel Publising Company, The Netherlands, 1984.
2. C. F. Kennel and H. E. Petschek, J. Geophys. Res., 1966, 71, 1.
3. P. A. Bespalov and V. Yu. Trakhtengerts, Rev. Plasma Phys., 1980, 10, 88.
4. E. Villalón and W. J. Burke, J. Geophys. Res., 1991, 96, 9655.
5. E. Villalón, W. J. Burke, P. L. Rothwell, M. B. Silevitch, J. Geophys. Res., 1989, 94, 15243.

6. H. C. Koons, Geophys. Res. Lett., 1975, 2, 281.

7. R. A. Kovrazhkin, M. M. Mogilevskii, O. A. Molchanov, Yu. I. Gal'perin, N. V. Dzhordzhio, Zh. M. Bosbe, and A. Rem, JETP Lett., 1984, 39, 228.

8. R. Gendrin, J. Atmospheric Terrest. Phys., 1968, 30, 1313.

9. M. G. Morgan, J. Geophys. Res., 1980, 85, 130.

10. D. R. Shklyar, Planet. Space Phys., 1986, 34, 1091.

11. R. A. Helliwell, J. Geophys. Res., 1967, 72, 4773.

12. K. B. Dysthe, J. Geophys. Res., 1971, 76, 6915.

Part IV

Nonlinear Optics

Optical Turbulence in Semiconductor Lasers

E. Abraham[1], *H. Adachihara*[1], *O. Hess*[1], *R.A. Indik*[2], *P. Jacobsen*[2], *J.V. Moloney*[2], and *P. Ru*[2]

[1]Department of Physics, Heriot-Watt University, Edinburgh, UK
[2]Department of Mathematics, University of Arizona, Tucson, AZ, USA

In this paper we study the dynamics of broad-area semiconductor lasers and arrays. In the case of the former, our numerical simulations show symmetry breaking as transverse modulation instabilities develop leading to optical turbulence: temporal chaos and random filamentation of the output beam. The spatio-temporal evolution of linear arrays of otherwise individually stable lasers also becomes turbulent for currents well above threshold and separation smaller than some critical value. Stability can be restored by injecting an external field of suitable amplitude and frequency.

The semiconductor laser is an interesting physical system that converts electric power into coherent light and plays a central role in modern technology [1]. The type we consider here is the double heterostructure laser which consists of a thin active region of GaAs sandwiched between two regions of GaAlAS of slightly lower refractive index (fig.1(a)).

Stimulated emission of radiation occurs in the active layer and remains confined due to the passive waveguiding. In the z direction, Fresnel reflection at the end faces, which can be reinforced by additional coating, provides the resonator for lasing. Finally, a current of holes and electrons in the y direction provide the gain in the active medium in which the the stimulated radiative electron-hole recombination takes place.

To study the dynamics we solve consistently Maxwell's wave equation and a semi-phenomenological equation for the electron-hole density [2]. An outline of the derivation goes as follows. For a linearly polarised electric field we can take the scalar wave equation

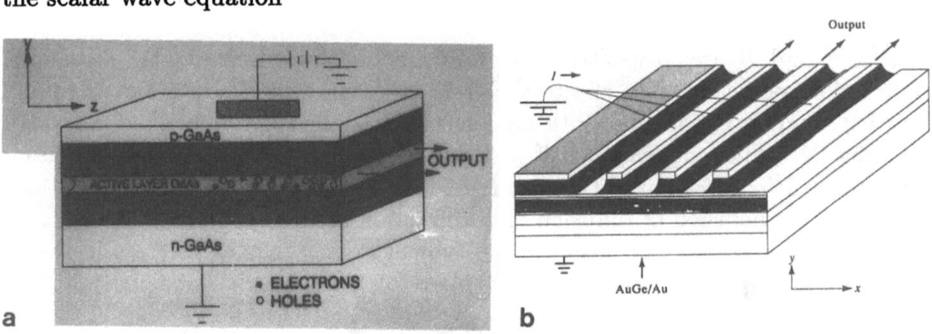

Figure 1: (a)Schematic of a double heterostructure laser. (b)Laser array.

$$\nabla^2 E(\vec{r},t) = \frac{1}{\epsilon_o c^2}\frac{\partial^2 D(\vec{r},t)}{\partial t^2} \tag{1}$$

where E is the electric field and D is the displacement vector given by

$$D(\vec{r},t) = \epsilon(\vec{r},t)E(\vec{r},t) \tag{2}$$
$$= (\epsilon_p(\vec{r}) + \epsilon_a(N))E(\vec{r},t) \tag{3}$$

where ϵ_p is the passive component of the dielectric 'constant' and ϵ_a is its active component which depends on the electron-hole density $N(\vec{r},t)$. Implicit in (3) is the assumption that the decay time of the polarisation is much shorter than any time scale in the problem and can therefore be adiabatically eliminated: except for plane waves, this approximation is the source of unphysical (numerical) instabilities that can be suppressed by adding a small imaginary part to the diffraction coefficient (defined below).

The electron-hole dynamics is governed by the reaction-diffusion equation,

$$\frac{\partial N}{\partial t} = D_f\frac{\partial^2 N}{\partial x^2} + \frac{\mathcal{J}}{qd} - \frac{N}{\tau} - \Gamma g(N)\frac{I_L}{\hbar\omega_0} \tag{4}$$

where D_f is the diffusion constant, $\mathcal{J}$ is the excitation (pumping) current density, e is the electron charge, d is the thickness of the active layer and τ is the electron-hole recombination time. The last term accounts for saturation effects caused by the laser intensity I_L inside the resonator; Γ is the radiation confinement factor, $g(N)$ is the gain and ω_0 the frequency of the emitted light.

After scaling variables and making the slowly-varying-envelope and paraxial approximations, we obtain the following non-dimensional system of pde's:

$$\frac{\partial F}{\partial t} + \frac{\partial F}{\partial z} = iD_p\frac{\partial^2 F}{\partial x^2} - i\Delta(x)F + \kappa\Gamma(x)[g(N) - iRN]F \tag{5}$$
$$\frac{\partial B}{\partial t} - \frac{\partial B}{\partial z} = iD_p\frac{\partial^2 B}{\partial x^2} - i\Delta(x)B + \kappa\Gamma(x)[g(N) - iRN]B \tag{6}$$
$$\frac{\partial N}{\partial t} = D_f\frac{\partial^2 N}{\partial x^2} + J(x) - \gamma N - \Gamma(x)g(N)(|F|^2 + |B|^2) \tag{7}$$

where F and B are, respectively, forward and backward envelopes, D_p is the diffraction coefficient, Δ is effectively a refractive index profile, J is the pumping current, and R is the linewidth enhancement factor [3] which plays an anti-guiding role. The x dependence of Δ, Γ and J, applies when eqns. (5-7) are used for arrays. It is to be noted that our treatment differs from previous approaches [4]-[7] in the inclusion of diffusion and diffraction in the dynamics. The boundary conditions at the end faces are the typical resonator ones whereas in the x direction they are assumed to be absorptive for both the electric fields and the electron-hole density.

Broad-area laser

The stripe width of a broad-area laser is of the order of 100 micrometres (μm) as opposed to 10 μm of a typical semiconductor laser. The object is to increase the output power by increasing the width of the active layer. Thus one can go from tens of milliwatts to a few watts [8]. However, from the viewpoint of our discussion, the broad-area laser is characterized by its ability to sustain several transverse modes.

To study the dynamics we take eqns.(5-7) with $\Delta(x) = 0$, i.e. no index guiding, and uniform confinement $\Gamma(x)$. In spite of the lack of passive waveguiding in x, a depression in the carrier density introduces a self-focussing nonlinearity by modulating the phase of the electric field through the line enhancement factor R in (5) and (6). The depression of the carrier density occurs via saturation which in turn increases the local refractive index thus acting as a radiation trap which is the origin of the filamentation of the beam as fig.2(a) shows. The temporal evolution is markedly aperiodic and the transverse profile asymmetric with a fairly rich structure.

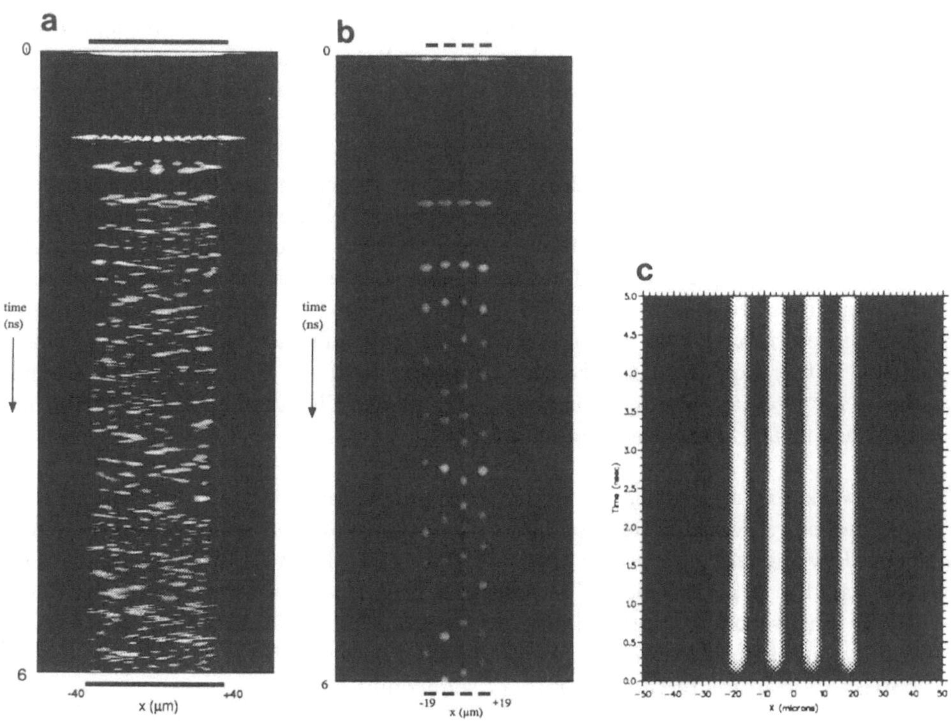

Figure 2: Optical turbulence in a broad-area laser (a) and in an array (b). The plots show the near-field in the transverse x direction and time. The brightness is proportional to the intensity of the field. (c) Same as (b) but stabilised via injection locking.

Previous time-independent, diffusion-free analysis suggested that the saturable nonlinearity for the cavity field stabilizes filament growth at a current value far above the minimum (threshold) for lasing. Our simulations predict the opposite trend in the presence of diffusion. Moreover, using experimental data and for a realistic range of currents our simulations do not exhibit a stable output.

Arrays

Consider a single element of fig.1(b). We can predict through a linear stability analysis and confirm numerically that the output will be stable after the characteristic relaxation oscillations have died down. However, if we bring in a second element, the dynamics will depend on their separation. This is because there is an evanescent coupling provided by the overlap of the fields and by R, which now also couples through diffusion the influence of the neighbour's carrier density to the phases of the fields in a particular element.

Our simulations of the twin-stripe system are illustrative since the interaction in our system is limited to nearest neighbours. They show that for a separation of the order of the carrier diffusion length, the phase difference taken at the centre of the beams is constant (phase locking) with a value $\sim \pi$. As the separation is decreased the phase difference becomes oscillatory and then chaotic. Fig.2(b) shows the typical turbulent dynamics of strongly coupled lasers with short spacing and large current.

It is possible to stabilise (fig.2(c)) an array by injecting into it an external field (injection locking) coming from another laser; each element thus becomes a forced nonlinear oscillator. The stability so achieved is concomitant with a strongly dominant single lobe in the far field which is essential for many applications. The optimum frequency offset, i.e. that for which the single lobe is obtained, can be determined from a Fourier transform both in space and time of the near-field. It turns out that the Fourier transform 'surface' is single-peaked around $k = 0$ for a certain frequency in the $(\omega - k)$ plane. The amplitude of the external field has to be determined more empirically.

Our work explains observed spatio-temporal instabilities, also showing up in commercially available semiconductor lasers, the cause of which was largely obscured by the lack of a proper theoretical framework. In addition to its intrinsic interest, the practical side of the work is that it also shows how to recover a stable single-lobed output. Finally we should point out that we are working on a fully quantum-mechanical many-body approach which we expect to give the correct description of the dynamics of semiconductor lasers.

Acknowledgements: We acknowledge financial support from SERC through the Computational Science Initiative (grant GR/G 13044) and from ACMS (grant AFOSR F4962086C0130).

References

[1] A. Yariv, *Optical Electronics*, 4th ed. (Saunders College Publishing, 1991)

[2] K. Hess, *Advanced Theory of Semiconductor Devices* (Prentice Hall, 1988)

[3] A. Yariv, *Quantum Electronics*, 3rd ed. (Wiley, 1989)

[4] S.S. Wang and H.G. Winful, Appl.Phys.Lett. **52**, 1774(1988)

[5] H.G. Winful and S.S. Wang, Appl.Phys.Lett. **52**, 1894(1988)

[6] K. Otsuka, Phys.Rev.Lett. **65**, 329(1990)

[7] D. Mehuys, R.J. Lang, M. Mittelstein, J. Salzman and A. Yariv, IEEE J. Quant. Electron. **QE-23**, 1909(1987)

[8] L. Goldberg and M.K. Chun, Appl.Phys.Lett. **53**, 1900(1988)

Chaotic Dynamics Due to Competition Among Degenerate Modes in a Ring-Cavity Laser

A. Aceves[1], *D.D. Holm*[2], *and G. Kovacic*[3]

[1]Department of Mathematics and Statistics, University of New Mexico, Albuquerque, NM 87131, USA

[2]Theoretical Division and Center for Nonlinear Studies, Los Alamos National Laboratory, MS B284, Los Alamos, NM 87545, USA

[3]Mathematical Sciences Department, Rennsalear Polytechnical Institute, Troy, NY 12181, USA

Competition between two degenerate spatial modes in a ring-cavity laser is shown to lead to chaotic time-evolution of their amplitudes. Analysis of this temporal chaos implies that the laser output will show randomly intermittent bursts of apparent spatial complexity due to interference between the two competing modes.

1 *Introduction*

The nonlinear dynamics of a ring cavity laser is generally characterized by competition among modes that describe the spatial structure of the fields [1-2]. This competition appears experimentally in the measurement of time-dependent transverse spatial patterns in the laser intensity output [3]. In some cases, the ring cavity resonator geometry can be adjusted, or tuned, to produce two degenerate modes, i.e. two modes with equal longitudinal wave numbers [1-2]. In the case of such a degeneracy, the dynamics of the mode competition simplifies to a system of ordinary differential equations for the time-dependent complex amplitudes of the two cavity modes (i.e., the transverse spatial structures of the fields). In this case, rigorous dynamical systems methods can be applied to exactly characterize the chaotic dynamics that results for the mode amplitudes. We show that this is a purely temporal type of chaos, although it manifests itself experimentally as chaotically-changing transverse spatial patterns in the laser intensity output.

The geometry for a ring-cavity laser experiment with degenerate modes consists of a resonant cavity with curved end mirrors, [1-3]. The ring-cavity laser dynamics is described by the Maxwell-Bloch equations for the slowly-varying complex envelopes E and P of the electric field, $\mathcal{E} = \mathrm{Re}[E e^{i(k_0 z - \omega_0 t)}]$, the polarizability of the medium, $\mathcal{P} = \mathrm{Re}[P e^{i(k_0 z - \omega_0 t)}]$, and the real level inversion, D. Here k_0 and ω_0 are the wave number and frequency of the incident light. The Maxwell-Bloch equations are given by [1-2]

$$-\frac{i}{4}\nabla_\perp^2 E + \frac{\partial E}{\partial z} + \frac{\partial E}{\partial t} = P, \quad \frac{\partial P}{\partial t} = ED - \delta P, \text{ and } \frac{\partial D}{\partial t} = -\text{Re}(E^*P) - \delta(D - D_0),$$

where $\nabla_\perp^2$ denotes the transverse Laplacian, longitudinal distance z is measured in units of Λ (the optical path length), time is measured in units of the round-trip time for light, Λ/c, and the small parameter $\delta = \gamma\Lambda/c << 1$ is the ratio of time scales: the round-trip time for light in the cavity, divided by the relaxation time of the medium. By choosing to scale time in units of Λ/c and setting $\delta = 0$, we address dynamics on the nondissipative time scale, i.e., on a time scale which is short compared to any dissipation time.

We assume separation of variables for the solution of the linearly polarized electric field in the resonant cavity. Because these cavity modes are taken to be degenerate in longitudinal wave number, we may assume a solution in the form $E(r,\theta,z,t) = \sum_l A_l(r,\theta,z)\,E_l(t)$, where $A_l(r,\theta,z)$ describes the spatial structure of the l-th empty-cavity field mode. That is, $A_l(r,\theta,z)$ is the l-th eigenfunction in the cavity of the spatial part of the linear operator in the first Maxwell-Bloch equation. (See, e.g., Ref. [2], appendix A). Similar modal expansions are also assumed for the fields P and D. Substituting these three modal expansions into the Maxwell-Bloch equations and projecting onto the l-th degenerate mode yields the following set of coupled Maxwell-Bloch equations, labeled by indices $l, m, n = 1, \ldots, L$,

$$\dot{E}_l = P_l, \qquad \dot{P}_l = \sum_{mn} \Gamma_{lmn} E_m D_n, \qquad \dot{D}_l = -\text{Re}\,(\sum_{mn} \Gamma_{lmn} E_m^* P_n). \qquad \text{(CMB)}$$

In these CMB equations, $\Gamma_{lmn} = \int_{\text{cavity}} A_l A_m A_n$. Consequently, the real arrays of constants Γ_{lmn} describing the coupling among the modes are totally symmetric tensors. That is, $\Gamma_{lmn} = \Gamma_{\sigma(lmn)}$ for every permutation σ of l, m, and n.

2 *Diagonalization of the Modal Equations*

The CMB equations derived above possess L constants of motion, $K_l = \frac{1}{2}\sum_{mn} \Gamma_{lmn} E_m E_n^* + D_l$, $l = 1, \ldots, L$. The conserved quantity K_l is the projection onto the l-th mode of the sum of the total field energy and atomic excitation energy, $\frac{1}{2}|E|^2 + D$. Since K_l is linear in D_l we may use it to eliminate D_l and thereby reduce the number of equations from 5L to 4L. After this elimination, we have,

$$\dot{E}_l = P_l, \qquad \dot{P}_l = \sum_{mn}\left[\Gamma_{lmn} E_m (K_n - \frac{1}{2}\sum_{pq} \Gamma_{npq} E_p E_q^*)\right].$$

These equations describe a set of L coupled complex Duffing oscillators. They may be expressed in Hamiltonian form as $\dot{E}_l = \{E_l, H\} = 2\partial H/\partial P_l^*$ and $\dot{P}_l = \{P_l, H\} = 2\partial H/\partial E_l^*$, with Hamiltonian function,

$$H = \sum_l \left[\frac{1}{2} |P_l|^2 + \frac{1}{2}(K_l - \frac{1}{2} \sum_{mn} \Gamma_{lmn} E_m E_n^*)^2 \right]. \qquad \text{(H)}$$

Now, $2H = \sum_l (|P_l|^2 + D_l^2)$, in terms of the original variables. Hence, conservation of H for the Duffing oscillator system represents preservation of the total unitarity of the system by the CMB dynamics (Holm, Kovacic, and Sundaram [4]). Invariance of the Hamiltonian H under the overall phase shift $(P_l, E_l) \rightarrow (e^{i\varphi}P_l, e^{i\varphi}E_l)$ also implies conservation of its infinitesimal generator, $J = \sum_l \mathrm{Im}(E_l^* P_l)$. This conserved quantity is the total self-interaction energy of the L modes.

Consider the situation in which the coupling coefficients Γ_{lmn} mutually commute, i.e., suppose $\sum_l (\Gamma_{lmn} \Gamma_{lpq} - \Gamma_{lmq} \Gamma_{lpn}) = 0$. Then the total Hamiltonian, H in equation (H), may be diagonalized. This diagonalization of H may be understood by regarding Γ_{lmn} as a set of L symmetric matrices, written $\Gamma^{(l)}{}_{mn}$. The commutation formula above means that the symmetric matrices $\Gamma^{(l)}$ and $\Gamma^{(j)}$ commute for all l and j. Hence, in this situation, a single orthogonal $L \times L$ matrix, O_{lm}, exists that simultaneously diagonalizes all the Γ – matrices. Because of the total symmetry of the tensors Γ_{lmn}, which is preserved under orthogonal transformations, the only nonzero entry remaining in the j-th tensor $\Gamma^{(j)}$ after such a diagonalization is $\Gamma^{(j)}{}_{jj} = \gamma_j$. Performing this orthogonal transformation on the fields E_l and P_l, and on the constants K_l, $(E_l, P_l, K_l) \rightarrow (E'_l, P'_l, K'_l) = (\sum_m O_{lm} E_m, \sum_m O_{lm} P_m, \sum_m O_{lm} K_m)$, yields the Hamiltonian for the system in the new variables (after dropping the primes) in diagonalized form,

$$H = \sum_l H_l = \sum_l \left[\frac{1}{2} |P_l|^2 + \frac{1}{2}(K_l - \frac{1}{2} \gamma_l |E_l|^2)^2 \right].$$

Thus, the Hamiltonian decouples into a sum of L complex-Duffing-oscillator Hamiltonians, provided the coupling coefficients Γ_{lmn} mutually commute as symmetric matrices. In this case, the conserved Hamiltonians H_l of the decoupled modes are the projections of the unitarity condition for the full system onto each mode.

This diagonalization renders each Hamiltonian H_l invariant under an independent phase change of the l-th fields: $(E_l, P_l) \rightarrow (e^{i\varphi_l} E_l, e^{i\varphi_l} P_l)$. Hence, the self-interaction energy for each mode, $J_l = \mathrm{Im}\,(E_l^* P_l)$, is separately conserved. These additional conservation laws imply that the dynamics of each decoupled Hamiltonian subsystem is completely integrable. Therefore, the entire system of L interacting ring-cavity modes is completely integrable when the coupling coefficients mutually commute.

3 *The Integrable Decoupled Model*

The Hamiltonian and the conserved self-interaction energy for a single decoupled mode are given by

$$H_l = \frac{1}{2}\,|P_l|^2 + \frac{1}{2}\,(K_l - \frac{1}{2}\gamma_l\,|E_l|^2)^2 \qquad \text{and} \quad J_l = \mathrm{Im}\,(E_l^* \, P_l),$$

where conservation of J_l follows from phase invariance of H_l. This phase invariance suggests the canonical transformation to polar coordinates given by $E_l = \mathcal{Q}_l\, e^{i\varphi_l}$ and $P_l = (\mathcal{P}_l + \frac{iJ_l}{\mathcal{Q}_l})\, e^{i\varphi_l}$. When expressed in these polar coordinates, the single-mode Hamiltonian is independent of φ_l,

$$H_l = \frac{1}{2}\mathcal{P}_l^{\,2} + \frac{J_l^2}{2\mathcal{Q}_l^2} + \frac{1}{2}\,(K_l - \frac{1}{2}\gamma_l \mathcal{Q}_l^2)^2,$$

and Hamilton's equations for $(\mathcal{Q}_l, \mathcal{P}_l, \varphi_l, J_l)$ are given by

$$\dot{\mathcal{Q}}_l = \{\mathcal{Q}_l, H_l\} = \mathcal{P}_l, \qquad \dot{\mathcal{P}}_l = \{\mathcal{P}_l, H_l\} = \gamma_l \mathcal{Q}_l (K_l - \frac{1}{2}\gamma_l \mathcal{Q}_l^2) + \frac{J_l^2}{\mathcal{Q}_l^3},$$

$$\dot{\varphi}_l = \{\varphi_l, H_l\} = \frac{J_l}{\mathcal{Q}_l^2}, \qquad \dot{J}_l = \{J_l, H_l\} = 0.$$

In the case $J_l = 0$ and $K_l > 0$, and only in this case, the solutions of this system have a homoclinic orbit in the $\mathcal{Q}_l$ - $\mathcal{P}_l$ phase plane. In the original variables, the homoclinic solution of equations (CMB) for $J_l = 0$ and $K_l > 0$ is given by

$$E_l = 2\sqrt{\frac{K_l}{\gamma_l}}\,\mathrm{sech}\left[\sqrt{K_l\gamma_l}\,(t - t_l)\right] e^{i\bar{\varphi}_l}, \quad P_l = -\,2\,K_l\,\mathrm{sech}\left[\sqrt{K_l\gamma_l}\,(t - t_l)\right]\tanh\left[\sqrt{K_l\gamma_l}\,(t - t_l)\right] e^{i\bar{\varphi}_l}$$

This solution connects the hyperbolic point at the origin, $E_l = 0 = P_l$, to itself. Geometrically the solution is a "pinched torus," i.e., the Cartesian product of a homoclinic loop in a plane $\varphi_l = \bar{\varphi}_l$ (which is constant for $J_l = 0$), times a circle parameterized by the value of phase angle $\bar{\varphi}_l$. For $J_l \neq 0$, all other solutions are periodic in the $\mathcal{Q}_l$ - $\mathcal{P}_l$ phase plane and, thus, are quasiperiodic with two periods in the original variables. See Ref [4] for further details.

4 *Recoupling the Modes*

In general, the coupling coefficients Γ_{lmn} do not mutually commute. Of particular interest is the situation in which they nearly commute, and are expressible as $\Gamma_{lmn} = \gamma_l\,\delta_{lmn} + \varepsilon\,\Delta_{lmn}$, where $\delta_{lmn} = 1$, if $l = m = n$, and vanishes otherwise. The coefficients Δ_{lmn} are assumed not to mutually commute with all the Γ_{lmn}, and ε is taken to be the small uniform coupling strength; ε

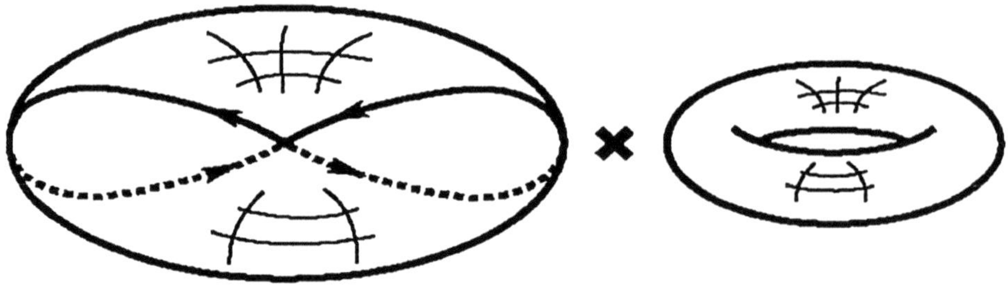

Figure 1: The homoclinic CMB solutions in the two-mode phase space can be regarded geometrically as a pinched torus (mode 1) crossed into a two-torus (mode 2).

<<1. We treat this situation as a nearly integrable problem, and ask what physical consequences arise when the coupling perturbation breaks integrability.

We consider the case of two coupled modes, $L = 2$, which is the case most frequently encountered in experiments. Suppose the unperturbed modes before coupling have $J_1 = 0$, $K_1 > 0$, and $J_2 \neq 0$, so we are coupling a homoclinic structure in mode 1 with the quasiperiodic structure of mode 2, see Figure 1. Except for the homoclinic orbits, most solutions in this case are quasiperiodic with four (i.e., 2L) frequencies. Thus, the pure second mode is a special solution, since it has only two frequencies. In the system of two modes, the solution corresponding to the pure second mode is hyperbolic; that is, it is unstable along the homoclinic orbits of the first mode. As a two-frequency quasiperiodic solution, the pure second mode can be thought of as a two-torus, and the unperturbed solutions homoclinic to it in the full phase space can be regarded geometrically as a pinched torus, crossed into the two-torus of the pure second mode, as shown in Figure 1. Therefore, the unperturbed orbits homoclinic to a particular pure second mode solution (for a particular choice of J_2 and K_2) form a four-dimensional homoclinic surface.

In the weakly coupled case, $\varepsilon \neq 0$, most of the special solutions with just two frequencies still persist and are $O(\varepsilon)$ close to the pure second-mode solutions. Moreover, the unstable directions of these perturbed solutions with only two frequencies are $O(\varepsilon)$ close to the homoclinic directions of the unperturbed pure first-mode solution. (The validity of these claims follows from the persistence theory of normally hyperbolic manifolds and the KAM theorem, see, e.g., Ref. [5].) Hence, orbits forward-asymptotic in time to any perturbed two-frequency quasiperiodic solution still form a smooth four-dimensional surface, which is $O(\varepsilon)$ close to the unperturbed homoclinic surface. This surface is called the stable manifold of the perturbed two-frequency quasiperiodic solution. Likewise, orbits backward-asymptotic to a two-frequency quasiperiodic solution form its smooth four-dimensional unstable manifold.

If the two-mode system were to remain integrable under the weak coupling, then the stable and unstable manifolds of the hyperbolic quasiperiodic solution would continue to coincide, as a slightly deformed four-dimensional homoclinic surface. However, when, as expected, the coupling perturbation destroys integrability, the stable and unstable manifolds will, in general, intersect. These intersections are the homoclinic orbits of the hyperbolic solution that survive after the coupling. These homoclinic intersection orbits form two-dimensional surfaces, since they are the intersections of four dimensional stable and unstable manifolds in the six-dimensional simultaneous level surfaces (lying in the eight-dimensional phase space) of the total unitarity H and self-interaction energy J.

5 *Showing Transverse Intersections*

In order to show that homoclinic intersections occur under weak coupling of the two modes, we use the standard Melnikov method [5-6]. Application of this method requires the computation of the distance between perturbed stable and unstable manifolds of the hyperbolic solution, to first order in ε. This first-order distance is given by a vector having two components measured along two of the normal directions to the unperturbed homoclinic surface. For these directions we choose ∇H_1 and ∇J_1, the eight-component gradients in the full phase space of the conserved quantities for the unperturbed first mode. Only two normal directions (out of the existing four) are needed, because both the perturbed stable and unstable manifolds of the hyperbolic solution lie on level surfaces of total H and total J (see [4,5]). These two components of the required first-order distance are given by

$$M_1 = \int_{-\infty}^{\infty} \nabla H_1 \cdot \mathbf{g}\, dt \qquad \text{and} \qquad M_2 = \int_{-\infty}^{\infty} \nabla J_1 \cdot \mathbf{g}\, dt,$$

where the integrals are taken along a particular unperturbed homoclinic orbit of mode 1, and the eight-component vector g is the linearization in ε of the perturbed flow. In terms of the field, we have

$$M_1 = \int_{-\infty}^{\infty} dt \sum_{m\,n} \Delta_{1mn} \left[(K_m - \tfrac{1}{2}\gamma_m |E_m|^2)\, \mathrm{Re}\,(E_n^*\, P_1) - \tfrac{1}{2}\gamma_1\, \mathrm{Re}\,(E_m E_n^*\, E_1^* P_1) \right],$$

$$M_2 = \int_{-\infty}^{\infty} dt \sum_{m\,n} \Delta_{1mn} \left[(K_m - \tfrac{1}{2}\gamma_m |E_m|^2)\, \mathrm{Im}\,(E_1^* E_n) \right],$$

where E_m is evaluated at $t - t_m$ and φ_m along the unperturbed homoclinic and quasiperiodic solutions.

We compute the Melnikov vector components (M_1, M_2), by considering the unperturbed motion in the $(\mathcal{Q}_2, \mathcal{P}_2)$ phase plane of the second mode and linearizing this motion in the vicinity of the center at $\mathcal{Q}_2 = \bar{Q}_2$ and $\mathcal{P}_2 = 0$. This linearized solution undergoes *harmonic* motion in the $(\mathcal{Q}_2, \mathcal{P}_2)$ phase plane. In the full phase space, these harmonic motions are given in terms of the complex envelope field variables (E_2, P_2) by

$$E_2 = \left[\bar{Q}_2 + A_2 \cos \Omega (t - t_2) - \frac{2i\omega A_2}{\Omega} \sin \Omega (t - t_2)\right] e^{i\omega(t - t_2) + i\bar{\varphi}_2}$$

$$P_2 = \left[i \bar{Q}_2 \omega + \frac{A_2}{\Omega}(2\omega^2 - \Omega^2) \sin \Omega (t - t_2) - iA_2\omega\cos\Omega(t - t_2)\right] e^{i\omega(t - t_2) + i\bar{\varphi}_2}$$

where we denote $\omega = J_2 / \bar{Q}_2^{\,2}$, $\Omega^2 = \gamma_2^{\,2} \bar{Q}_2^{\,2} + 4\ \omega^2$, and t_2 and $\bar{\varphi}_2$ are constant parameters.

Substituting into the integrands of the Melnikov vector the expressions for the unperturbed homoclinic mode 1 solution (E_1, P_1) given earlier and the linearized quasiperiodic mode 2 solution (E_2, P_2) given above yields, after considerable algebra and evaluation of integrals,

$$M_1 = a_1 \sin (\omega\Delta t - \Delta\bar{\varphi}) + a_2 \sin [(\Omega + \omega)\Delta t - \Delta\bar{\varphi}] + a_3 \sin [(\Omega - \omega)\Delta t - \Delta\bar{\varphi}] + a_4 \sin (\Omega\Delta t),$$

$$M_2 = b_1 \sin (\omega\Delta t - \Delta\bar{\varphi}) + b_2 \sin [(\Omega + \omega)\Delta t - \Delta\bar{\varphi}] + b_3 \sin [(\Omega - \omega)\Delta t - \Delta\bar{\varphi}],$$

where $\Delta t = t_1 - t_2$, $\Delta\bar{\varphi} = \bar{\varphi}_1 - \bar{\varphi}_2$, and the coefficients a_i and b_j are given in Aceves et al. [7].

The Melnikov vector components M_1 and M_2 have simultaneous zeros, when $\Delta t = 0$ and $\Delta\bar{\varphi} = 0$, or π. (Since the Jacobian $(\partial M_1, \partial M_2)/(\partial \Delta t, \Delta\bar{\varphi})$ is nonvanishing, these Melnikov zeros are simple.) Hence, the consequence of coupling small-amplitude oscillations of mode 2 to mode 1 is to produce transverse intersections of the stable and unstable manifolds of the hyperbolic quasiperiodic solutions of the coupled system. These transverse intersections are sketched in Figure 2.

The formulas for M_1 and M_2 show that the transverse intersections caused by mode coupling are two-dimensional and parameterized either by $t - t_2$ and $\bar{\varphi}_2$, or, equivalently, by $t - t_1$ and $\bar{\varphi}_1$. As shown in the next section, these transverse intersections imply chaotic dynamics for phase points in their vicinity via a Smale horseshoe construction.

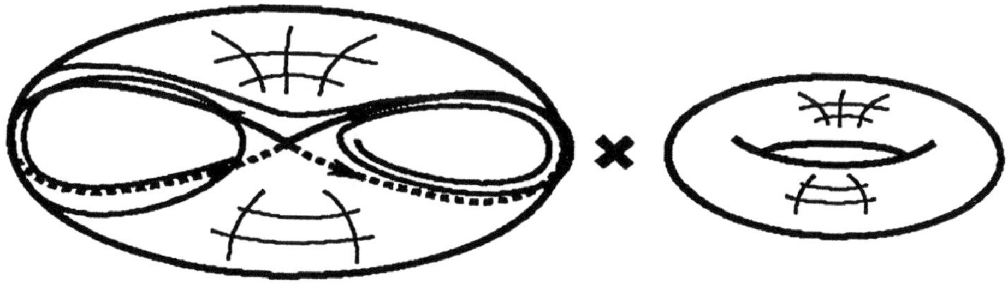

Figure 2: The weak-coupling perturbation causes transverse intersections of the stable and unstable manifolds of the homoclinic solutions to occur near $\Delta t = 0$ and $\Delta\bar{\varphi} = 0$, or π. A typical orbit near these transverse intersections switches randomly from side to side in the vicinity of the homoclinic periodic orbit and near the plane containing $\Delta\bar{\varphi} = 0$ and $\Delta\bar{\varphi} = \pi$.

6 *Generalized Smale Horseshoe Construction*

We consider the Poincaré map Φ obtained by taking a Poincaré section at $t - t_2 = 0$, modulo 2π. In the φ_2-reduced phase space (Q_2, P_2) for the linearized harmonic mode 2 motion, this Poincaré section is the positive Q_2 axis. In the Poincaré section the quasiperiodic mode 2 solution is a circle, and the Poincaré map Φ takes this circle into itself. This circle is hyperbolic in the phase space for the coupled system and, according to the Melnikov calculation just given, its stable and unstable manifolds intersect transversely. These transverse intersections are two-dimensional surfaces in the continuous-time dynamics, and are circles in our Poincaré section. The angle parameterizing each intersection circle in the Poincaré section is $\bar{\varphi}_2$ or, equivalently, φ, the angle conjugate to the total interaction energy, J_{tot}. Because J_{tot} is conserved, the dynamics for φ decouples from the rest of the system and may now be factored out by Hamiltonian reduction. The transverse intersections on the resulting *reduced* Poincaré section form the usual homoclinic tangle of points, as guaranteed by the Poincaré-Birkhoff-Smale theorem. That is, an iterate of the reduced Poincaré map results in a Smale horseshoe construction, thereby leading to an invariant Cantor set of points in the reduced Poincaré section on which the dynamics is topologically conjugate to a Bernoulli shift [5,6]. Adding back the angle φ now yields an invariant Cantor set of *circles*, whose dynamics under the reconstructed Poincaré map is extremely sensitive to initial conditions.

7 *Physical Implications*

Given that the dynamics in the vicinity of the homoclinic tangle is chaotic, what implications are there for ring-cavity laser experiments? For phase points in the vicinity of the homoclinic tangle, the time series for the amplitude of the first mode $|E_1|$ will show intermittent excursions away from zero in the form of the homoclinic solution,

$$|E_1| = 2\sqrt{\frac{K_1}{\gamma_1}}\,\mathrm{sech}\left[\sqrt{K_1\gamma_1}\,(t - t_\mu)\right],$$

where t_μ is now indistinguishable from a *random* variable, because of the extreme sensitivity to initial conditions for this motion. These intermittent bursts in the first mode will interfere with the quasiperiodic evolution of the second mode, in the total output intensity of the laser,

$$|E|^2 = \left|E_1(t)\,A_1(r, \theta, z) + E_2(t)\,A_2(r, \theta, z)\right|^2,$$

where $A_1(r, \theta, z)$ and $A_2(r, \theta, z)$ represent the two transverse spatial patterns of the competing degenerate cavity modes. Hence the resulting transverse spatial pattern of the total resulting laser output will show a chaotic sequence of bursts of interference. During these bursts, the transverse output will appear spatially complex. Thus an effect which might appear to be spatio-temporal chaos is actually due to temporal chaos alone, arising from the competition between two degenerate spatial modes.

Acknowledgement

This work was partially supported by the U.S. Air Force Office of Scientific Research, contract numbers AFOSR-91-0009 and AFOSRISSA900024. The authors would like to thank Roberto Camassa, Tasso Kaper, and Vittorio Penna for their constructive comments and encouragement.

References

[1] L. A. Lugiato, G. L. Oppo, J. R. Tredicce, L. M. Narducci and M. A. Pernigo, J. Opt. Soc. of America B 7, 1019 (1990); M. Brambila, F. Battipede, L. A. Lugiato, V. Penna, F. Prati, C. Tamm and C. O. Weiss, Phys. Rev. A 43, 5090 (1991).

[2] C. Tamm and C. O. Weiss, J. Opt. Soc. of America B 7, 1034 (1990).

[3] G. Giusfredi, J.F. Valley, R. Pon, G. Khitrova and H. M. Gibbs, J. Opt. Soc. of America B 5, 1181 (1988).

[4] D. D. Holm, G. Kovacic and B. Sundaram, Phys. Lett. A 154, 346 (1991).

[5] S. Wiggins, *Global Bifurcations and Chaos: Analytical Methods*, Springer Verlag: New York (1988).

[6] J. Guckenheimer and P. J. Holmes, *Nonlinear Oscillations, Dynamical Systems, and Bifurcations of Vector Fields*, Springer-Verlag, New York (1983).

[7] A. Aceves, D.D. Holm and G. Kovacic, Phys. Lett. A (in press).

Quantum Self Phase Modulation in Optical Fibres

K.J. Blow[1], *R. Loudon*[2], *and S.J.D. Phoenix*[1]

[1]BT Laboratories, Martlesham Heath, Ipswich, IP5 7RE, UK
[2]University of Essex, Colchester, UK

We discuss various aspects of quantum propagation effects in optical fibres. Self phase modulation is shown to give quadrature squeezing for small values of the classical nonlinear phase shift. At high values of the phase shift this squeezing is destroyed and the coherent state develops into a state which does not have minimum uncertainty.

Quantum theory, as normally formulated, describes systems enclosed in an artificial box through its time evolution. In some systems, such as a laser, the box may be a physical element such as a cavity. In optical fibre communications the situation is somewhat different. The optical field is continuously generated by a source such as a laser. Once generated the field propagates along an optical fibre and is detected in some form of receiver at the far end. Quantum theory, with its reliance on Hamiltonian systems necessarily conserves the total energy in the system whereas fibre communication is better described as a balance between sources and sinks. In this paper we describe an approach to the study of quantum propagation in optical fibres. The discussion will be based on the description of single mode weakly guiding fibres so that a one dimensional scalar wave equation can be used to describe the electric field. This formalism is also applied to self phase modulation where the quadrature squeezing is calculated.

Previous work has concentrated on approximate methods such as linearizing the quantum operators about a semiclassical pump to describe quantum modulation instability [1] and soliton squeezing [2] or use of Hartree states to describe solitons [3] or self phase modulation [4]. There has also been considerable work in the single mode case [5], [6].

In the absence of a quantization box the modes of the system are continuous in both the time and frequency domain which enables us to work directly with continuum operators rather than with their discrete counterparts. The positive frequency part of the electric field is written [7], for a narrowband excitation,

$$\hat{E}^{+} = i\left(\frac{\hbar\omega_0}{4\pi\varepsilon_0 cA}\right)^{\frac{1}{2}} \int d\omega\, \hat{a}(\omega)\exp\{-i\omega(t - \frac{z}{c})\}, \tag{1}$$

where A is the mode area and ω_0 is the central frequency of the wavepacket. The equation of motion for the electric field is obtained from the quantised version of Maxwell's equations suitably processed to treat a single mode optical fibre. The normally ordered Poynting vector is defined as

$$\hat{S}(z,t) = \frac{1}{\mu_0}\left\{\hat{E}^{-}(z,t)\hat{B}^{+}(z,t) + \hat{B}^{-}(z,t)\hat{E}^{+}(z,t)\right\} \tag{2}$$

and is conveniently expressed in terms of the Fourier transform $\hat{a}(t)$ of the destruction operator $\hat{a}(\omega)$

$$A\hat{S}(z,t) = \hbar\omega_0 \hat{a}^\dagger(t - \frac{z}{c})\hat{a}(t - \frac{z}{c}). \tag{3}$$

This has a particularly simple physical interpretation as the energy per photon multiplied by an operator representing the flux of photons per unit time. We note that this operator is finite for a CW excitation whereas the total energy would be infinite.

In quantum optics we are concerned with the evaluation of matrix elements of coherent states defined by

$$\hat{a}(t) \mid \{\alpha\}> = \alpha(t) \mid \{\alpha\}> \tag{4}$$

with $\alpha(t)$ representing the time dependent output field from a laser. In order to detect the presence of squeezing in a light beam it is necessary to perform homodyne detection where a coherent local oscillator is mixed with the signal. This leads us to calculation of second order moments of the two field operators $\hat{a}$ and $\hat{a}^\dagger$. The problem of the 1D impenetrable Bose gas [8] is similar to this problem in many ways but the main difference is that the ground state of the system is an N particle state which, in quantum optics terminology, is a photon number state.

In the study of χ_3 processes it will be necessary to evaluate operators of the form $\exp(\hat{O})$, where

$$\hat{O} = \int dt\, g(t)\hat{a}^\dagger(t)\hat{a}(t). \tag{5}$$

We have proved the following continuous version of a single mode theorem [9],

$$\exp\{\int dt\, g(t)\hat{a}^\dagger(t)\hat{a}(t)\} = :\exp\left\{\int dt\, [e^{g(t)} - 1]\hat{a}^\dagger(t)\hat{a}(t)\right\}: \tag{6}$$

where the colons denote normal ordering.

The equation of motion for the annihilation operator can be obtained from a spatial Heisenberg formalism [10] and for an instantaneous nonlinear response it is

$$i\frac{\partial\hat{a}}{\partial z} = \kappa\hat{a}^\dagger\hat{a}\hat{a} \tag{7}$$

which has the following operator solution

$$\hat{a}(z,t) = e^{i\kappa z\hat{a}^\dagger(0,t)\hat{a}(0,t)}\hat{a}(0,t). \tag{8}$$

The problem of quantum self phase modulation is therefore exactly of the form of equation 5. The pure Kerr part of the nonlinear response is normally treated as instantaneous, obtained by taking $g(t) = \delta(t)$ in equation 6 and giving the operator solution of equation 8, but this leads to non-integrable singularities in the continuous mode quantum theory [11]. These singularities can be removed if the physical response time of the medium is included. In silica the response time of the Kerr effect is a few femtoseconds and we are interested in describing the properties of pulses in the picosecond regime. Thus, we can model the response function in a particularly simple way

$$\begin{aligned} g(t) &= \frac{1}{\tau_0} \quad t < \tau_0 \\ &= 0 \quad t > \tau_0 \end{aligned} \tag{9}$$

where τ_0 is the response time.

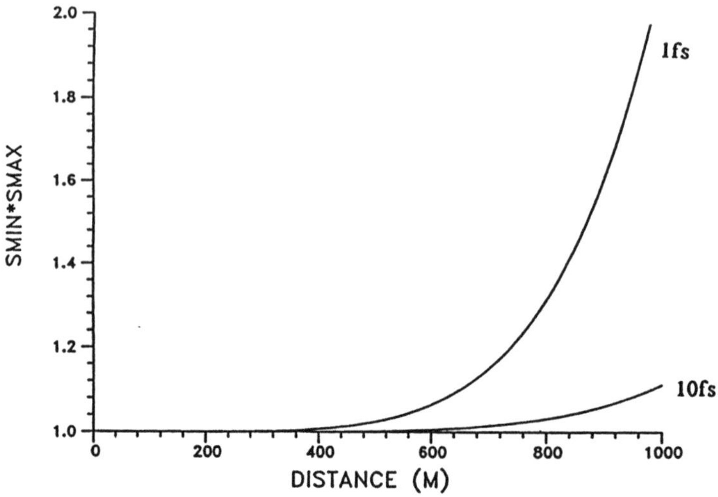

Figure 1. Evolution of the product of the coherent state variances as a function of distance along the fibre.

Using the theorem in equation 5, we have been able to calculate the squeezing spectrum of an arbitrary coherent state pulse propagating in a lossless fibre as a function of the classical phase shift of the pulse and the vacuum phase shift associated with the medium response. In the regime of interest the exact expressions can be expanded to second order in the vacuum phase shift. In particular we have obtained an expression for the product of the two quadrature noise variances as follows,

$$S_{\min} S_{\max} = 1 + \left(\frac{40}{3}\phi_c^3 + \frac{16}{3}\phi_c^5\right)\phi_v - \left(\frac{82}{3}\phi_c^4 + \frac{32}{3}\phi_c^6\right)\phi_v^2 \tag{10}$$

where ϕ_c and ϕ_v are the classical and vacuum phase shifts respectively. In a typical optical fibre experiment at $1.5\mu m$ these phase shifts are given by

$$\begin{aligned} \phi_c &= 3.33L \\ \phi_v &= 4.4\times10^{-4}\frac{L}{\tau_0} \end{aligned} \tag{11}$$

where L is the distance in kilometres and the medium response time is in femtoseconds. The product in equation 10 is plotted in figure 1 as a function of distance for a typical optical fibre. There are two curves for the two response times of 1fs and 10fs showing the crucial dependence on this parameter. As we can see from equation 10 the development of the uncertainty product above the minimum increases rapidly with the classical phase shift. Thus, this effectively represents an additional noise source at high values of the classical phase shift which may be present in a recent soliton squeezing experiment [12]. In this experiment a classical phase shift of several π occurred since the nonlinear loop mirror used was many soliton periods long.

REFERENCES

1. M. J. Potasek and B. Yurke, Squeezed-light Generation in a Medium governed by the Nonlinear schrodinger Equation, Phys Rev A 35 3974-7 (1987)
2. P. D. Drummond, S. J. Carter and R. M. Shelby, Time Dependence of Quantum Fluctuations in Solitons, Optics Letts 14 373-5 (1989)
3. Y. Lai and H. A. Haus, Quantum Theory of Solitons in Optical Fibres I. Time-dependent Hartree Approximation, Phys Rev A 40 844 (1989)
4. E. M. Wright, Quantum Theory of Self Phase Modulation, J Opt Soc Am B 7 1142-6 (1990)
5. R. Tanas and S. Kielich, Self-squeezing of Light Propagating Through Nonlinear Optically Isotropic Media, Opt Commun, 45 351 (1984)
6. M. Kitagawa and Y. Yamamoto, Number-Phase Minimum-uncertainty State with Reduced Number Uncertainty in a Kerr Nonlinear Interferometer, Phys Rev A 34 3974-88 (1986)
7. K. J. Blow, R. Loudon, S. J. D. Phoenix and T. J. Shepherd, Continuum Fields in Quantum Optics, Phys Rev A 42 4102-14 (1990)
8. V. E. Korepin and N. A. Slavnov, The Time Dependent Correlation Function of an Impenetrable Bose Gas as a Fredholm Minor. Commun Math Phys 129 103-13 (1990)
9. W. H. Louisell, Quantum Statistical Properties of Radiation (Wiley, New York 1973)
10. I. Abram, Quantum Theory of Light Propagation: Linear Medium, Phys Rev A 35 4661-72 (1987)
11. K. J. Blow, R. Loudon and S. J. D. Phoenix, Exact Solution for Quantum Self Phase Modulation, to be published J Opt Soc Am B (1991)
12. M. Rosenbluh and R. M. Shelby, Squeezed Optical Solitons, Phys Rev Lett 66 153-6 (1991)

Quantum Groups: Q-Boson Theories of Integrable Models and Applications in Non-Linear Optics

R.K. Bullough[1] and *N.M. Bogoliubov*[2]

[1]Department of Mathematics, UMIST,
P.O. Box 88, Manchester, M60 1QD, UK
[2]On leave from Steklov Mathematical Institute,
Fontanka 27, 191011 St. Petersburg, Russia

This short report sketches new results in the *application* of the 'quantum groups'[1,2] in 1 + 1 dimensions. They concern two aspects:- (i) generation of new quantum (and classical) integrable models in 1 + 1 - we report solution of a new "q-boson" lattice model which is quantum integrable; (ii) the very *physical* question of whether there are experimentally realisable systems where quantisation by a quantum group ("q-quantisation") could be distinguished from ordinary canonical quantisation. We suggest quantum optical experiments involving Rydberg atoms in microwave cavities (cavity q.e.d.[3,4]) where our predictions are different in the two cases. In contrast we show that for the integrable quantum *field* theories predictions are identical in the two cases. Thus ordinary q.e.d. based on the quantised Maxwell equations is apparently unchanged by q-quantisation. We show that the new "q-boson" lattice model exemplifies similar features.

The Fig. 1 attached is updated from that in Ref. 4: new features are the connection between the classical KP-I equations in 2 + 1 dimensions and 2D quantum gravity reported in Refs. 5 and 6 and the several new connections from the 'box' "Hopf Algebra $\Delta T = T \otimes T$ Quantum Groups". The notation used in Ref. 2, $\Delta T = T \dot{\otimes} T$, with the dot, for the *co-multiplication* of the elements of the quantum monodromy matrices T which form the quantum group is now introduced: $\Delta(T_{ij}) = T \dot{\otimes} T \equiv \sum_k T_{ik} \otimes T_{kj}$, and this is matrix multiplication in which $\otimes$ is the Kronecker product tensoring between two spaces. Ref. 1 *interpreted* $\Delta T = T \otimes T$ in this way.

The main result of q-quantisation, marked in Fig. 1, is the commutation relations for the elements of the monodromy matrices $T(\lambda)$ (called $\hat{T}(\lambda)$ in this note)

$$R(\lambda,\mu)\ \hat{T}(\lambda) \otimes \hat{T}(\mu) = \hat{T}(\mu) \otimes \hat{T}(\lambda)\ R(\lambda,\mu) \tag{1}$$

in which spectral parameters $\lambda, \mu \in \mathbb{C}$ are introduced: $\otimes$ is Kronecker product, and $R(\lambda,\mu)$ is the quantum R-matrix which can be found by solving the Yang- Baxter equations

$$(I\otimes R(\lambda,\mu))(R(\lambda,\nu)\otimes I)(I\otimes R(\mu,\nu)) = (R(\mu,\nu)\otimes I)(I\otimes R(\lambda,\nu))(R(\lambda,\mu)\otimes I). \quad (2)$$

For particular solutions R of (2) the elements of $\hat{T}(\lambda)$ for $\lambda \to \pm \infty$ form a non-commutative and non-co-commutative quasi-triangular Hopf algebra - a quantum group[7,8]. The idea of the quantum groups can be extended to the elements of $\hat{T}(\lambda)$ for *finite* $\lambda \in \mathbf{C}$: these form a co-algebra, loosely called a quantum group in this note. From (2) the commutation relations (1) lead to particular quantum integrable models by choosing representations $\hat{T}(\lambda)$ of the quantum group. The matrix trace of (1) means $[\hat{\Delta}(\lambda), \hat{\Delta}(\mu)] = 0$ with $\hat{\Delta}(\lambda) \equiv \mathrm{Tr}\ \hat{T}(\lambda)$, a quantum integrability condition (a 'quantum Liouville theorem'): then $\ln \hat{\Delta}(\lambda)$ is a generator of integrable Hamiltonians $\hat{H}$.

This route involves q-quantisation: canonical quantisation is not used. Thus our new q-quantised "q-boson" lattice shows the new features reported below. In contrast the quantum inverse method (QIM) reaches exactly the commutation relations (1) by imposing canonical quantisation on the local transition matrices $\hat{L}_n(\lambda)$. These satisfy

$$R(\lambda,\mu)\ \hat{L}_n(\lambda) \otimes \hat{L}_n(\mu) = \hat{L}_n(\mu) \otimes \hat{L}_n(\lambda)\ R(\lambda,\mu) \quad (3)$$

and $\hat{T}(\lambda) = \overset{\curvearrowleft}{\prod_n} L_n(\lambda)$ induces (1) from (2). For integrable quantum field theories (e.g. the bose gas, for which calculation of the correlation functions was reported in Ref.9), one can use (1) as constraints on the quantum partition function (Fig. 1 top right) $Z = \mathrm{Tr}\int \mathcal{D}\mu \exp S[p]$ in which $S[p]$ is the classical action of the integrable model expressed in action-angle variables[10] : the *measures* $\mathcal{D}\mu$ are evaluated under constraints derived from (1). Since (1) is used, results are the same for field theories by either mode of quantisation, and the physical output, the free energy $F = -\beta^{-1} \ln Z$, β^{-1} = temperature, is the same. There are equivalent fermion *and* boson descriptions for F.[10] This fermi-bose equivalence is apparently vested in two equivalent representations of the algebra $g\ell(\infty)$[11]. We suggest that the q-deformed algebra $g\ell_q(\infty)$ (a quantum group) may be more correctly involved; but the representation theory of $g\ell_q(\infty)$ is not developed (and that of $g\ell(\infty)$ is (apparently) itself still very incomplete).

The equivalence of q-quantisation and canonical quantisation exemplified by the partition function Z for the bose gas is apparently generic for quantum integrable field theories in 1 + 1. The situation can be different for discrete systems with a finite number of degrees of freedom : thus we look[3] to cavity q.e.d. to find experimentally realisable systems where the two quantisations lead to different predictions. *Microwave* cavity q.e.d. can involve single, discrete, electromagnetic field modes. For these we introduce q-bosons instead of ordinary bosons and

"SOLITONS"

U, V and Ψ are $N \times N$ Matrices $\left\{ \begin{array}{c} sl(N,C) \\ gl(N,C) \end{array} \right\} = g \ni U, V$

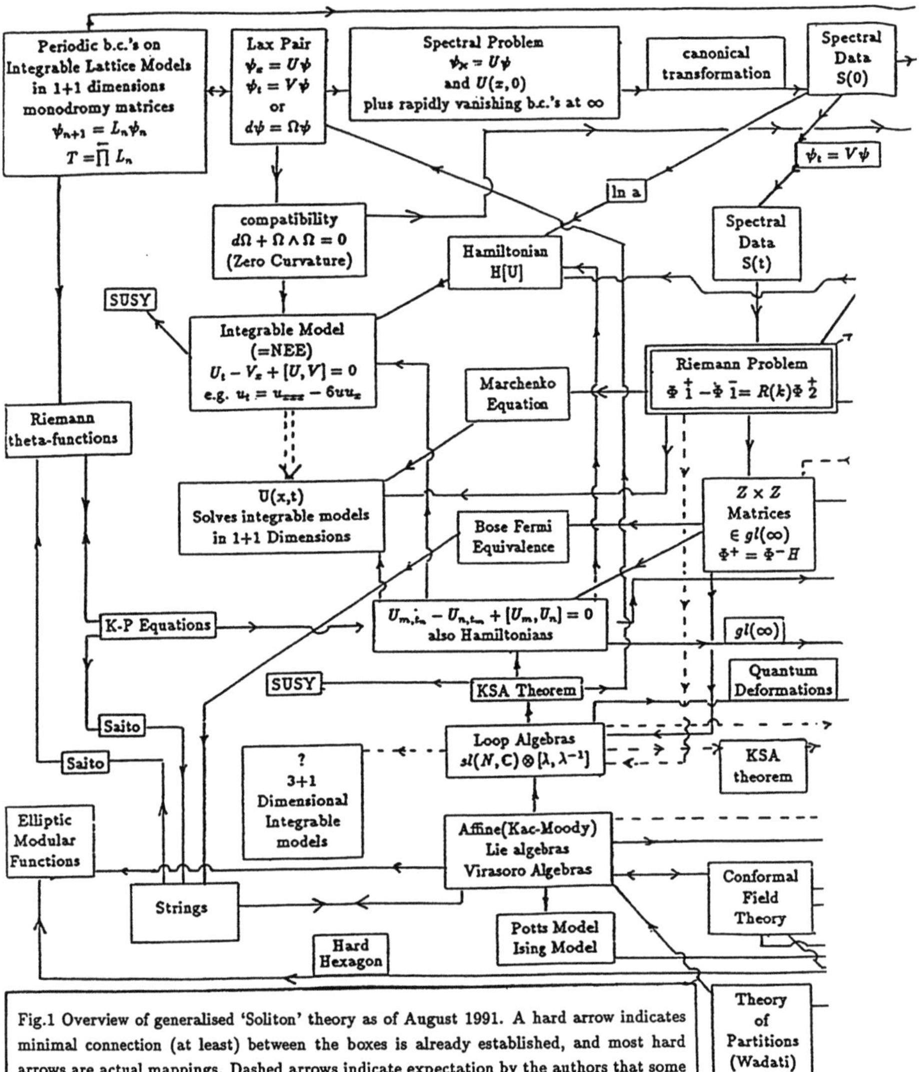

Fig.1 Overview of generalised 'Soliton' theory as of August 1991. A hard arrow indicates minimal connection (at least) between the boxes is already established, and most hard arrows are actual mappings. Dashed arrows indicate expectation by the authors that some such minimal connection can be achieved or stronger. New features compared with Ref.4 include the connections to the "Hopf Algebras" and the experimental output through q-bosons to the microwave cavity q.e.d.; also the 2D-Quantum Gravity.

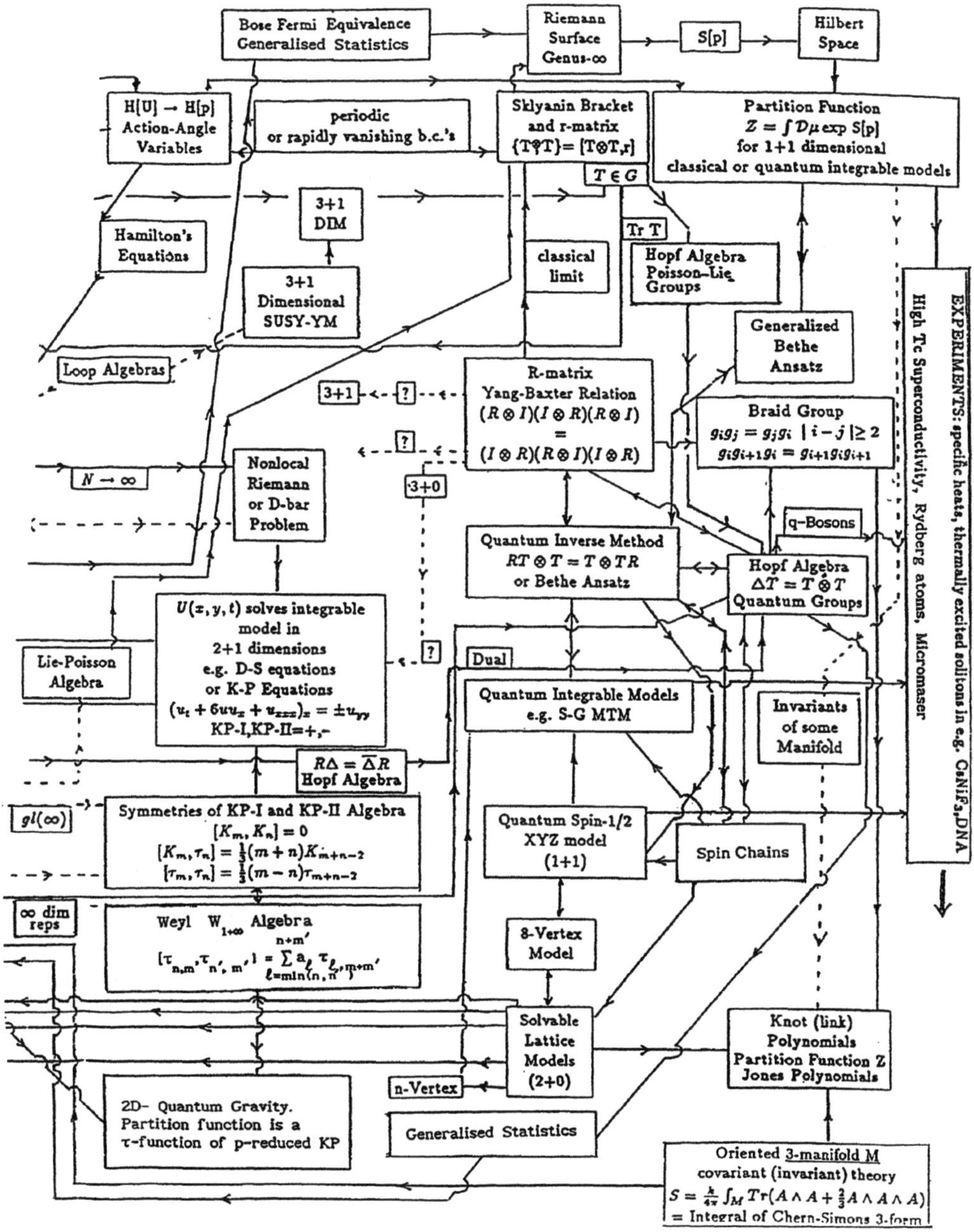

Bose Fermi Equivalence
Generalised Statistics
Riemann Surface Genus-∞
S[p]
Hilbert Space
H[U] → H[p]
Action-Angle Variables
periodic or rapidly vanishing b.c.'s
Sklyanin Bracket and r-matrix
Partition Function
for 1+1 dimensional classical or quantum integrable models
3+1 DIM
Hamilton's Equations
Tr T
classical limit
Hopf Algebra Poisson-Lie Groups
3+1 Dimensional SUSY-YM
Loop Algebras
Generalized Bethe Ansatz
3+1
?
R-matrix Yang-Baxter Relation
Braid Group
N → ∞
Nonlocal Riemann or D-bar Problem
3+0
q-Bosons
Quantum Inverse Method
or Bethe Ansatz
Hopf Algebra
Quantum Groups
solves integrable model in 2+1 dimensions
e.g. D-S equations
or K-P Equations
KP-I,KP-II=+,-
Lie-Poisson Algebra
Dual
Quantum Integrable Models
e.g. S-G MTM
Invariants of some Manifold
Hopf Algebra
Symmetries of KP-I and KP-II Algebra
Quantum Spin-1/2
XYZ model
(1+1)
Spin Chains
∞ dim reps
Weyl Algebra
8-Vertex Model
Solvable Lattice Models
(2+0)
Knot (link) Polynomials
Partition Function Z
Jones Polynomials
n-Vertex
2D- Quantum Gravity. Partition function is a τ-function of p-reduced KP
Generalised Statistics
Oriented 3-manifold M
covariant (invariant) theory
= Integral of Chern-Simons 3-form
EXPERIMENTS: specific heats, thermally excited solitons in e.g. CsNiF3,DNA
High Tc Superconductivity, Rydberg atoms, Micromaser

develop a q-boson cavity q.e.d. from this : some of the observable consequences are reported below.

For a single mode the q-boson algebra has the three elements $a, a^\dagger, N = N^\dagger$ satisfying[3,12,13]

$$[a^\dagger, N] = -a^\dagger, \quad [a, N] = a, \quad a\,a^\dagger - q\,a^\dagger a = q^{-N}. \tag{4}$$

This reduces to the Heisenberg-Weyl algebra when the deformation parameter $q(= e^{\hbar\gamma}, \gamma \in \mathbb{C}$ and $\hbar = 1) \to 1$(i.e. $\gamma \to 0$). One finds (4) by appropriate generalisations of the Primakov-Holstein or Jordan-Schwinger transformation on the quantum group $su_q(2)$[12], a q-deformation of the universal enveloping algebra of su(2)[12]. Our generalisation of the Primakov-Holstein tranformation is

$$S^+ = [\alpha - N]^{\frac{1}{2}} a, \quad S^- = a^\dagger [\alpha - N]^{\frac{1}{2}}, \quad S^z = \tfrac{1}{2}\alpha - N \tag{5a}$$

for $\alpha \in \mathbb{C}$: $S^\pm, S^z$ satisfy $su_q(2)$, namely

$$[S^\pm, S^z] = \mp S^\pm, \quad [S^+, S^-] = [2S^z] \equiv (q^{2S^z} - q^{-2S^z})/(q - q^{-1}) \tag{5b}$$

if, and only if, $a, a^\dagger, N$ satisfy (4).

Deformation (by q) is distinct from q-quantisation. From

$$\{a, a^*\} = i \;, \quad \{N, a\} = -\,ia, \quad \{N, a^*\} = ia^* \tag{6a}$$

we derive by q-deformation

$$\{a, a^*\} = i \cosh \gamma N, \quad \{N, a\} = -\,ia, \quad \{N, a^*\} = ia^*, \tag{6b}$$

with $a\,a^* = (\sinh \gamma N)/\gamma$ and $\{\ldots\}$ a Poisson bracket. Thus $\{a, a^*\} = i\sqrt{1 + \gamma^2 (a\,a^*)^2}$, a *classical* nonlinear oscillator. We use the semiclassical limit of (5a) to construct $S^+(S^-) = \sqrt{\sinh \gamma(\alpha - N)/\gamma}\; a(a^*), S^z = \frac{1}{2}\alpha - N (\alpha \in \mathbb{C})$. This forms the 'Poisson-Lie group'[2], $\{S^-, S^+\} = -i(\sinh 2\gamma S^z)/\gamma$, $\{S^z, S^\pm\} = \pm i\, S^\pm$, the semiclassical limit of $su_q(2)$, which is a *commutative* and co-commutative Hopf algebra. This can now be quantised in the fashion of Ref.8 to the *non* commutative Hopf algebra $su_q(2)$ providing $a \to a$, $a^* \to a^\dagger$, $N \to N$ satisfying (4) with $q = e^{i\hbar\gamma}$. Moreover the classical bosons (6a) can be q-quantised directly to the q-bosons (5). In this picture q-quantisation of (6a) yields the q-bosons; canonical quantisation of (6a) yields ordinary bosons.

The q-bosons have a complete set of number states $|n\rangle$ satisfying $N|n\rangle = n|n\rangle$, $n \in Z_+$, and $a|0\rangle = 0$. Consequently $|n\rangle = ([n]!)^{-\frac{1}{2}}(a^\dagger)^n|0\rangle$, where $[n]! \equiv [1].[2]. \ldots .[n]$ and $[n] \equiv (q^n - q^{-n})/(q - q^{-1})$. On this Fock space $a^\dagger a = (q^N - q^{-N})/(q - q^{-1}) \equiv [N]$. Cavity q.e.d. can now be based on (4) and (5) instead of ordinary bosons, and experiment must confirm or otherwise the alternative choices.

Quantum integrable lattices also provide discrete integrable systems. Particular representations of the quantum group formed by the elements of the $\hat{T}(\lambda)$, $\lambda \in \mathbb{C}$, are determined by choosing the $\hat{L}_n(\lambda)$, equation (3). For our new quantum integrable lattice model we choose

$$\hat{L}_n(\lambda) = \begin{bmatrix} [i\lambda + \tfrac{1}{2}\alpha_n - N_n] & -i\, a_n^{\dagger}\, [N_n - \alpha_n]^{\frac{1}{2}} \\ i[N_n - \alpha_n]^{\frac{1}{2}}\, a_n & [i\lambda - \tfrac{1}{2}\alpha_n + N_n] \end{bmatrix} \tag{7}$$

in which α_n is a complex number depending on the lattice site n, and a_n, $a_n^{\dagger}$, N_n satisfy the q-boson algebra (5) at each lattice site, operators all commuting at different sites. This $\hat{L}_n(\lambda)$ is evidently a q-deformed form of $\hat{L}_n(\lambda)$ for the lattice NLS(= LNLS or lattice bose gas model) of Ref.15. Likewise the R-matrix is the q-deformed R-matrix of the LNLS model (Ref.15 equation (1)) in which $f(\mu,\lambda) = [i(\lambda-\mu) + 1]\ g(\mu,\lambda)$; $g(\mu,\lambda) = [i(\lambda-\mu)]^{-1}$. The parameter $q = e^{i\gamma}$ and $[x] = \sinh \gamma x/\sinh \gamma$ in the 'box' notation used for the q-bosons. When $q \to 1$ $(\gamma \to 0)$ $f(\mu,\lambda)$ and $g(\mu,\lambda)$ become these quantities for the LNLS and NLS models. We therefore call the new quantum lattice model the q-deformed LNLS (or "q-boson" lattice) model.

We note that by the appropriate form of the Primakov-Holstein transformation on (7) we reach $-\sigma^3\ \hat{L}_n^{XXZ}(\lambda)$ (σ^3 is the Pauli matrix) for the XXZ model of arbitrary spin. The elements instead of (7) are now $-[i\lambda + S_n^z]$, $-\ S_n^-$, S_n^+, $[i\lambda - S_n^z]$ and $S_n^{\pm}$, S_n^z form a representation of $su_q(2)$. The R-matrix is (in terms of γ) exactly that for the XXZ model of arbitrary spin.[14]

We find also that there is a *classical* integrable nonlinear lattice with the Poisson-Lie group structure which corresponds to (7). From this classical lattice the q-boson lattice model is reached by the deformation just described in which the commutative Hopf algebra is quantised by $q = e^{\gamma h}$ to the non- commutative and non-co-commutative algebra. Results for the classical lattice model are still be to be worked out. However, we have completely solved the quantum q-boson lattice model using the method of algebraic Bethe ansatz (QIM)[12,13]. Results, which are substantially different from the canonically quantised LNLS model, are reported below. We have shown that a continuum limit of this q-quantised model is the ordinary bose gas model. This is also a continuum limit of the canonically quantised LNLS model[15]. Thus these results illustrate the difference between discrete and continuum models concerning q-quantisation and canonical quantisation. The R-matrix of the q-bose gas model is also that of the quantum lattice sine-Gordon models.[15] We therefore expect an alternative continuum limit which is the quantum s-G model: this is being investigated. Finally we note that the XXZ models of arbitrary spin are *not* q-quantised by $su_q(2)$: apparently γ acts only as the coupling constant of the XXZ models. This illustrates the intricate relationship between the *physics* of quantum groups (q-quantisation) and their *mathematics.*

We turn to the cavity q.e.d. We find for Hamiltonian $\hat{H} = \hbar\omega N$ for q-bosons in equilibrium with a heat bath at temperature β^{-1} that $\bar{N} \equiv \langle N\rangle = \sum_n \langle n|Ne^{-\beta\hbar\omega N}|n\rangle = (e^{\beta\hbar\omega}-1)^{-1}$, the Planck function for ordinary bosons. But, although $\bar{N}e^{\beta\hbar\omega} = \bar{N} + 1$ and $\langle a^\dagger a\rangle\, e^{\beta\hbar\omega} = \langle a\, a^\dagger\rangle$, $\langle a^\dagger a\rangle = (e^{\beta\hbar\omega}-1)/(e^{\beta\hbar\omega}-q)(e^{\beta\hbar\omega}-q^{-1})$ and has a maximum at $\beta = \beta_0 = \omega^{-1}\,\ell n(1 + 2(1-\cos\gamma\hbar)) > 0$ for $q = e^{i\gamma\hbar}(\gamma>0)$ (For $q = e^{\gamma\hbar}$ there is a singularity at $\beta = \gamma\,\omega^{-1}$). If $\langle a^\dagger a\rangle$ can be measured it will depend on $q \neq 1$.

We have derived the master equation for the density operator ρ_a for $N_A \geq 1$ Rydberg (2-level) atoms in a heavily damped ("low Q", Q is the cavity quality factor[3,16]) microwave cavity at finite β^{-1} assuming the resonant cavity mode is occupied by q-bosons. The cavity is resonant with the atomic frequencies ω. N_A such 2-level atoms form an $N_A + 1$ representation of $su_q(2)$ if q-bosons are involved: this is the *q-deformed* Dicke model[3,4,16]. We find that the mean number $N_+ = \frac{1}{2}N_A + \mathrm{Tr}(\rho_a S^z)$ of atoms in their upper states in thermal equilibrium is given by

$$N_+ = X\left\{1 - (N_A+1)X^{N_A} + N_A X^{N_A+1}\right\}\left\{1-X-X^{N_A+1} + X^{N_A+2}\right\}^{-1} \qquad (8)$$

in which $X = \langle a^\dagger a\rangle_{th}/\langle a\, a^\dagger\rangle_{th} = e^{-\beta\hbar\omega}$: $\langle\ \rangle_{th}$ is the thermal average of the q-boson occupations of the cavity mode. When $N_A = 1$, $N_+ = \langle a^\dagger a\rangle_{th}/\langle a\, a^\dagger + a^\dagger a\rangle_{th}$, but when $N_A \to \infty$, $N_+ = \langle a^\dagger a\rangle_{th}/\langle a\, a^\dagger - a^\dagger a\rangle_{th}$. For ordinary thermal bosons in the cavity the corresponding results are $N_+ = \bar{n}_{th}/(1+2\bar{n}_{th})$ when $N_A = 1$ and $N_+ = \bar{n}_{th}$ (simply) when $N_A \to \infty$[4,16]. But it is obvious that *in both cases* these results, as well as the formula (8), are identical and have $\bar{N}_{th} = \bar{n}_{th} = (e^{\beta\hbar\omega}-1)^{-1}$, the Planck function. Thus the asymptote $\bar{N}_{th} = \bar{n}_{th}$ which has been observed[17] does *not* distinguish q-bosons from ordinary bosons. However the experiment[17] is also an observation of bose-fermi equivalence (for ordinary bosons) or of q-boson-q-fermion equivalence (for q-bosons)[4]. For $N_A = 1$ the single atom is one fermion (or one q-fermion). Or it is equivalently one boson (or one q-boson). For $N_A \to \infty$ the system is one giant boson (one giant q-boson). The experiment of Ref. 17 does not distinguish q-bosons from ordinary bosons but the giant oscillator is q-deformed in the q-boson case : if[3] $S^+ \to \bar{S}^- = [2S]^{-\frac{1}{2}} S^+ q^{\frac{1}{2}N}$, $S^- \to \bar{S}^+ = [2S]^{-\frac{1}{2}} q^{\frac{1}{2}N} S^-$, then

$$\bar{S}^+\bar{S}^- - q\,\bar{S}^-\bar{S}^+ = q^{-N} \qquad (9)$$

and $\langle \bar{S}^+ \, \bar{S}^- \rangle = (e^{\beta\hbar\omega}-1)/(e^{\beta\hbar\omega}-q)(e^{\beta\hbar\omega}-q^{-1})$. (This is the usual group contraction now extended to the q-bosons).

Observational evidence for the q-bosons could be found in the super-radiant emission which we are calculating for a comparison with this in Ref.16 and as observed[16]. It could also be found in the micromaser[3]. The micromaser is essentially a quantum integrable system with two degrees of freedom[3,4], but it is 'pumped' by an input of inverted resonant Rydberg atoms. We have solved for the energy eigenstates and eigenvalues of the q-quantised model of one atom and one cavity field mode[3]. The energy eigenspectrum is q-dependent and this could be observed in an observation of the fluorescence spectrum[3]. We have calculated the quantum revivals of the atomic inversion in the micromaser[3]. In principle the actual observations[3] of these revivals puts a bound on $q = e^{\gamma\hbar}$ (i.e. on γ) if the bosons are q-quantised. This bound is being investigated.

The main results for the new q-boson lattice model are:- the Hamiltonian $\hat{H}$ is local involving direct interactions between sets of four lattice sites; the energy spectrum $h(\lambda)$ has both an acoustic *and* an optical branch with gap $\Delta_0 > 0$; the optical branch has quantum breather-like bound states involving strings of a finite length determined by γ. In the appropriate continuum limit the spectrum becomes the spectrum $h(\lambda) = \lambda^2$ of the bose gas : this has the single acoustic branch and no bound states. Likewise $\hat{H}$ becomes that for the bose gas. We have also used conformal field theory to calculate the long distance asymptotics of the correlation functions. These coincide in form with the corresponding asymptotics for the LNLS model[18] e.g. $\langle a_n \, a_1^\dagger \rangle \sim A \; n^{-1/\Theta}$, and this is because these asymptotics do not distinguish continuum and lattice models. But the critical exponent Θ is now q-deformed[12,18]. Connections to the corresponding correlation functions for the XXZ and lattice s-G and quantum s-G models have still to be worked out.

References

1. R.K.Bullough *et al.*, NATO, ASI Series B, 1990, Vol **245**, Plenum Press, New York.

2. Quantum Groups, ed. H. Doebner and J. Henning, Springer Lect.Notes in Phys. 1990, **370**, Springer-Verlag, Berlin.

3. R.K.Bullough *et al.*, Proc. NATO Adv. Res. Workshop "Quantum Measurements in Optics", 1991, Plenum, New York. In press.

4. R.K.Bullough and J.Timonen in "Diff. Geom. Methods in Theor. Phys" eds. C. Bartocci, U. Bruzzo and R. Cianci, Springer Lect. Notes in Phys., 1991, **375**, Springer-Verlag, Berlin, p.69.

5. Y. Kodama, "A Generalized Sato's Equation of the KP Theory and Weyl Algebra", This volume.

6. I. Krichever, "Algebraic Geometrical Perturbation Theory of Integrable Equations", This meeting.

7. V.E. Drinfeld, Proc. Intl. Congr. Math. Berkely, 1986, p. 798.

8. L.A. Takhtajan, p. 3 of Ref. 2.

9. V.E. Korepin, "Differential Equations for Quantum Correlation Functions", This meeting.

10. R.K. Bullough, Y-z. Chen and J. Timonen in Proc. IV Intl. Workshop (Kiev IV) "Nonlinear World", eds. Bar'yakhtar *et al* ., 1990, **2**, World Scientific, Singapore, p. 1377.

11. M. Jimbo and T. Miwa in "Integrable Systems in Stat. Mechanics" eds. G.M. D'Ariano, A. Montorsi and M.G. Rasetti, 1985, World Scientific, Singapore, p. 65, especially p. 113.

12. R.K. Bullough and N.M. Bogoliubov, Proc. XXth Intl. Meeting on Diff. Geom. Methods in Theor. Phys., Baruch College, NYC, June 1-7, 1991, eds. S. Catto and A. Rocha, 1992, World Scientific, Singapore and references.

13. N.M. Bogoliubov and R.K. Bullough. To be published (and references).

14. A.N. Kirillov and N.Yu. Reshetikhin, J.Phys A:Math.Gen., 1987, **20**, 1565.

15. A.G. Izergin and V.E. Korepin, Nucl. Phys. B, 1982, **205** [FS 5], 401.

16. G.P. Hildred *et al.*, J. Phys. B: At. Mol. Phys., 1984, **17**, L535-539.

17. J.M. Raimond *et al.*, Phys. Rev. Lett., 1982, **49**, 117.

18. N.M. Bogoliubov *et al.*, J. Phys. A: Math. Gen., 1987, **20**, 5361.

Perturbation Method and Optical Solitons

Y. Kodama

Department of Mathematics, Ohio State University,
Columbus, OH 43210, USA

We develop a perturbation method based on an infinite dimensional extension of the Lie transform of the vector fields, and discuss several applications to optical soliton problems. The applications include soliton propagation under rapidly oscillating perturbations [1], and higher order effects on multi-soliton interactions [2].

1. A. Haswgawa and Y. Kodama: Opt. Lett. 15(1990) 1443; Phys.Rev. Lett. 66 (1991) 161.
2. Y. Kodama and D. Muraki, "Dynamical Interaction of Optical Solitons" (in preparation).

Additional references on this topic

3. Y. Kodama and A. Hasegawa, "Theoretical Foundation of Optical-Soliton Concept" appear in "Progress in Optics" ed. by E. Wolf (Elsevier Science Publishers B.V. 1992) Chapter IX.
4. Y. Kodama and A. Hasegawa, "Generation of Asymptotically Stable Optical Solitons and Suppression of he Gordon-Haus Effect" to appear in Opt. Lett. (1992, Jan. 1 issue).
5. Y. Kodama and S. Wabnitz, Elect. Lett. 27 (1991) 1931.
6. Y. Kodama and A. Hasegawa, IEEE J.Q.E. QE-23 (1987) 510.
7. M. Nakazawa et.al, IEEE J.Q.E. QE-26 (1990) 2095.
8. L. F. Mollenauer et. al, J. Lightwave Tech. LT-9 (1991) 194.
9. L. F. Mollenauer et. al, Opt. Lett. 15 (1990) 1203.

Birefringent Optical Fibers: Modulational Instability in a Near-Integrable System

D.J. Muraki, O.C. Wright, and D.W. McLaughlin[1]

Princeton University, Program in Applied and Computational Mathematics, Princeton, NJ 08544, USA

Introduction

The propagation of light pulses in a nonlinear optical fiber operated near the zero-dispersion point is governed by the nonlinear Schrödinger (NLS) equation. For fibers having *normal* dispersion, the NLS equation is of de-focussing type so that the simplest exponential plane waves are stable.

In nearly-axisymmetric fibers, there are two co-propagating modes, or *polarizations* — both governed by NLS-type equations which are nonlinearly coupled. When there is a slight asymmetry in the linear propagation characteristics of these two polarized modes (*birefringence*), the possibility arises for a modulational instability of the plane waves due to the cross-coupling. *This instability occurs even when the underlying NLS equation is de-focussing.*

The observation of this instability has been recently reported in laboratory experiments using this phenomenon for the generation of optical oscillations at terahertz frequencies [1].

The following outlines some highlights of work in progress which (we hope) will eventually lead to a modulation theory for this instability based upon a treatment of the birefringence as a perturbation to an exactly-integrable vector NLS system.

Basic Equations

The equations describing co-propagation of light pulses in a nonlinear fiber with group velocity birefringence are

$$\begin{aligned} i\,(u_t+\delta u_z) &+ \tfrac{1}{2}u_{zz} &- (|u|^2+B|v|^2)u &= 0 \\ i\,(v_t-\delta v_z) &+ \tfrac{1}{2}v_{zz} &- (B|u|^2+|v|^2)v &= 0 \end{aligned} \qquad (1)$$

where u, v denote orthogonal linear polarizations of the electric field components; and t, z are the time & (longitudinal) space coordinates.

The asymmetric derivative coefficient δ generates a difference in the linear group velocities of the two fields and represents the (strong) birefringence.

The nonlinear coupling coefficient B represents the difference in the self- and cross-phase modulations. The usual Kerr nonlinearity for glass requires $B = 2/3$,

[1]funded by AFOSR-90-0161 and NSF DMS 8922717 A01

however, *when $B = 1$ the system is exactly-integrable via the inverse scattering transform* [2]. Thus, by perturbing B from unity to the natural Kerr value, the equations for the birefringent fiber can be considered as a near-integrable system.

More detailed explanations of the birefringent fiber system in terms of their asymptotic reduction from the Maxwell equations and physical relevance appear in the recent books of Hasegawa [3] and Agrawal [4]. In addition, some of the optical ramifications of the dual polarized fiber were discussed by Wai in a previous presentation [5].

Plane Wave Instability

Upon inspection, one exact solution to these coupled PDEs has the form of an exponential plane wave

$$u = a\, e^{i(k_u z - \omega_u t)} \qquad \text{and} \qquad v = b\, e^{i(k_v z - \omega_v t)} \tag{2}$$

where the frequencies ω are required to satisfy the amplitude-dependent dispersion relations

$$\begin{aligned} \omega_u &= \delta k_u + \tfrac{1}{2}k_u^2 + |a|^2 + \tfrac{2}{3}|b|^2 \\ \omega_v &= -\delta k_v + \tfrac{1}{2}k_v^2 + \tfrac{2}{3}|a|^2 + |b|^2. \end{aligned} \tag{3}$$

It is assumed that the amplitudes a, b of the fields are constant, thereby defining a wave having fixed *polarization* — that is, a wave whose division of intensity between the two components u & v is uniform in both time and space.

Earlier works have documented the observation that a linearized stability analysis determines these plane waves to be unstable to low wavenumber fourier disturbances even in the normal dispersion regime where the usual Benjamin-Feir instability is absent [6, 7].

Modulation Equations

Alternatively, the above modulational instability can be illustrated via a nonlinear WKB approach [8, 9] similar to that used by Levermore [10] in his earlier discussion of the de-focussing NLS. A modulating plane wave is assumed which takes the more general exponential form

$$u = a\; exp\{\int^z \alpha\} \qquad \text{and} \qquad v = b\; exp\{\int^z \beta\} \tag{4}$$

where the parameters a, α, b, β are all allowed to be slowly-varying in z. Assembling these parameters into the vector

$$\vec{v} = (|a|^2, |b|^2, \alpha, \beta)^T, \tag{5}$$

the slow variations in time & space are obtained by direct substitution into the birefringent PDEs (1).

For this nonlinear problem, quite unlike the linear WKB, the usual eikonal and leading order transport equations do not decouple from each other and a set of *modulation equations* for the slow-scale variations is obtained which is most conveniently written in the quasi-linear form

$$\vec{v}_t + \begin{bmatrix} \delta+\alpha & 0 & |a|^2 & 0 \\ 0 & -\delta+\beta & 0 & |b|^2 \\ 1 & B & \delta+\alpha & 0 \\ B & 1 & 0 & -\delta+\beta \end{bmatrix} \vec{v}_z = \vec{0}. \tag{6}$$

With initial data specified for all z at $t = 0$, this system is well-posed only as long as it remains strictly hyperbolic — or equivalently, as long as the Riemann eigenspeeds of the matrix are real.

Direct calculation of the eigenvalues of the matrix coefficient demonstrates that for certain values of the parameters in $\vec{v}$ there can be complex eigenspeeds so that the above system can become elliptic. *The existence of a region in z-t space for which the modulation equations (6) lose hyperbolicity is indicative of instability to a more complex wave form.*

Generally, ellipticity of the modulation equations is indicative of a failure in the basic assumptions of the nonlinear WKB solution ansatz. In linear WKB, the breakdown of the exponential solution at turning points is resolved through the introduction of the Airy function and its higher-order cousins. In the nonlinear case, the determination of an analogous hierarchy of waveforms which resolve the ellipticity motivates the study of the nearby integrable limit.

It is important to note that for sufficiently small amplitudes, the modulation equations above can be shown to remain hyperbolic — this differs considerably from the focussing NLS for which there is always a modulational instability.

Integrable Theory

When the coefficients $B = 1$ and $\delta = 0$, the integrability of the coupled equations (1) is obtained via the Lax pair representation

$$\vec{\psi}_z = \mathbf{L}\,\vec{\psi} \qquad \text{and} \qquad \vec{\psi}_t = \mathbf{B}\,\vec{\psi}. \tag{7}$$

The operators $\mathbf{L}, \mathbf{B}$ contain an eigenvalue E and are defined as

$$\mathbf{L} = iE\,\mathbf{D} + \mathbf{N} \qquad \text{and} \qquad \mathbf{B} = iE\,\mathbf{L} + \frac{1}{2}\mathbf{D}(\mathbf{N}_z + \mathbf{N}^2) \tag{8}$$

based on a constant diagonal matrix $\mathbf{D}$ and a scattering matrix $\mathbf{N} = \mathbf{N}(z,t)$ whose explicit forms are given by

$$\mathbf{D} = \begin{bmatrix} -1 & 0 & 0 \\ 0 & 1 & 0 \\ 0 & 0 & 1 \end{bmatrix} \qquad \text{and} \qquad \mathbf{N} = \begin{bmatrix} 0 & u & v \\ u^* & 0 & 0 \\ v^* & 0 & 0 \end{bmatrix} \tag{9}$$

where the superscript * denotes complex conjugation. The compatibility of the mixed partial derivatives in z & t requires that the scattering matrix $\mathbf{N}$ satisfies the PDE

$$i\mathbf{N}_t - \frac{1}{2}\mathbf{D}\mathbf{N}_{zz} + \mathbf{D}\mathbf{N}^3 = 0, \tag{10}$$

which is the (integrable) vector NLS equation for u & v (recall, $B = 1$ and $\delta = 0$).

Quasi-Periodic Theory

To generate quasi-periodic solutions, it proves more convenient to reformulate the Lax pair by utilizing a *squared eigenfunction* approach similar to those found for the scalar NLS and sine-Gordon equations [11]. In the case of the vector NLS equation, an analogous construction begins with the left eigenmatrices and negative right eigenmatrices for the matrix operators

$$\Psi_z = \mathbf{L}\,\Psi \qquad \text{and} \qquad \Phi_z = -\Phi\,\mathbf{L} \tag{11}$$

(likewise for the temporal operator $\mathbf{B}$) from which the simple matrix product $\mathbf{Q} \equiv \Psi\Phi$ results in an equivalent *commutator Lax pair*

$$\mathbf{Q}_z = [\mathbf{L}, \mathbf{Q}] \qquad \text{and} \qquad \mathbf{Q}_t = [\mathbf{B}, \mathbf{Q}] \tag{12}$$

where the vector NLS equation is again obtained through the requisite compatibility of the mixed partials.

Just as the special soliton solutions of the focussing NLS are associated with the reflectionless eigensolutions, *the fundamental solutions of the quasi-periodic NLS are generated through special commutator* $\mathbf{Q}$ *whose representations are polynomials in the scattering eigenvalue* E — these are more commonly referred to as *N-phase waves.* For the scalar NLS, the 0-phase waves are simply the exponential plane waves.

Our present working hypothesis is that the N-phase waves for the coupled NLS equation provide the natural resolution for the breakdown of the nonlinear WKB analysis of the birefringent fiber system.

Future Thinking

The aim of current efforts is to catalog the lowest orders of N-phase waves. At present we have established the existence of quasi-periodic solutions whose spatial structure do not have uniform polarization like the plane waves (2), but display a time-dependent *rippling* of wave intensity between field components.

Determined via the solution of an integro-differential hamiltonian system, these waves are suspected to be of the 1-phase variety and seem to be a likely candidate for the primary mode of the modulational instability. They also bear a strong resemblence to the oscillatory instability observed in the optical experiments.

References

[1] J.E. Rothenberg, Opt Lett **16**, 18 (1991)

[2] S.V. Manakov, ZETP **65**, 505 (1973)

[3] A. Hasegawa, Optical Solitons in Fibers, Springer-Verlag (1989)

[4] G.P. Agrawal, Nonlinear Fiber Optics, Academic Press (1989)

[5] A. Wai, this proceedings

[6] G.P. Agrawal, Phys Rev Lett **59**, 880 (1987)

[7] J.E. Rothenberg, Phys Rev A **42**, 682 (1990)

[8] R.M. Miura & M.D. Kruskal, SIAM J Appl Math **26**, 376 (1974)

[9] G.B. Whitham, Proc Roy Soc A **283**, 238 (1965)

[10] C.D. Levermore, this proceedings

[11] M.G. Forest & D.W. McLaughlin, J Math Phys **23**, 1248 (1982)

Rigorous Results in Quantum Theory of Stimulated Raman Scattering

V.I. Rupasov

Institute of Spectroscopy, USSR Academy of Sciences,
142092 Troitsk, Moscow Region, Russia

1. Introduction. The modern theory of stimulated Raman scattering (SRS) of light in resonant media is based on the investigations of appropriate integrable models of the classical field theory[1–3] by means of the inverse problem method. But, strictly speaking, Raman scattering is a pure spontaneous process and, hence, it is necessary to take into account a quantum nature of the phenomenon. Moreover, there are some questions and problems, for example, the problem of scattered photons statistics, which can be studied only within the framework of the quantum field theory.

We have developed an exact quantum theory of SRS for the case of point-like geometry of resonant media (two-level atoms or harmonic oscillators) of the radius $r \ll \lambda_0$, where λ_0 is the typical wavelength of the light, but all our results are also valid for the case of short extended medium of the length $L \ll l_p$ (l_p is the typical size of pulses) when the spatially homogeneous approximation is valid.

The Hamiltonian of an effectively one-dimensional (within the framework of dipole resonance approximation) quantum model of the SRS problem can be written as

$$H = -\int_{-\infty}^{\infty} \mathrm{d}x[i\epsilon_\sigma^+(x)\partial_x\epsilon_\sigma(x) + J\delta(x)\epsilon_\sigma^+(x)(\sigma_{\sigma\mu}^+R^- + \sigma_{\sigma\mu}^-R^+)\epsilon_\mu(x)] \tag{1}$$

Here we combine the operators of 'slow' envelopes of dipole components of laser (E_L) and Stokes (E_S) fields into an isotopic spinor

$$\epsilon_\sigma = \begin{pmatrix} E_L \\ E_S \end{pmatrix} \tag{2}$$

with commutation relations

$$[\epsilon_\sigma(x), \epsilon_\mu^+(y)] = \delta_{\sigma\mu}\delta(x-y), \tag{3}$$

where isospin variable σ takes on two possible values: $\sigma = L, S$; J is a coupling constant and σ^+, σ^- are the Pauli matrices. First term of the Hamiltonian describes a free propagation of the fields and second term corresponds to an interaction of the fields with a point-like resonance medium which is described in the case of two-level atom system by a spin operator

$$\mathbf{R} = \sum_{a=1}^{a=M} \mathbf{r}_a = (R^x, R^y, R^z), \qquad R^\pm = R^x \pm iR^y \tag{4}$$

with commutation relations

$$[R^i, R^j] = e^{ijk}R^k. \tag{5}$$

In the case when resonant medium can be described as a system of harmonic damping oscillators the Hamiltonian of the problem is given by

$$H = -\int_{-\infty}^{\infty} dx[\quad i\epsilon_\sigma^+(x)\partial_x\epsilon_\sigma(x) + \delta(x)\epsilon_\sigma^+(x)(\sigma_{\sigma\mu}^+ B + \sigma_{\sigma\mu}^- B^+)\epsilon_\mu(x) + i\gamma B^+ B] \tag{6}$$

where $B^+(B)$is an operator of creation (annihilation) of an excitation in an oscillator system with commutation relations

$$[B, B^+] = M, \tag{7}$$

M is the number of oscillators in a system and γ is a damping constant.

2. One-particle problem. Before we take up the solution of general many-body scattering problem, we will examine the one-particle problem, in which a photon is propagating along the x axis and scattered by m-fold excited system of two-level atoms. The initial state of the field+atoms system is

$$| \text{in} >= [\frac{(M-m)!}{m!M!}]^{1/2} a_k^+ (R^+)^m \mid 0 > \tag{8}$$

where

$$a_k^+ = \int dx \phi(x) e^{ikx} E_L(x)$$

and $\phi(x)$is some arbitrary (normalized) wave function of an incident laser photon. The vacuum state is defined as the state in which there are no particles and all of the atoms are in the ground state:

$$\epsilon_\sigma(x) \mid 0 >= 0, \quad R^- \mid 0 >= 0; \quad R^z \mid 0 >= -(M/2) \mid 0 >$$

The final state is evidently given by

$$| \text{out} >= S \mid \text{in} > \tag{9}$$

where the S-matrix of the particle scattering by atomic system is found by solving a one-particle Schrödinger equation with Hamiltonian (1). The result is

$$S = A + \sigma^- R^+ C \tag{10}$$

$$A = \cos(J\sqrt{R^- R^+}) \tag{11}$$

$$C = \frac{i \sin(J\sqrt{R^- R^+})}{\sqrt{R^- R^+}} \tag{12}$$

where A and C are the operators of the amplitudes of elastic and inelastic scattering of a photon. Here, it is necessary to note that the S-matrix does not depend on a momentum of photon and, hence, the shape of incident laser pulse plays a completely insignificant role.

3. Many-particle problem. To avoid tedious intermediate calculations in the derivation of the physical observables of scattered field, one can transform from a continuous description of the field to a discrete model using the fact that S-matrix is independent on a particle momentum. Then, the final state of the many-particle scattering problem is given by

$$| \text{out} >= (\prod_{j=1}^{j=N} S_{j0}) \mid \text{in} >,$$

where N is the number of incident laser photons. Using the method of recurrent relations for physical observables calculations and taking into account the higher spherical harmonics of the field which do not interact in the dipole approximation with point-like resonant medium we found the following exact expressions for the intensities of laser (unshifted) and Stokes components of scattered light:

$$
\begin{aligned}
I_L(t) &= 4I_D(t)\sin^4\left[g(m(t))/2\right] &(13)\\
I_S(t) &= I_D(t)\sin^2\left[g(m(t))\right] &(14)
\end{aligned}
$$

where I_D is the intensity of dipole component of incident laser field and a function $m(t)$ is a solution of the equation

$$\frac{\mathrm{d}m}{\mathrm{d}t} = I_D(t)\sin^2\left[g(m)\right] \tag{15}$$

with the asymptotic condition $m(t \to -\infty) \to 0$. And a function $g(m)$ is given by

$$g(m) = J\sqrt{m(M-m)} \tag{16}$$

4. Critical phenomena. The nature of the scattering process and the time evolution of the population of an atomic system are determined completely by the quantity

$$G \equiv g_{\max}(M) = JM/2$$

We first consider the case $G \ll 1$. The solution of Eq.(15) in this case is trivial and we obtain the soliton-like solutions for the intensities

$$
\begin{aligned}
I_L(t) &= (I_D/64)(JM)^4\cosh^{-4}\left[(t-t_0)/\tau_0\right] &(17)\\
I_S(t) &= (I_D/4)(JM)^2\cosh^{-2}\left[(t-t_0)/\tau_0\right] &(18)
\end{aligned}
$$

where$t_0 = (\tau_0/2)\ln M$, $\tau_0 = 2/I_D J^2 M$. The intensity of the unshifted component of scattered light is proportional to M^4, but this result applies specifically to a point-like atomic system. In the scattering process the system of atoms becomes completely inverted, i.e. the maximum value of population $\mu \equiv m(t \to +\infty)$ is equal to M.

A qualitative change in the scattering occurs at the value $M = M_{\text{cr}}$, where $G(M_{\text{cr}}) = \pi$: $M_{\text{cr}} = 2\pi/J$. For $M > M_{\text{cr}}$ the atomic system is not inverted completely at any intensity and duration of incident laser pulse. For $M \gg M_{\text{cr}}$ the value μ and, along with it the total number of Stokes photons, N_S, created in the course of scattering, decrease in inverse proportional to the number of atoms, M:$\mu \propto M_{\text{cr}}^2/4M$. At the point $M = M_{\text{cr}}$ the function $\mu(M)$ is discontinuous:

$$\mu(M_{\text{cr}} - 0) - \mu(M_{\text{cr}} + 0) = M/2.$$

Thus, the rigorous analysis of the quantum model allows us to predict the critical phenomena in the dynamical process of Raman scattering of light by a system of two–level atoms.

5. Bose–system. The method of solution described above can be also used in the case when a resonant medium is described within the framework a model of a system of damping harmonic oscillators with the Hamiltonian (6). Repeating all of the calculations we obtain

$$I_S(t) = I_D(t)\exp\left[-2\gamma m(t)\right]\sin^2 J\sqrt{Mm(t)} \tag{19}$$

and the following equation for the function $m(t)$:

$$\frac{dm}{dt} = I_D(t)\exp\left[-2\gamma m(t)\right]\sin^2 J\sqrt{Mm(t)} \tag{20}$$

It is easy to see that a dissipation does not play the crucial role for a creation of Stokes soliton in the course of stimulated Raman scattering, but only can help this process if the value $\gamma > (\pi/J\sqrt{M})^2$. At $\gamma = 0$ the function $m(t)$ changes from zero at $t \to -\infty$ to the value $(\pi/J\sqrt{M})^2$ and the Stokes component of the scattered light has a soliton–like form.

6. Statistical properties. The exact solution of the quantum models of the SRS allow us to investigate the statistical properties of Stokes photons and a role of quantum fluctuations in the course of scattering. For initial period of scattering process the probability distribution function $P(n)$ of a creation of n Stokes photons as a result of scattering of N_L incident laser photons is given by

$$P(n) = \exp(-\xi)[1 - \exp(-\xi)]^n, \tag{21}$$

where $\xi = J^2MN_L$. It is easy to see the fluctuations of the number of Stokes photons are very large at the initial stage of scattering and, hence, the Stokes component intensity should strongly randomly oscillates at the beginning of the pulse.

7. Conclusions. The following main results of the quantum theory of SRS have been obtained:

- The critical phenomena in SRS process when the number of atoms M in atomic system is equal to some critical value $M_{cr} = 2\pi/J$ (where J is the coupling constant) are predicted[4]. When $M < M_{cr}$ the total number of Stokes photons created in the course of scattering N_S is, naturally, proportional to M and an atomic system can be completely inverted. When $M > M_{cr}$ the value of N_S decreases in inverse proportion to M: $N_S \propto M_{cr}^2/M$, and an atomic system can not be completely inverted for any duration and form of an incident laser pulse. At the point $M = M_{cr}$ the function $N_S(M)$ is discontinuous: $N_S(M_{cr} - 0) - N_S(M_{cr} + 0) = M/2$.

- It was shown that in the case of SRS by a system of damping harmonic oscillators the dissipation does not play the crucial role in Stokes solitons creation but it only promotes soliton creation[5]. This contradicts the results of the classical SRS-theory for long extended resonant medium[3].

- The exact solution of the quantum models allows us to study the statistics of scattered photons and the role of quantum fluctuations in SRS process[6]. It was shown that the fluctuations of the number of Stokes photons are very large at the initial stage of scattering and, hence, the Stokes component intensity should strongly randomly oscillates at the beginning of the pulse.

In the case of long extended media, when spatially homogeneous approximation is not valid it was shown[4] that the SRS problem can be formulated and investigated within the framework of the so-called six-vertex model of 2D statistical mechanics[7].
A creation process of anti-Stokes photons can be included[4] into our consideration but this does not change drastically the main results.

References

1.Steudel H. Physica, 1983, 6D, 155.
2.Kaup D. J. Physica, 1983, 6D, 143.
3.Kaup D. J. Physica, 1986, 19D, 125.
4.Rupasov V. I. Sov. Phys. JETP, 1988, 67, 2003.
5.Ogievetsky E. I. and Rupasov V.I., to be published.
6.Rupasov V. I., to be published.
7.Baxter R. J. Exactly Solved Models in Statistical Mechanics, 1982, 'Academic Press', London.

Bäcklund Transformations as Physical Equations

H. Steudel

Central Institute of Optics and Spectroscopy,
W-1000 Berlin, Fed. Rep. of Germany

We give some examples where dynamical equations of certain physical systems take the form of Bäcklund transformations. By this way we discover structural relations between several equations of nonlinear optics.

1 Simple Examples

(i) Stimulated Raman scattering (SRS) under "real amplitude" (r.a.) conditions can be described by

$$\psi_T = \sin\phi, \phi_X = \sin\psi \tag{1}$$

where X is a properly scaled space variable and T is a "nonlinear" retarded time variable. ψ and ϕ are functions describing atoms and field respectively. Now it is easy to recognize (1) as the Bäcklund transformation to sine-Gordon equation. Introducing $\beta_{1,0} = \phi \pm \psi$ in (1) we see that β_1 becomes the Bäcklund tranform of β_0 with the usual Bäcklund parameter fixed as $a = 1$.

(ii) Resonant two-photon propagaion (TPP) or "two-photon SIT" under analogous conditions leads to [1]

$$\psi_T = \sinh\phi, \phi_X = \sin\psi \tag{2}$$

Defining $\Phi = \psi + i\phi$ in (2) we find that both Φ and its conjugate complex Φ^* fulfill the complex sine-Gordon equation and that Φ^* is the Bäcklund tranform of Φ with $a = 1$.

(iii) Degenerate TPP under corresponding assumptions gives [2]

$$\psi_T = \exp\phi, \phi_X = \sin\psi \tag{3}$$

With $\Phi = \phi + i\psi$ from (3) we find that Φ and Φ^* fulfil the complex Liouville equation and that Φ^* is a Bäcklund transform of Φ.

2 The SRS/TPP System

SRS and TPP may be described by [3]

$$\partial_T \mathbf{r}_+ = i(gs_3\mathbf{r}_+ + \mathbf{r}_3 s_+), \mathbf{r}_\pm = \mathbf{r}_1 \pm \mathbf{r}_2$$

$$\partial_x s_+ = i(g\mathbf{r}_3 s_+ + \epsilon s_3 \mathbf{r}_+), s_\pm = s_1 \pm is_2 \tag{4}$$

$$\partial_T \mathbf{r}_3 = -\partial_x s_3 = (\mathbf{r}_+ s_- - \mathbf{r}_- s_+)$$

with ϵ being +1 (SRS) or -1 (TPP) and g is a real constant. $\mathbf{r} = (\mathbf{r}_1, \mathbf{r}_2, \mathbf{r}_3)$ is the Bloch vector $(\mathbf{r}_1^2 + \mathbf{r}_2^2 + \mathbf{r}_3^2 = 1)$ while $s = (s_1, s_2, s_3)$ describes the field and fulfills

$$\epsilon(s_1^2 + s_2^2) + s_3^2 = \frac{1}{2}(1 + \epsilon) \tag{5}$$

In case of SRS we distinguish SRS1 ($g < 1, f$ real) and SRS2 ($g > 1, f$ imaginary). (4) are the integrability conditions of the linear system[3,4]

$$U = U(\lambda, f) = \begin{pmatrix} -1\lambda\mathbf{r}_3 & (if + \lambda)\mathbf{r}_+ \\ \epsilon(if - \lambda)\mathbf{r}_- & i\lambda\mathbf{r}_3 \end{pmatrix} \tag{6}$$

$$V = V(\lambda, f) = \frac{i}{2}\ g\sigma_3 + \frac{1}{2\lambda + g} \begin{pmatrix} -i\epsilon\lambda s_3/2 & \epsilon(if + \lambda)s_+ \\ (if - \lambda)s_- & i\epsilon\lambda s_3/2 \end{pmatrix}$$

with

$$f^2 = (\epsilon - g^2)/4 \tag{7}$$

Introducing the matrix

$$\Lambda = \lambda I - if\sigma_3 \tag{8}$$

and letting the letter W stand for U or for V, it is easy to verify the symmetry relations

$$\mathcal{C} : \sigma_2 W^*(\lambda^*, f)\sigma_2 = W(\lambda, f^*) \tag{9}$$

$$\Lambda W(\lambda, f)\Lambda^{-1} = W(\lambda, -f) \tag{10}$$

3 Two Hierarchies of Dynamical Systems

Let us define the dynamical systems $\sum, \sum_D$ by starting from two linear systems of AKNS type,

$$v_t = Vv, \quad v_x = Uv \tag{11}$$

with

$$V = \begin{pmatrix} -i\zeta & q \\ r & i\zeta \end{pmatrix}, \quad U = \frac{i}{4\zeta}\begin{pmatrix} a & -b \\ -c & -a \end{pmatrix} \tag{12}$$

$$a^2 + bc = \{ \begin{matrix} 1, & \sum \\ 0, & \sum_D \end{matrix} \tag{13}$$

By the restrictions

$$r = -q^*, \quad c + b^*, \quad a = real \tag{14}$$

$\sum$ and $\sum_D$ are reduced to (sharp-line) SIT and second harmonic generation (SHG) respectively. Further restriction to real q, b leads to sine-Gordon and Liouville equations.

4 Kaup's Gauge Transformation and Bäcklund Transformations

Kaup applied a gauge transformation

$$\tilde{u} = Ku, \quad (UV) \rightarrow (\tilde{U}, \tilde{V}) \tag{15}$$

to the system (6) in order achieve an AKNS form, see Kaup[4] for details. In the case of SRS1 the resulting system (U, V) has the same structure as the linear system associated with SIT. Let us define as another transformation

$$u' = \Lambda u, \quad W' = \Lambda W \Lambda^{-1} \tag{16}$$

which according to (10), has the property to change the sign of f. Now we combine both mappings as seen in Fig. 1.

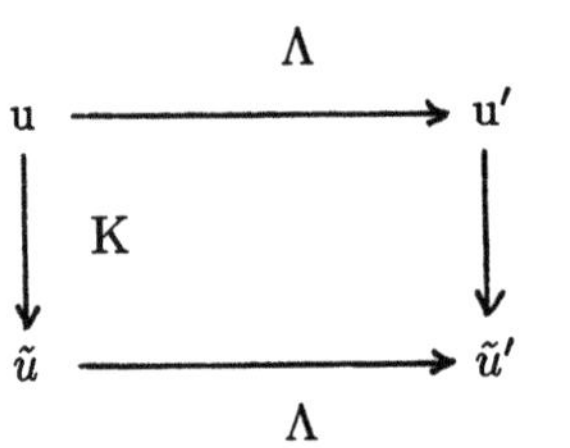

Fig. 1

For the resulting gauge transformation

$$\tilde{u}' = A\tilde{u}, \quad A = K\Lambda K^{-1}, \quad (\tilde{U}, \tilde{V}) \rightarrow (\tilde{U}', \tilde{V}') \tag{17}$$

it holds that

a) A is a linear function of the spectral parameter λ with the coefficient of λ being the unity matrix, and

b) the system $(\tilde{U}, \tilde{V})$ is form invariant.

These properties characterize a BT, see, e.g., Neugebauer and Meinel[5]. By this way our example (i) is generalized. Examples (ii) and (iii) may be generalized in quite an analogous way. With $\mathcal{C}$ denoting the discrete symmetry (9) and C standing for complex conjugation we may summarize our results in a symbolic notation.

$$SRS1 = BT(SIT) \tag{18a}$$

$$SRS1(g = 0, r.a.) = BT(sine - Gordon) \tag{18b}$$

$$SRS2 \ or \ TPP = (BT = \mathcal{C})\sum \tag{19a}$$

$$TPP(g = 0, r.a.) = (BT = C) \ complex \ \ sine - Gordon \tag{19b}$$

$$deg.\ TPP = (BT = \mathcal{C}) \sum_D \tag{20a}$$

$$deg.\ TPP(g = 0, r.a.) = (BT = C)\ complex\ Liouville \tag{20b}$$

The three relations indicated by "b" are specifications of those indicated by "a".

Finally we would like to mention that Kaup's gauge transformation also leads to a space-time symmetry of (sharp-line) SIT which subsequently could be established in a direct way[6].

References

[1] H. Steudel, Exp. Techn. Physik 20 (1972) 409.

[2] H. Steudel, "Resonant Two-Photon Propagation-An Application of the Liouville Equation and of its Auto-Bäcklund Transformation" preprint and communication at the 2nd Soliton Workshop, Jadwisin/Poland 1979.

[3] H. Steudel, Physica 6D (1983) 155.

[4] D.J. Kaup, Physica 6D (1983) 143.

[5] G. Neugebauer and R. Meinel, Phys. Lett. 100 (1984) 467.

[6] H. Steudel, "Space-Time Symmetry of Self-Induced Transparency and of Stimulated Raman Scattering", to appear in Phys. Lett.

Level Splitting and Band Formation of Dark Soliton Eigenvalues

G.A. Swartzlander Jr.

Code 6546, Naval Research Laboratory, Washington, DC 20375-5000, USA

Dark soliton "bands" are found in the soliton eigenvalue spectrum for periodic amplitude and phase gratings. A generalized solution is also determined for the eigenvalues when the initial condition has three discrete regions, each having a uniform, but arbitrary, complex transmittance.

Dark solitons were discovered as particular solutions [1,2] of the (1+1)-D nonlinear Schrodinger equation (NSE) (see Eq. (1) below), and have been observed in the temporal regime [3]. Recently, spatial dark soliton (SDS) stripes and grids [4] have been discovered in the cross-section of a laser beam. This novel phenomenon can be described by the (2+1)-D NSE, i. e., with two transverse degrees of freedom. Analytic solutions are yet unknown, although experimental evidence shows that dark solitons stripes are almost identical to (1+1)-D or slab solitons. Contrary to expectations of significant modulational instabilities, we find that (2+1)-D high contrast SDS's are remarkably stable.

Spatial and temporal (1+1)-D solitons, and to some extent (2+1)-D spatial soliton stripes [5], can be described in terms of the normalized (1+1)-D NSE,

$$i\partial u/\partial z + \partial^2 u/\partial x^2 - 2|u|^2 u = 0 \tag{1}$$

where u, x and z are the normalized electric field, and transverse and longitudinal coordinates, respectively. Equation (1) can be completely solved for an arbitrary initial condition, u(x,z=0), using the inverse scattering technique (see Ref [1]). The intensity profile of a single soliton is given by

$$|u|^2 = 1 - \nu^2\, sech^2\left[\nu(x - x_0 - 2\lambda z)\right] \tag{2}$$

where the parameter λ determines the visibility or contrast ($\nu^2 = 1-\lambda^2$), width (ν^{-1}), and angular direction ($\tan^{-1}(2\lambda)$) of the soliton, and x_0 is the position of the soliton at $z=0$.

Spatial dark solitons can be formed by passing, for example, a plane wave through a diffracting screen and into a nonlinear medium. Given the transmitted field, $u(x,z=0)$, the soliton parameters, λ_i, are determined by solving the "direct scattering" problem. A practical and versatile soliton generator can be composed of discrete regions of uniform transmittance. Such a screen allows solitons to be easily directed by changing either the beam power, or by actively changing the transmittance function.

The basic unit cell, or building block for a prototypical dark soliton generator is composed of a diffracting plate with three regions, each having uniform transmission and phase retardation, and step discontinuities at the interfaces. The field in each of these regions will be expressed as

$$u_j(x,z=0) = \gamma_j\, exp\,(i\,\phi_j), \quad , x_j < x < x_{j+1} \tag{3}$$

where $x_1 = -\infty$, $x_2 = 0$, $x_3 = +L$, $x_4 = +\infty$, and $L = X_L/X_{NL}$ is the normalized width of the central region of the diffraction plate which increases with both the beam intensity and the physical width of the opaque region. Following Ref [1], we set $u(x=\pm\infty)=1$ (i. e., $\gamma_1=\gamma_3=1$).

According to the direct scattering method, the values of λ are determined by satisfying the continuity of two sets of waves at the diffracting interfaces:

$$
\begin{aligned}
&A_j\exp(p_jx_{j+1})+C_jg_j\exp(-i\phi_j-p_jx_{j+1}) \\
&\quad =A_{j+1}\exp(p_{j+1}x_{j+1})+C_{j+1}g_{j+1}\exp(-i\phi_{j+1}-p_{j+1}x_{j+1}) \\
&A_jg_j\exp(i\phi_j+p_jx_{j+1})+C_j\exp(-p_jx_{j+1}) \\
&\quad =A_{j+1}g_{j+1}\exp(i\phi_{j+1}+p_{j+1}x_{j+1})+C_{j+1}\exp(-p_{j+1}x_{j+1})
\end{aligned}
\tag{4}
$$

where $g_j=\gamma_j^{-1}\left[\lambda-ip_j\right]$, and $p_j\equiv(\gamma_j^2-\lambda^2)^{1/2}$ if $\gamma_j^2>\lambda^2$; otherwise $p_j\equiv-i\lambda(1-\gamma_j^2/\lambda^2)^{1/2}$. Note that as $\gamma_j\to0$, $g_j\to0$ for nonzero values of λ. Physical solutions require finite amplitudes at $|x|\to\infty$; thus $C_1=A_3=0$. A transcendental function of λ is found by setting the determinant of the remaining coefficients to zero, resulting in the transcendental relation:

$$
\begin{aligned}
&\left[g_1e^{+i\phi_1}-g_2e^{+i\phi_2}\right]\left[g_2e^{-i\phi_2}-g_3e^{-i\phi_3}\right]e^{-p_2L} \\
&\qquad =\left[g_1g_2e^{i(\phi_1-\phi_2)}-1\right]\left[1-g_2g_3e^{i(\phi_2-\phi_3)}\right]e^{+p_2L}
\end{aligned}
\tag{5}
$$

where it has been assumed that L is finite; on the other hand, as $L\to\infty$, the continuum $\lambda^2<1$ are eigenvalues. The eigenvalues of Eq. (6), λ_i can be numerically computed. Several special cases are shown in Fig. 1 (see [6] for other interesting cases). Both $\pm\lambda_i$ are solutions.

When an opaque "wire" of width L is placed at the input face of the nonlinear material ($\gamma_2=0$, $\phi_1=\phi_3=0$), Eq (6) reduces to $\nu=\lambda\tan(\lambda L)$ (the expression $\lambda=\pm\cos(\lambda L)$ [1,6] is incomplete). As shown in Fig. 1(a), only a single pair of solitons are formed when $L<\pi$; successive pairs are formed with π increments of L. Although "black" solitons are not formed (except as $1/L\to0$), high contrast is achieved above $L=1$, where the visibility is $\nu>0.67$.

When a phase difference exists across the null region, Eq. (6) reduces to $\lambda=\nu\cot(\lambda L-\phi_3/2)$. Fig. 1(c) shows that when $\phi_3=\pi$ ($\gamma_2=0$), a single "black" dark soliton ($\lambda=0$) is formed for all values of L, and that, unlike the case $\phi_3=0$, the first *pairs* of solitons do not occur until $L>\pi/2$. Again, additional pairs of solitons form with π increments of L. The lack of paired solitons between $L=0$ and $\pi/2$ means that radiation transmitted through the screen can not couple into additional nonlinear modes, and is thus "diffracted" away.

The most important case, from an application point of view, is when a <u>lossless</u> mask is used, i.e., $\gamma_{1,2,3}=1$. When $\phi_1=\phi_3=0$, Eq. (6) reduces to $\lambda^2=\cos^2(\phi_2/2)+\sin^2(\phi_2/2)\exp(-2\nu L)$. For all values of L, only a single pair of solitons can form (unless $\gamma_2<1$). A nearly degenerate pair of "black" solitons which propagate practically parallel to each other can form when $\phi_2=\pi$ and $L>\pi$, as shown in Fig. 1(b). Alternatively, two converging (diverging) solitons form when $-\pi<\phi_2<0$ ($0<\phi_2<\pi$); they intersect at $z\approx L/4\lambda$. For large values of L, the asymptotic value of λ is identical to that of a single phase discontinuity, and the solitons originate in an independent manner. It is clear that a lossless screen with N regions of π phase retardation, with widths and separations $>\pi$, should produce $2N$ nearly parallel high contrast dark solitons. For example, at a wavelength of 1 μm, these black solitons can be packed as close as 100 μm without cross-talk in a nonlinear medium with $\Delta n/n_0=10^{-4}$.

This opens the question of what happens when more than three regions of constant transmittance are used. Eq. (4) can be solved with numerical techniques to obtain meaningful

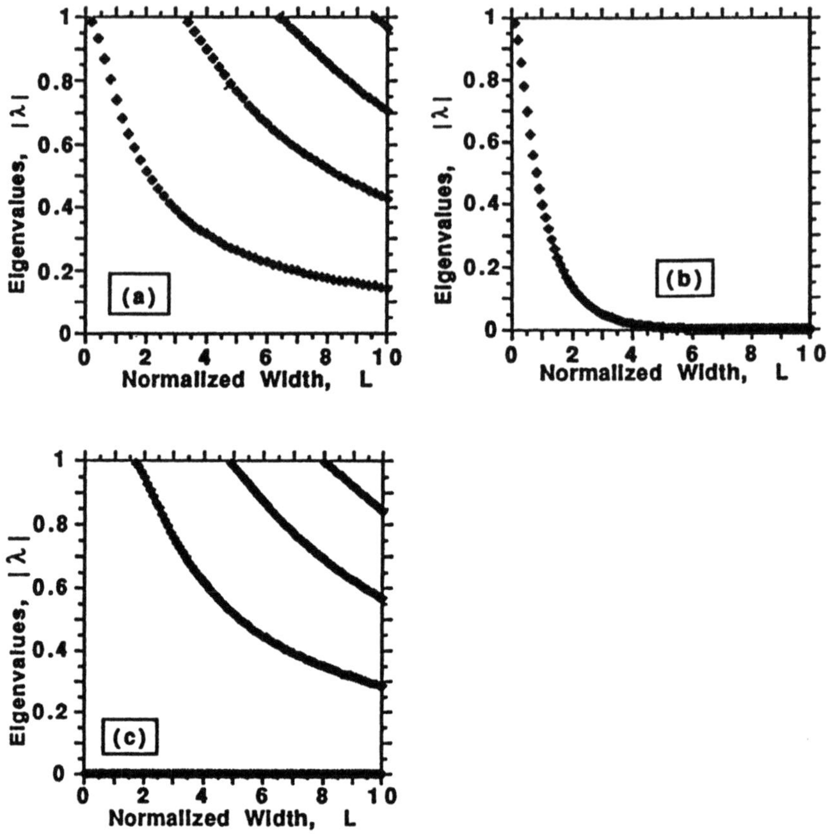

Figure 1 Dark solitons eigenvalues for the three-region problem. L is the width of the central region.

(a) $\gamma_1=\gamma_3=1,\ \gamma_2=0,\ \phi_1=\phi_2=\phi_3$

(b) $\gamma_1=\gamma_2=\gamma_3=1,\ \phi_1=\phi_3=0,\ \phi_2=\pi$

(c) $\gamma_1=\gamma_3=1,\ \gamma_2=0,\ \phi_1=0,\ \phi_3=\pi$

solutions. Fig. 2(a) shows the eigenvalue $|\lambda|$ for as many as N=10 amplitude wells ($\gamma_{j=even}=0$, $\gamma_{j=odd}=1$) for the case $L_{j=even,odd}=2$. For $N=1$ there is only one value of $|\lambda|$. For $N=2$, there are two values, owing to "tunneling" of waves from neighboring $\gamma=0$ regions. This is similar to level splitting in quantum mechanics for two coupled wells. As the "wells" are brought further apart, the eigenvalues become degenerate ($\nu=\lambda\tan(\lambda L)$). On the other hand, as the wells are brought together, $L\rightarrow 2L$ in this equation. In this case we see that for any number of wells, N, there are N values of $|\lambda|$. It is clear from Fig. 2(a) that as $N\rightarrow\infty$, the solitons appear to be constrained within a band of width $\Delta\lambda\approx0.2$.

A similar phenomenon is seen for a screen with many phase regions. Fig. 2(b) shows the case of $\phi_{j=even}=\pi$, $\phi_{j=odd}=0$, with $L_{j=even,odd>2}=2$. As discussed above, when two regions with a π phase difference exists, a black soliton forms ($\lambda=0$). It can be seen in Fig. 2(b) that this solu-

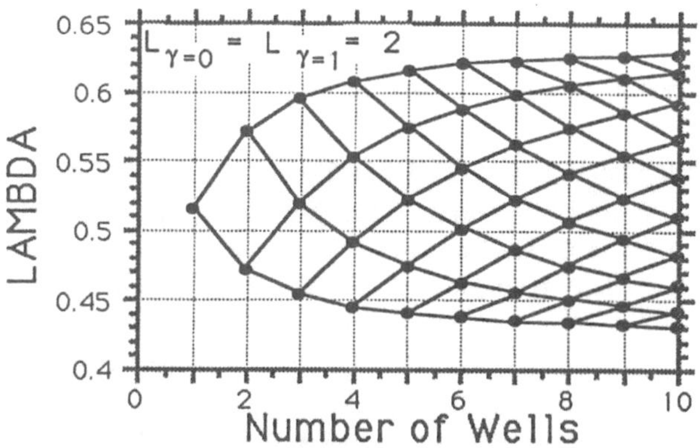

Figure 2(a) Dark soliton eigenvalues showing level splitting and band formation as the number of amplitude "wells" of zero transmittance is increased.

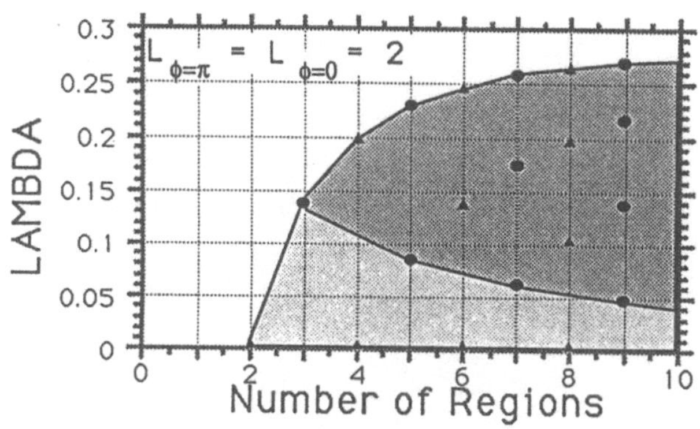

Figure 2(b) Level splitting for "phase well" case as the number of phase regions is increased.

tion occurs only when an even number of regions is present (triangles). Level splitting is observed when either an even or odd (dots) number of regions exists. For *M=even* regions, there are M/2 values of $|\lambda|$. For *M=odd* regions, there are (M-1)/2 values of $|\lambda|$. Band formations are evident in both cases.

One physical meaning of these bands is found in the relation between λ and the angular direction of the solitons, $\theta_i=\tan^{-1}(2\lambda_i)$, e. g., the angular distribution of solitons will be characterized by a high density of solitons within narrow regions, with no solitons outside these regions. In real space, however, these solitons may be well isolated from each other, owing to their mutual repulsion, as well as their difference origins, $x_{0,i}$. The dynamical interaction of these "band-solitons" is the subject of continuing interest.

I would like to thank A. E. Kaplan for valuable discussions regarding level-splitting, D. R. Andersen for detailed conversations regarding dark solitons, and C. T. Law for expert advice on numerical computations. This work was supported by DARPA, ONR, and the Pittsburgh Supercomputing Center. G. Swartzlander is presently an ONT postdoctoral fellow at the Naval Research Laboratory.

References

[1] V. E. Zakharov and A. B. Shabat, Zh. Eksp. Teor. Fiz. 64, 1627 (1973) [Sov. Phys.-JETP 37, 823 (1973)].

[2] A. Hasegawa and F. Tappert, Appl. Phys. Lett., 23, 171 (1973).

[3] P. Emplit, J. P. Hamaide, F. Reynaud, C. Froehly, and A. Barthelemy, Opt. Comm., 62, 374 (1987); D. Krökel, N. J. Halas, G. Giuliani, and D. Grischkowsky, Phys. Rev. Lett. 60, 29 (1988); A. M. Weiner, J. P. Heritage, R. J. Hawkins, R. N. Thurston, E. M. Kirschner, D. E. Leaird, and W. J. Tomlinson, Phys. Rev. Lett. 61, 2445 (1988).

[4] G. A. Swartzlander, Jr., D. R. Andersen, J. J. Regan, H. Yin, and A. E. Kaplan, Phys. Rev. Lett. 66, 1583 (1991).

[5] D. R. Andersen, D. E. Hooton, G. A. Swartzlander, Jr., and A. E. Kaplan, Opt. Lett., 15, 783 (1990).

[6] S. A. Gredeskul, Yu. S. Kivshar, and M. V. Yanovskaya, Phys. Rev. A 41, 3994 (1990).

Soliton Propagation in a Random Medium

P.K.A. Wai, C.R. Menyuk†*, and H.H. Chen*

University of Maryland, College Park, MD 20742, USA
†Electrical Engineering Department, University of Maryland, Baltimore, MD 21228, USA

The effects of randomly varying birefringence in a single mode fiber on soliton propagation are studied. It is shown analytically that the evolution equation can be reduced to the nonlinear Schrödinger equation if the variation length is much shorter than the soliton period. The soliton does not split at high values of the average birefringence, but does undergo spreading and loss of polarization. The effects of randomly varying birefringence on soliton interactions are also investigated. We found that for initial separations of less than ten pulsewidths, the phase dependent, short range interaction dominates. For separations larger than ten pulsewidths, the soliton interacts through the dispersive radiation which they generate. This interaction is too weak to explain the phase independent, long range soliton interaction observed experimentally.

Summary

Because of the presence of linear birefringence, a single-mode fiber is really bimodal. It has been shown[1,2] that in the case of constant birefringence, walk-off can be avoided by modulation; the two fractional pulses in each polarization shift each other's frequency such that any initial difference in group velocity is eliminated. As a result, they are self-trapped. In real fibers, the orientation of the birefringence tends to shift randomly, scattering light from one state of polarization to another in the process. This polarization dispersion has potentially detrimental effects in a long distance optical communication system.

For a single-mode fiber with constant linear birefringence, the equation of motion is given by [3],

$$\begin{aligned} i\frac{\partial U}{\partial z} + i\delta\frac{\partial U}{\partial t} + \frac{1}{2}\frac{\partial^2 U}{\partial t^2} + (|U|^2 + \frac{2}{3}|V|^2)U + \frac{1}{3}V^2U^*\exp(iR\delta z) &= 0, \\ i\frac{\partial V}{\partial z} - i\delta\frac{\partial V}{\partial t} + \frac{1}{2}\frac{\partial^2 V}{\partial t^2} + (\frac{2}{3}|U|^2 + |V|^2)V + \frac{1}{3}U^2V^*\exp(-iR\delta z) &= 0, \end{aligned} \tag{1}$$

where U and V are the pulse amplitudes in the two polarizations, δ is the normalized time delay, and we are assuming that the fiber is linearly birefringent. The quantity $R\delta$ equals 4×(soliton period)/(beat length) which is very large in case of practical interest for communications system.[3,4] It follows that the state of polarization of the electric field is rapidly changing, corresponding to rapid rotation on the Poincaré sphere.[5] Hence, the terms with the factor $\exp(\pm iR\delta z)$ are rapidly varying, will average to zero, and can be dropped.

In the following, the fiber is assumed to periodically undergo a sudden, random rotation θ. This angle is uniformly spaced over the interval $[0, 2\pi]$. In between each rotation the fiber is assumed to maintain a constant linear birefringence, so that the state of polarization at each point in time is steadily and rapidly changing. The state of polarization is characterized by the angle $\tan^{-1}[\mathrm{Im}(E_1E_2^*)/\,\mathrm{Re}((E_1E_2^*)]$, where E_1 and E_2 are the complex field amplitudes in the two polarizations. We will assume that when θ changes, a random phase factor ϕ is added to the phase difference between the pulse envelopes U and V, corresponding to E_1 and E_2. We do so because in real fibers, the state of polarization changes with great rapidity, and slight variations in the birefringence over the sections which are being treated as constant will lead to complete randomization of the state of polarization. The overall change in the evolution equation is thus given by the following transformation,

$$\Psi' = \mathcal{U}\Psi, \tag{2}$$

where,

$$\mathcal{U} = \begin{pmatrix} \cos\theta & \sin\theta\, e^{i\phi} \\ -\sin\theta\, e^{-i\phi} & \cos\theta \end{pmatrix} \tag{3}$$

and $\Psi = (U, V)^t$ is the polarization state envelope vector. The matrix, $\mathcal{U}(\theta, \phi)$ in Equation (3) is unitary and its determinant equals 1. It is a member of the mathematical group $SU(2)$ and its action is equivalent to an arbitrary spatial rotation on a sphere.[6] Using the Poincaré sphere representation for the polarization state vector, Eq. (3) is a rotation of the corresponding Stokes vector. From another point of view, the effect of the sudden change in the birefringence axes can be viewed as rotation of a Poincaré sphere on which the current state of polarization is the North pole. If the angles θ and ϕ are uniformly distributed random variables, the representative point on the Poincaré sphere is also uniformly distributed. Applying Eq. (3) to the coupled nonlinear Schrödinger equation with linear birefringence and dropping the prime, the resulting equation of motion is given by

$$i\frac{\partial\Psi}{\partial z} + i\delta\sigma\frac{\partial\Psi}{\partial t} + \frac{1}{2}\frac{\partial^2\Psi}{\partial t^2} + \frac{5}{6}|\Psi|^2\Psi + \frac{1}{6}(\Psi^\dagger\sigma\Psi)\sigma\Psi = 0, \tag{4}$$

where $\Psi^\dagger = (U^*, V^*)$, and the matrix σ is given by,

$$\sigma = \begin{pmatrix} \cos 2\theta & -\sin 2\theta\, e^{i\phi} \\ -\sin 2\theta\, e^{-i\phi} & -\cos 2\theta \end{pmatrix}. \tag{5}$$

In subsequent analysis, the angles θ and ϕ are assumed to vary randomly at intervals of z_h, which is typically of length 100 meters or so. For the 50 psec solitons which would be used in a long distance communication system, the nonlinear length scale z_0 is on the order of thousands of kilometers. Thus, the Poincaré sphere would rotate many times in the course of a soliton period. The rotations are so fast that the fiber appears to be isotropic to the lowest order of approximation in a perturbation series. That is, pulses launched with different polarization state evolve identically. The equation describing the averaged evolution of the pulse can be obtained by averaging Eq. (4) on the Poincaré sphere. Using the pulse amplitudes U and V explicitly instead of Ψ, we have,

$$\begin{aligned} i\frac{\partial U}{\partial z} + \frac{1}{2}\frac{\partial^2 U}{\partial t^2} + \frac{8}{9}(|U|^2 + |V|^2)U &= 0, \\ i\frac{\partial V}{\partial z} + \frac{1}{2}\frac{\partial^2 V}{\partial t^2} + \frac{8}{9}(|U|^2 + |V|^2)V &= 0. \end{aligned} \tag{6}$$

Manakov[7] has shown that Eq. (6) can be solved by the spectral transform method and that it possesses soliton solutions. If the pulses are initially in the same polarization as a function of time, *i.e.*, $U(z=0,t) = q(t)\cos\chi$ and $V(z=0,t) = q(t)\sin\chi\exp(i\alpha)$, where χ and α are constants, then Eq. (6) can be further reduced Eq. (6) reduces to,

$$i\frac{\partial q}{\partial z} + \frac{1}{2}\frac{\partial^2 q}{\partial t^2} + \frac{8}{9}|q|^2 q = 0, \tag{7}$$

to the nonlinear Schrödinger equation. Hence the nonlinear Schrödinger equation remains an excellent description of nonlinear wave propagation in a birefringent fiber, provided that the principal axes vary rapidly in the course of a soliton period.

We now choose χ and α to be zero with no loss of generality, and determine the first order correction to Eq. (7) using a simple multiple length scale expansion.[8] We thus obtain,

$$\begin{aligned} U(z,t) &= U^{(0)} + i\frac{1}{12}\Big[\int^z dz'(\cos 4\theta - \frac{1}{3})\Big]\left|U^{(0)}\right|^2 U^{(0)}, \\ V(z,t) &= \delta\Big[\int^z dz' \sin 2\theta \exp(-i\phi)\Big]\frac{\partial U^{(0)}}{\partial t} - i\frac{1}{12}\Big[\int^z dz' \sin 4\theta \exp(-i\phi)\Big]\left|U^{(0)}\right|^2 U^{(0)}, \end{aligned} \tag{8}$$

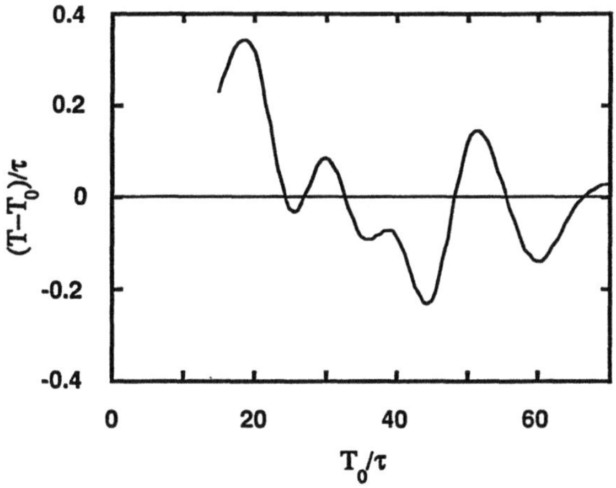

Figure 1: Change in pulse separation $(T - T_0)/\tau$ as a function of in put separation T_0 for a distance of 40 soliton periods

where $U^{(0)}(z,t)$ is the soliton solution to Eq. (7),

$$U^{(0)} = \sqrt{\frac{9}{8}} A \operatorname{sech} A\Big(t - \delta \int^{z} dz' \cos 2\theta\Big) \exp(-iA^2 z/2). \tag{9}$$

Randomly varying birefringence leads at this order to modification of the soliton delay, deformation of the soliton shape, and creation of shadows. The shadow is the fractional pulse in the V-polarization. The distribution function of these random coefficients can be shown to be Gaussian with zero mean value and a variance equal to $\pi^2 z_h Z/8z_0^2$, where Z is the fiber length.

Equation (4) is then integrated numerically, it is found that the results agree with Eq. (8) and (9) very well. The pulse stays bounded in all cases investigated. No pulse splitting is observed although a significant amount of dispersive wave radiation is generated. Next, the evolution of a pair of 1-solitons initially T_0 apart are studied numerically up to 40 z_0. The pulses separation T_0 ranges from 2τ to 70τ, where τ is the initial pulsewidth of the soliton. For $T_0 \lesssim 10\tau$, the phase dependent short range interaction dominates. If the solitons are in phase, they initially attract each other and then separate after one oscillation due to the randomly varying birefringence. If the solitons are out of phase, they repel each other.

For $T_0 \gtrsim 10\tau$, the solitons interact through the dispersive wave components they generate. In Fig. 1, the change in pulse separation is plotted as a function of T_0 for a fiber length of 40 z_0. The variation does not display any periodic behavior. Therefore, although the randomly varying birefringence causes a well-separated pair of solitons to interact, its variation and magnitude does not account for the long-range interaction observed experimentally.[9] Recently, Dianov, *et. al.* [10] have shown that the soliton long range interaction could be the result of an electrostrictive excitation of acoustic oscillations by a periodic sequence of solitons.

In summary, we have shown analytically that in a birefringent fiber of rapidly and randomly varying orientation, the underlying evolution equation is the nonlinear Schrödinger equation if the variation length is much shorter than a soliton period. This shows the robustness of the solitons and the usefulness the nonlinear Schrödinger equation in fibers. Using numerical simulations, we further show that the soliton does not split even at high values of δ, although large amount of dispersive wave radiation may be generated. We have also studied numerically the effects of a randomly oriented birefringence on solitons interaction. Single solitons initially less than 10τ apart are dominated by the phase dependent short range interaction. The random variation causes solitons with the same initial phase to separate. For $T_0 > 10\tau$, the solitons interact through the dispersive waves which they generate, but the resulting variation in pulse separation cannot account for the long range interaction between solitons which has been observed experimentally.

The work was supported by the National Science Foundation and the Department of Energy. Computational work is supported by National Energy Research Supercomputer Center and San Diego Spercomputer Center.

Reference

1. C. R. Menyuk, Opt. Lett., **12**, 614 (1987).
2. C. R. Menyuk, J. Opt. Soc. Am. B, **5**, 392 (1988).
3. C. R. Menyuk, IEEE J. Quantum Electron., **QE-25**, 2674 (1989).
4. L. F. Mollenauer, K. Smith, J. P. Gordon, and C. R. Menyuk, Opt. Lett. **14**, 1219 (1989).
5. Born and Wolf, *Principles of Optics,* 4th ed. (Pergamon, Oxford, 1970).
6. G. Arfken, *Mathematical methods for Physicists,* 2nd ed, (Acdemic Press, London, 1970).
7. S. V. Manakov, Zh. Eksp. Teor. Fiz. **65**, 505 (1973) [Sov. Phys. JETP **38**, 248 (1974)].
8. P. K. A. Wai, C. R. Menyuk, H. H. Chen, and Y.C. Lee, IEEE J. of Quantum Electron., **QE-24**, 373 (1988).
9. K. Smith and L. F. Mollenauer, Opt. Lett. **14**, 1284 (1989).
10. E. M. Dianov, A. V. Luchnikov, A. N. Pilipetskii, and A. N. Starodumov, Sov. Lightwave Commun. **1**, 37 (1991).

Part V

Hydrodynamics and Turbulence

Length Scales and the Navier-Stokes Equations

M. Bartuccelli[1], *C.D. Doering*[2], *J.D. Gibbon*[1], *and S.J.A. Malham*[1]
[1]Department of Mathematics, Imperial College, London SW7 2BZ, UK
[2]Department of Physics, Clarkson University, Potsdam, NY 13676, USA

Despite the lack of a regularity proof for the 3d Navier-Stokes equations, it is an interesting and important question whether it can be shown that large fluctuations or excursions away from temporal and spatial averages can occur. If it can be demonstrated that the Navier-Stokes equations allow such fluctuations away from averages, then these must have narrow spatial and temporal bandwidths and the width of these will give information about the smallest scale in the flow. Any numerical scheme must 'resolve' these spikes to get an accurate representation of the flow. The question of the smallest length scale is the topic of this paper.

In the theory of attractors in PDEs [1, 2, 3, 4, 5, 6, 7, 8, 9, 10], it has become conventional to measure the smallest scale using upper bounds on the Lyapunov dimension of the universal attractor. This upper bound $\mathcal{N}$, is usually interpreted as the number of degrees of freedom in the PDE under investigation. On a domain $[0, L]^d$ of volume L^d (d spatial dimensions), $\mathcal{N}$ is related to the minimum scale l by $\mathcal{N} = (L/l)^d$. In other words, $\mathcal{N}$ represents the number of eddies of volume l^d which are contained in the box of volume L^d. The only problem that exists with this approach is whether computing a lower bound on l from an upper bound of the Lyapunov dimension is always correct. In [11] it was shown that this identification is not quite so simple as one might hope. For the complex Ginzburg-Landau (CGL) equation on periodic boundary conditions,

$$A_t = RA + (1 + i\nu)\Delta A - (1 + i\mu)|A|^2 A \tag{1}$$

it was shown analytically that when $d = 2$ (but not when d = 1), very spiky fluctuations (interpreted as strong turbulence) can occur in regions of the (μ, ν) plane but not in the rest of this plane. The spiky fluctuations occur on scales $l \geq R^{-|\nu|}$, whereas in the rest of the plane, where only weak turbulence can occur, $l \geq R^{-1/2}$. The Lyapunov dimension of the attractor in both regions was found to be uniform giving $l \geq R^{-1/2}$. In some sense this result is not surprising as the method of computing the evolution of volumes elements of the universal attractor through which one finds an upper bound on the Lyapunov dimension, requires a time on low norms of the velocity field. It contains no information on higher norms, the energy in which may be appreciable during the interval of the spike. This average takes account only of dynamics on scales larger than the spikes. In this sense, this method is certainly good for weak turbulence but is less sure for strong turbulence. It therefore seems necessary to find an alternative definition of a length scale which takes account of the possibility of large spikes or fluctuations and which, for the Navier-Stokes equations, gives sensible results.

The conventional scales in 3d turbulence is the Kolmogorov scale which is usually defined as $\lambda_{KG} = (\nu^3/\epsilon)^{1/4}$ where ϵ is the energy dissipation rate. Another scale is the Kraichnan scale, more appropriate in 2d, which is defined as $\lambda_{KR} = (\nu^3/\eta)^{1/6}$ where η is the enstrophy dissipation rate. Both these scales can be found via the Lyapunov dimension of the 2d and 3d Navier-Stokes attractor: in the latter case it is necessary to make certain assumptions because of the lack of a regularity proof. Below we show how to obtain a set of length scales which give the Kolmogorov scale for flows which represent homogeneous decaying turbulence. For more general flows we show that the length scales are much smaller than Kolmogorov, as they come out to be proportional to λ_{KG}^4.

We take the Navier-Stokes equations on a periodic domain $\Omega = [0, L]^d$ in d dimensions, with ν as viscosity and a C^∞ forcing function $\mathbf{f}$. In the velocity form they are

$$u_t + (u \cdot \nabla)u = \nu\Delta u - \nabla\mathcal{P} + \mathbf{f} \qquad\qquad div\ u = 0 \tag{2}$$

In the vorticity notation these assume the following form:

$$\omega_t + (u \cdot \nabla)\omega = \nu\Delta\omega + (\omega \cdot \nabla)u + F \tag{3}$$

$$\omega = curl(u), \qquad\qquad F = curl(\mathbf{f}) \tag{4}$$

The vortex strechting term $(\omega \cdot \nabla)u$ vanishes for $d = 2$, but not for $d = 3$. In this latter case it has a strong influence on the evolution of the flow.

Now we define a set of quantities in d dimensions ($d = 2, 3$)

$$H_N = \sum_{i=1}^{d} \sum_{|n|=N} \int (D^n u_i)^2 dx \equiv \|D^N u\|_2^2 \tag{5}$$

where $D^N u_i = \frac{\partial^{n_1+\cdots+n_d}}{\partial x_1^{n_1} \ldots \partial x_d^{n_d}}$ and $n_1 + \cdots n_d = N$.

Inserting the H_N in equation (2) we prove [12] that for $N \geq 1$ and $d = 2, 3$ they satisfy the two differential inequalities (ladder inequalities)

$$\frac{1}{2}\dot{H}_N \leq -\frac{\nu}{2}\frac{H_N^2}{H_{N-1}} + \frac{b_N}{\nu} H_N \|u\|_\infty^2 + H_N^{1/2} \sum_{i=1}^{d} \|D^N f_i\|_2 \tag{6}$$

$$\frac{1}{2}\dot{H}_N \leq -\nu H_N^2/H_{N-1} + c_N H_N \|Du\|_\infty + H_N^{1/2} \sum_{i=1}^{d} \|D^N f_i\|_2. \tag{7}$$

b_N, c_N are constants and $\|u\|_\infty = \sup_{x\in\Omega} |u(x)|$. $\|g\|_p^p := \int_\Omega |g(x)|^p dx < +\infty, \quad p \geq 1.$

When $d = 2$, it is possible to find long time upper bounds for all the H_N making no assumptions. This proves the existence of a C^∞ attractor. When $d = 3$, we are forced to make some a priori assumptions in order to obtain upper bounds for all the H_N, such as taking H_1 bounded.

From the $\|Du\|_\infty$ version of the ladder inequalities (7), we can compute a lower bound on the length scale (in what follows we take the leading order term only: the forcing produces terms of lower order)

$$\ell^{-2} \leq \ < \frac{H_N}{H_{N-1}} > \ \leq \ c\nu^{-1} < \|Du\|_\infty > \tag{8}$$

where $< g >= limsup \lim_{t\to\infty} \left(\frac{1}{t}\int g(s)ds\right)$.

Then we can prove the following vorticity inequalities ($d = 3$) [13]

$$\overline{\|\omega\|_\infty} \leq \ c\nu^{-3}\overline{\|\omega\|_2^4} \tag{9}$$

$$< \|\omega\|_\infty > \ \leq \ c_1\nu^{-3} < \|\omega\|_2^4 > \tag{10}$$

The overbar means limsup over all smooth initial conditions as $t \to \infty$ and the c's are dimensionless absolute constants.

For weak turbulence, when $\sup_t \|Du\|_\infty \approx \sup_t \|Du\|_2 L^{-3/2}$ with $\epsilon = 2\nu \sup_t H_1 L^{-3/2}$ we get

$$\ell \geq c\lambda_{KG}. \tag{11}$$

This result has been obtained also by Henshaw, Kreiss and Rejna [14]. More generally, with $\epsilon = 2\nu < H_1^2 >^{1/2} L^{-3}$ we get [13]

$$\ell \geq c\lambda_{KG}^4 L^{-3}. \tag{12}$$

References

[1] C. Foias and G. Prodi. Sur le comportement global des solutions non-stationnaires des equations de navier-stokes en dimension 2. *Rendiconti di Padova*, 1:1, 1967.

[2] R. Temam. *Navier-Stokes Equations, Theory and Numerical Analysis.* North-Holland, Amsterdam, second edition, 1979.

[3] R. Temam. *The Navier-Stokes Equations and Non-linear Functional Analysis.* CBMS-NSF Regional Conference Series in Applied Mathematics. SIAM, 1983.

[4] R. Temam. *Infinite Dimensional Dynamical Systems in Mechanics and Physics*, volume 68 of *Springer Applied Mathematics Series.* Springer, Berlin, 1988.

[5] P. Constantin, C. Foias, R. Temam, and B. Nicolaenko. *Integral and Inertial Manifolds for Dissipative PDE's*, volume 70 of *Springer Applied Mathematics Series.* Springer, Berlin, 1988.

[6] P. Constantin, C. Foias, O.P. Manley, and R. Temam. Determining modes and fractal dimension of turbulent flows. *J. Fluid Mech.*, 150:427–440, 1985.

[7] C. Foias, O.P. Manley, R. Temam, and Y.M. Treve. Asymptotic Analysis of the Navier-Stokes Equations. *Physica 9D*, pages 157–188, 1983.

[8] P. Constantin, C. Foias, and R. Temam. On the Dimension of the Attractors in Two-Dimensional Turbulence. *Physica 30D*, pages 284–296, 1988.

[9] P. Constantin and C. Foias. *The Navier-Stokes Equations*. Chicago University Press, 1989.

[10] P. Constantin, C. Foias, and R. Temam. Attractors Representing Turbulent Flows. *Memoirs of AMS*, 1985.

[11] M. Bartuccelli, P. Constantin, C.R. Doering, J.D. Gibbon, and M. Gisselfält. On the Possibility of Soft and Hard Turbulence in the Complex Ginzburg-Landau Equation. *Physica D*, 44:421–444, 1990.

[12] M. Bartuccelli, C. Doering, and J.D. Gibbon. Ladder Theorems for the 2d and 3d Navier-Stokes Equations on a Finite Periodic Domain. *Nonlinearity*, 4:531–542, 1991.

[13] M.V. Bartuccelli, C.R. Doering, J.D. Gibbon, and S.J.A. Malham. Vorticity Fluctuations in the 3d Incompressible Navier-Stokes Equations. In Preparation.

[14] W.D. Henshaw, H.O. Kreiss, and L.G. Reyna. Smallest Scale Estimates for the Incompressible Navier-Stokes Equations. Preprint.

Local and Nonlocal Transfer of Motion Integrals in Wave Turbulence

G.E. Falkovich[1] *and M.D. Spector*[2]

[1]Department of Nuclear Physics, Weizmann Institute of Science, Rehovot 76100, Israel

[2]Department of Fluid Mechanics and Heat Transfer, Tel Aviv University, Ramat Aviv, 69978 Tel Aviv, Israel

Introduction

The picture of cascade turbulence suggested by Richardson, Kolmogorov and Obukhov is based on the concept of interaction locality [1–5]. That means that those modes (vortices or waves) effectively interact which are of comparable scales only. The question naturally arise: whether a locality property should be satisfied on the steady Kolmogorov-like spectrum only or on the slightly differing distributions as well? Proceeding from continuity-like speculations, one might suppose that in general case interaction locality for Kolmogorov distribution leads to that for close ones. Such a supposition is, however, incorrect since Kolmogorov spectrum usually possesses higher degree of symmetry (for example, being isotropic) than arbitrary yet close distributions. A stationary locality does not mean thus an evolutionary locality as it was stated in [4,6].

What could we say about the properties of the system with local Kolmogorov-like spectrum and with perturbations interacting nonlocally? Our report deals with that matter. We demonstrate here that nonlocality could lead to either the growth or the decreasing of perturbation, depending on the number of angular harmonic. The growth of perturbation means the instability of stationary isotropic spectrum with respect to that harmonic. Nevertheless, we demonstrate the possibility of Kolmogorov-like spectrum to still alive in average which means that, despite the steady spectrum is anisotropic, angle averaging spectrum decreases with k-modulus by the Kolmogorov law. The energy is thus transferred locally. As far as momentum is concerned, it can be transferred nonlocally in $\vec{k}$-space.

1. Stationary and evolutionary locality

If the level of wave excitation is small enough to provide the weakness of nonlinear effects comparing with those of dispersion, the turbulence is regarded as weak and can be described with the help of kinetic equation for quantities $n(\vec{k})$, called wave occupation numbers in the $\vec{k}$-space. Under small level of nonlinearity the three wave processes are sufficient to be restricted with, which are allowed if the dispersion law $\omega(\vec{k})$ admits the decay equality $\omega(\vec{k}+\vec{k}') = \omega(\vec{k}) + \omega(\vec{k}')$. In this case, the kinetic equation reads as following

$$\begin{aligned}\frac{\partial n(\vec{k},t)}{\partial t} = I(\vec{k},t) = \int U_{k12}\left(n_1 n_2 - n_k n_1 - n_k n_2\right) d\vec{k}_1 d\vec{k}_2 \\ -2\int U_{1k2}\left(n_k n_2 - n_k n_1 - n_1 n_2\right) d\vec{k}_1 d\vec{k}_2 \ .\end{aligned} \tag{1}$$

Here $U_{k12} = V_{k12}^2 \delta(\vec{k}-\vec{k}_1-\vec{k}_2)\delta(\omega_k - \omega_1 - \omega_2)$ and V_{k12} is the matrix element of the three wave interaction. In the presence of the scale invariance of wave system implying that

$$\omega(\mu k) = \mu^\alpha \omega(k) , \quad V(\mu k, \mu k_1, \mu k_2) = \mu^m V(k, k_1, k_2) ,$$

a stationary solution of equation (1) turns out to be power one, $n(k) = Ak^{-\nu}$, with the index $\nu = m + d$. Distributions of such type correspond to constancy of the energy flux in the k-space and are used to be called Kolmogorov-like spectra of weak wave turbulence by the direct analogy with those for turbulence of incompressible fluid. Such distributions are stationary solutions of the kinetic equation provided that the inegrals involved converge. If the matrix element has the asymptotics $V^2(k, k_1, k_2) \propto k_1^{m_1} k^{2m-m_1}$ under $k_1 << k$, then the convergnce condition of the collision integral in (1) defined on the power distributions $n(k) \propto k^{-s}$ takes the following form:

$$m_1 + d - 1 + 2\alpha \equiv s_1 > s > s_2 \equiv 2m - m_1 + d + 1 - 2\alpha \ ,$$

Therefore, the interval of indices (s_1, s_2) exists, which provides an interaction locality for isotropic spectra, if

$$m_1 - m + 2\alpha - 1 > 0 \ . \tag{2}$$

It is curious to note, that the Kolmogorov index lies quite in the middle of the locality interval: $\nu = (s_1 + s_2)/2$.

Let us turn to the problem of interaction locality for small perturbations of Kolmogorov-like spectrum, $n(\vec{k}, t) = n(k) + \delta n(\vec{k}, t)$, $\delta n(\vec{k}, t) \ll n(k)$. The linearized kinetic equation necessiates the following conditions of convergence and locality interval existence:

$$m_1 + d + \alpha \equiv s_3 > s > s_4 \equiv 2m - m_1 + d - \alpha \ . \tag{3}$$

$$m_1 - m + \alpha > 0 \ . \tag{4}$$

The latter one is more restrictive than (2) since the frequency index $\alpha > 1$ (to allow three waves processes).

If the condition (4) is satisfied, then interaction is local for perturbations which decrease faster than k^{-s_4} at $k \to \infty$ and increase slower than k^{-s_3} at $k \to 0$. If the condition (4) is invalid for given system, then the interaction with largest and smallest motions effects on the behavior of quite arbitrary small perturbation. Such a property is refered to as evolutional non-locality [4,6].

Let us turn to the consideration of specific system for which inequality (2) is satisfied while (4) is not.

2. The turbulence of capillary waves on a shallow water

The well-known expression for the dispersion law of surface waves

$$\omega^2(k) = \tanh kh(gk + \sigma k^3/\rho)$$

for the intermediate wave numbers $\sqrt{\rho g/\sigma} \ll k \ll h^{-1}$ can be written as follows:

$$\omega^2 = \frac{\sigma h}{\rho} k^4 \ . \tag{5}$$

Here ρ, σ, h are the density, the surface tension and the depth of the fluid. The local Kolmogorov-like stationary solutions,

$$n(k) = P^{1/2} k^{-4} * (\frac{\rho h \pi^2}{\sigma)}, \tag{6}$$

, and the coefficient of triple interaction (allowed by (5)), $V(k, k_1, k_2) \propto k^4$, were found in [7] with indices $\alpha = 2$, $m = 2$, $m_1 = 0$ and $d = 2$.

Let us turn to the stability problem for the Kolmogorov-like spectrum(6). Linearized kinetic equation can be considered separately for each angular harmonic, $\delta n(\vec{k},t) = \sum n_l(k,t)e^{il\theta}$. It can be verified directly that for even harmonics the divergences in the linearized collision integral disappear for $n_l(k) \propto k^{-s}$ with $3 < \mathrm{Re}\, s < 5$, which agrees with (2). Due to the presence of locality interval, for solving stability problem one can use the theory developed in [4,6], exploiting the Mellin transform

$$n_l(k,t) = \int_\gamma a_l(s) e^{W(s)t} k^{-s}\, ds\ .$$

Temporal behavior of perturbations is governed by function $W(s)$ which is the Mellin image of the operator of linearized kinetic equation. It can be computed by direct integration. For example, for $l = 0$ it is

$$W_0(s) = B\left(\frac{1-s}{2}, -\frac{3}{2}\right) - B\left(\frac{1-s}{2}, \frac{1}{2}\right) + B\left(\frac{s-5}{2}, \frac{1}{2}\right) + B\left(\frac{s-1}{2}, -\frac{3}{2}\right)\ .$$

Here $B(x,y)$ is Euler beta function. As one can see, $W_0(4) = 0$ which corresponds to the neutral stability of Kolmogorov-like spectrum with respect to the variation of the flux value. The derivative $dW_0(s)/ds$ at $s = 4$ yields the group velocity of perturbations [8] which is positive in this case. So the isotropic perturbations drift to the large k's (damping region) at the same direction as an energy flux carried by spectrum, which, according to [4], is stable with respect to isotropic perturbations. Stability of any even harmonic can be proved in a similiar way.

Let us turn to odd harmonics. The absence of locality strip for index s can be seen yet from (4) since $m_1 - m + \alpha = 0$ in this case. Indeed, considering the main divergencies in the collizion integral, we get for $n_l(k,t) \propto k^{-s}$ and for $l = 2j+1$:

$$\frac{\partial n_l}{\partial t} \propto (-1)^j l k^{-4} \left(\int^\infty k_1^{3-s}\, dk_1 + \int_0 k_1^{3-s}\, dk_1 \right)\ .$$

Whatever the $\mathrm{Re}\, s$, the linearized collision integral diverges for the perturbations of power type, $n_l(k,t) \propto k^{-s}$ with $l = 2j + 1$. That is not only some mathematical difficulty arising from the use of Mellin transform. It can be shown that the time derivatives of perturbation moments

$$M_l(s,t) = \int_0^\infty k^{s-1} n_l(k,t)\, dk$$

turn into infinity or, strictly speaking, take very large values defined by cutting off on the left k_0 (wave source) and right k_m (sink) boundaries of the inertial interval, where Kolmogorov-like spectrum could take place:

$$\frac{\partial M_l(s,t)}{\partial t} = (-1)^j l M_l(4,t) \left(\int^{k_m} k^{s-5}\, dk + 2\int_{k_0} k^{s-5}\, dk \right)\ . \tag{7}$$

One can see the twice odd moments (with $l = 2j+1$ and $j = 2i+1$) to be quickly diminishing with time.

On the contrary, moments with $l = 4i + 1$ quickly grow that corresponds to the instability of isotropic spectrum (6) The turning of some moment into infinity (or into large value defined by cut off scales) means the arising of power asymptotics at $k \to 0, \infty$ (or at $k \to k_0, k_m$). Since every moment is proportional to $M(4)$ it is easy to see that the asymptotics of perturbation arises in the form $n_l(k,t) \to k^{-4}$ at $k \to 0, \infty$. Indeed, to have the asymptotics k^{-4} is the only way to satisfy the moment relations

$$M(s,t)/M(4,t) \propto k_m^{s-4}, \quad s > 4$$
$$M(s,t)/M(4,t) \propto k_0^{s-4}, \quad s < 4 ,$$

following from (7).

Therefore, strongly changing the angular shape of the spectrum, instability preserves the Kolmogorov-like k-dependence corresponding to the cascade transfer of energy. The nonlocal character of interaction of anisotropic perturbations means that momentum is nonlocally transferred from source to sink.

References

1. A.N.Kolmogorov: Dokl. Akad. Nauk SSSR **30**, 299 (1941)
2. A.M.Obukhov: Izv. Akad. Nauk SSSR **4**, 453 (1941)
3. A.S.Monin, A.M.Yaglom: *Statistical Fluid Mechanics* vol.2 (MIT Press, Cambridge 1975)
4. V.E.Zakharov, V.S.Lvov, G.E.Falkovich: *Kolmogorov Spectra of Developed Turbulence* vol.1 (Springer Verlag, Heidelberg 1991)
5. M.Lesieur: *Turbulence in Fluids* (Kluwer, London 1990)
6. A.M.Balk, V.E.Zakharov: in *Integrability and Kinetic Equations for Solitons* (Naukova Dumka, Kiev 1990) [in Russian]
7. A.V.Kats, V.M.Kontorovich: Zh. Prikl. Meh. Teh. Fiz. **6**, 97 (1974) [English translation: J. Appl. Mech. Tech. Phys.]
8. G.E.Falkovich: Sov. Phys. JETP **66**(1), 97 (1987)

Weak and Strong Turbulence in the CGL Equation

J.D. Gibbon[1], *M.V. Bartuccelli*[1], *and C.R. Doering*[2]

[1]Department of Mathematics, Imperical College, London SW7 2BZ, UK
[2]Department of Physics, Clarkson University, Potsdam, NY 13676, USA

To many fluid dynamicists, the only real turbulence is the fine scale 3-dimensional turbulence which occurs at high Reynolds numbers, with an energy cascade and an inertial subrange. The number of degrees of freedom in 3d strong turbulence is clearly many orders of magnitude greater than in such phenomena as convection in a box (see references in [1]) where perhaps only a few spatial modes govern the dynamics.

Only in 2d are the incompressible Navier Stokes equations understood analytically in the sense that there is a rigorous proof of the existence of a finite dimensional global attractor [2,3,4]. Computational methods are generally good enough to resolve the smallest scale in a 2d flow (see refs in [5]) and, for 2d homogeneous decaying turbulence, the vorticity obeys a maximum principle. No such maximum principle is known to exist in 3d and regularity remains to be proved. Numerical resolution of the smallest scale in a fully turbulent 3d flow is still a long way off.

There is yet one more important phenomenon which, as yet, lies in the realm of the unknown & that is the question of finite time singularities in the Euler equations [6]. These are the inviscid limit of the Navier Stokes equations and it is generally believed that, starting from smooth initial data, the 3d Euler equations do blow up in finite time but do not in 2d. However, there is no proof of either of these conjectures. Studies in Zakharov's equations have also done much to highlight the role played by finite time singularities in plasma turbulence [7,8].

In order to attempt to get a better grip on the tantalising phenomena displayed by the Navier Stokes equations, it is a useful exercise to see whether it is possible to mimic some limited features of the 3d Navier Stokes equations with a different PDE system which displays similar functional properties but in a lower spatial dimension. This exercise, however, must obviously be limited by the fact that simpler models in lower dimensions cannot display the vortex stretching properties displayed by the 3d Navier Stokes equations, although the lowering of the spatial dimension does make it easier to compute the dynamics.

One equation which we will show has some of the desired properties is a version of the d dimensional complex Ginzburg Landau (CGL) equation

$$A_t = RA + (1+i\nu)\Delta A - (1+i\mu)A|A|^{2q} \tag{1.1}$$

on the periodic domain $[0,1]^d$. R, μ and ν are real parameters with $R \geq 0$ and μ and ν of either sign. It is not our intention here to treat it in its physical context: indeed we shall divorce it from that entirely and treat it purely as an abstract model. In this sense, we do not pretend that it describes true fluid turbulence. Our intention in using it is to try and mimic limited features of the Navier Stokes equations with an equation over which we have more analytical control. Specifically we would like to have the following properties:

(i) A finite time singularity in the inviscid limit.
(ii) An area of the parameter space, going out to the inviscid limit, where strong turbulence occurs.
(iii) A transition boundary from strong turbulence to a much weaker form of turbulence as dissipation increases.
(iv) The possibility that these phenomena may occur in 2d or even 1d thereby making numerical studies much easier.
(v) The occurrence of an inertial subrange in the strong region.

What we have loosely called 'strong' and 'weak' turbulent behaviour needs some explanation. We define strong or hard turbulence to be a phenomenon where large fluctuations away from temporal & spatial averages occur. If this occurs, then solutions must be spatially & temporally narrow (i.e. spiky) with a consequent cascade of energy to high k. The spiking of the solution would rationally have its source in the finite time singularity in the inviscid limit trying to force the solution to blow up, only for the dissipation to finally control the system and bring it back down again. In contrast, weak or soft turbulence maybe chaotic but would display no great fluctuations away from spatial & temporal averages. In such a system, one naturally expects strong turbulent behaviour to occur near the inviscid limit where dissipation is small and the weak turbulent behaviour where dissipation is stronger.

We note straight away that condition (i) is fulfilled in the case of the CGL equation. The inviscid limit ($R \to 0$, $|\nu|, |\mu| \to \infty$), produces the NLS equation in d dimensions

$$iA_t = \nu \Delta A - \mu A|A|^{2q} \tag{1.2}$$

which is well known to blow up in finite time [9] provided $qd \geq 2$ and provided the energy, a constant of the motion

$$E = \int [\,|\nabla A|^2 + \frac{\mu}{2\nu}|A|^{2(q+1)}] \tag{1.3}$$

is negative which can only be true provided μ and ν are of opposite sign. It is particularly the so-called "critical case" $dq = 2$ which interests us here because it allows us to to look at the $d = 2$ (with $q = 1$) and the $d = 1$ (with $q = 2$) cases together. $dq > 2$ is the supercritical case.

As a summary of references [10,11], it is possible to construct a set of functionals of rank 2N when $q = 1$ (see [10] for $q = 1$ and [11] for $q \geq 1$)

$$F_N = \int [\,|\nabla^{N-1}A|^2 + \alpha_N|A|^{2N}]dx \tag{1.4}$$

where the F_N satisfy a set a ladder theorem

$$\dot{F}_N \leq (2NR + c_N\|A\|_\infty^2)F_N - b_N F_N^2/F_{N-1} \tag{1.5}$$

with $b_N = 2\min(1, \alpha_{N-1}/\alpha_N)$ & $c_N \geq 0$. Using an interpolation inequality to connect $\|A\|_\infty$ with the two halves of F_N

$$\|A\|_\infty^2 \leq c\left[F_N^{\frac{2}{2N-d}} + F_N^{\frac{1}{N}} \right] \qquad 2N > d + 2 \tag{1.6}$$

Since $2N > d + 2$, this allows a construction of absorbing balls for each of the F_N provided upper bounds on the "bottom rung" can be found. For d

≥ 2 this bottom rung is F_2. These bounds turn out to have an interesting dual structure in the μ-ν plane [10]. When d = 1, upper bounds on F_2 and $\|A\|_\infty^2$ are well controlled everywhere in the μ-ν plane. For d = 2, however, two regions in the μ-ν plane show widely differing behaviour. In one region, the so-called "weak" region,

$$\overline{\lim_{t\to\infty}}\, F_2 \leq cR^2 \qquad \langle F_2 \rangle \leq cR^2 \tag{1.7a}$$

$$\overline{\lim_{t\to\infty}}\, \|A\|_\infty^2 \leq cR^2 \qquad \overline{\lim_{t\to\infty}}\, \|A\|_2^2 \leq R \tag{1.7b}$$

where $\langle . \rangle$ is a time average. Hence large time supnorms are close to temporal & spatial averages. In contrast, there are boundaries in the 2nd and 4th quadrants [10,12], which are hyperbolae, which distinguishes between the weak region above and the rest of the μ-ν plane which goes out to the inviscid limit ($\pm\,\mu,\ \mp\,\nu \to \infty$), where the NLS equation has a finite time blow-up singularity, there is a strong difference between large time supnorms & spatial & temporal averages.

$$\overline{\lim_{t\to\infty}}\, F_2 \leq cR^{1/2+|\nu|} \qquad \langle F_2 \rangle \leq cR^2 \tag{1.8a}$$

$$\overline{\lim_{t\to\infty}}\, \|A\|_\infty^2 \leq cR^{1/2+|\nu|} \qquad \overline{\lim_{t\to\infty}}\, \|A\|_2^2 \leq R \tag{1.8b}$$

For large $|\nu|$, this means that fluctuations which achieve, or nearly achieve, the supnorm must be both spatially & temporally narrow i.e. spiky fluctuations must occur and can be thought of as a form of hard turbulence. It is not difficult to demonstrate that the spikes are homoclinic loops in the global attractor which can associated with the exact rotating wave solutions of the CGL equation [14]. There is also preliminary evidence [15] that there is an inertial range exists, but this has yet to be solidly confirmed. Certainly, the connection between the 2d CGL equation and the finite time singularity in the NLS equation (the bounds $R^{|\nu|+1/2}$ obviously fail in the inviscid limit) is what fluid dynamicists would like to establish between the Navier Stokes equations and the Euler equations. Although the situation is very different, such transitions gave been seen in gaseous Helium [13].The 2d CGL equation can therefore be thought of as a simple 'toy' model for studying transitions from weak to strong turbulence with a transitional boundary.

For general q, what of the supercritical case qd > 2? It turns out that solutions are regular everywhere in the region but not beyond where only weak solutions can be found [11].

$$|\nu|^2 + 1 < \left[\frac{qd}{qd-2}\right]^2 \tag{1.9}$$

Finally. what of the navier Stokes equations? A ladder theorem similar to (1.5) has indeed been proven [16] which reproduces many of the well known navier Stokes results found by classical methods [17]. It also produces the correct natural length scale in the flow in agreement with [18].

[1] Libchaber A.: Proc. Royal Soc. London <u>413A</u>, 63, 1987 and *"From chaos to turbulence in Benard convection"*, published in *"Dynamical chaos"*, p63, Princeton University Press 1987.

[2] Constantin P., Foias C. & Temam R.; Physica <u>30D</u>, 284, 1988.

[3] Constantin P., Foias C.; *The Navier Stokes equations,* Chicago University Press 1989.
[4] Temam R.; *Infinite dimensional dynamical systems in mechanics & physics.* Springer Applied Math Series 68, 1988.
[5] Weiss N. O.; Proc. Royal Soc. London 413A, 71, 1987 and *"From chaos to turbulence in Benard convection"*, published in *"Dynamical chaos"*, p71, Princeton University Press 1987.
[6] Majda A.; Comm. Pure & Appl. Math. 39, 187-220, 1986.
[7] Zakharov V.E.; Sov. Phys. JETP, 35, 908, 1972.
[8] Goldman M.; Rev. Mod. Phys. 56, 709-35, 1984.
[9] Landman M.J., Papanicalou G.C., Sulem C. & Sulem P.L.; Phys. Rev. A38, 3837-43, 1988. K. Rypdal, I Rasmussen & K. Thomsen; Physica 16D, 339, 1985. I. Rasmussen & K. Rypdal; Phys. Scr. 33, 481, 1986.
[10] Bartuccelli M., Constantin P., Doering C., Gibbon J. D. & Gisselfält M.; Physica D, 44, 421-444, 1990.
[11] Doering C.R., Gibbon J.D. & Levermore D.: *"Weak and strong solutions of the CGL equation"* preprint 1991.
[12] Levermore C. D.; private communication.
[13] Heslot F., Castaing B. & Libchaber A.; Phys. Rev. A36, 5870-3, 1987.
[14] Doering C., Gibbon J. D., Holm D.D. & Nicolaenko B.; Nonlinearity 1, 279-309, 1988 *and* Phys. Rev. Letts. 59, 2911-4, 1987.
[15] Doering C.D., Gibbon J.D., Holm D.D., Hyman J.M., Levermore D.; in preparation 1990.
[16] Bartuccelli M., Doering C. R., & Gibbon J. D.; *"Ladder theorems for the 2D & 3D Navier Stokes equations on a finite periodic domain"* Nonlinearity, 4, 531-542, 1991.
[17] R. Temam: *"The Navier Stokes equations & nonlinear functional analysis"*, CBMS-NSF Regional Conference Series in Appl. Math., SIAM 1983.
[18] W.D. Henshaw, H.O. Kreiss & L.G. Reyna: *Smallest scale estimates for the incompressible Navier Stokes equations*; preprint 1990.
W.D. Henshaw, H.O. Kreiss & L.G. Reyna: *On the smallest scale for the incompressible Navier Stokes equations*; J. Theor. Appl. Fluid Mech., 1, 1-32, 1989.

Numerical Test of a Weak Turbulence Approximation for an Electromagnetically Driven Langmuir Turbulence

A. Hanssen and E. Mjølhus

Institute of Mathematical and Physical Sciences, University of Tromsø, 9000 Tromsø, Norway

Abstract. In ionospheric radio modification experiments, manifestations of excited Langmuir turbulence are observed by means of VHF or UHF radars. Such experiments are performed in Arecibo, Puerto Rico,[1] and at Tromsø, Northern Norway.[2] A weak turbulence theory involving parametric cascade of Langmuir waves, has earlier dominated the theoretical understanding of these experiments.[3] This has recently been challenged, both from a theoretical[4] and an experimental[5] point of view, and a theory of strong Langmuir turbulence, involving a large number of nucleation – collapse – burnout cycles has been advocated.[4] A version of the Zakharov model including damping and parametric driving, contains both of these scenarios, the crucial parameter being $\Delta\Omega = \omega - \omega_{pe}$ where ω is the applied frequency and ω_{pe} the plasma frequency.[6] This model allows the construction of a weak turbulence wave kinetic equation. In the present work[6,7] spectra obtained from full wave solutions of the one dimensional Zakharov model are compared with saturation spectra of the wave kinetic model. The results can be described as follows: (i) for large values of $\Delta\Omega$, cascades are formed, and the number of cascades increases with the strength of the driver E_0; (ii) the number of cascades found in the full wave solution is smaller than that obtained from the wave kinetic equation; (iii) when E_0 becomes sufficiently large, the narrowly peaked cascade structure of the full wave spectrum breaks down, and a broad spectrum comes instead; (iv) this breakdown comes far before the cascade sequence has reached the Langmuir condensate; thus the Langmuir condensate plays no role in this process.

At smaller values of $\Delta\Omega$, the turbulence is characterized by caviton nucleation[4] resulting in broad wave number spectra. Also a coexistence range is found at intermediate values of $\Delta\Omega$, in which a few cascade lines ride upon a broad cavitation type spectrum, similar to what has also recently been reported in two dimensional simulations.[8] A detailed report on this work will be sought published elsewhere.[7]

References

1. J.A. Fejer, M.P. Sulzer, and F.T. Djuth, *J. Geophys. Res.*, Sept. 1991.

2. P. Stubbe, H. Kohl, and M.T. Rietveld, subm. to *J. Geophys. Res.*, 1991.

3. F.W. Perkins, C. Oberman, and E.J. Valeo, *J. Geophys. Res.*, **79**, 1478, 1974.

4. D.F. Dubois, H.A. Rose, and D. Russell, *J. Geophys. Res.*, **95**, 21221, 1990.

5. P.Y. Cheung et al., *Phys. Rev. Lett.*, **62**, 2676, 1989.

6. A. Hanssen and E. Mjølhus, *AGARD Conf. Proc.*, **485**, 9–1, 1990.

7. A. Hanssen, E. Mjølhus, D.F. Dubois, and H.A. Rose, subm. to *J. Geophys. Res.*, Oct. 1991.

8. D.F. Dubois, H.A. Rose, and D. Russel, *Phys. Rev. Lett.*, **66**, 1970, 1991.

Drift Wave Vortices in Inhomogeneous Plasmas

W. Horton, X. Su, and P.J. Morrison
Institute for Fusion Studies and Department of Physics,
The University of Texas at Austin, Austin, TX 78712, USA

Drift wave turbulence in weakly driven or decaying states possesses strong correlations requiring the concept of a weakly correlated vortex gas. Recent progress on the effects of inhomogeneities on the structure, stability and life-time of the vortices is reviewed. In particular, two cases (i) of a finite temperature gradient, and (ii) of a shearing of the magnetic field across the vortex structure are analyzed. A new formulation of drift wave turbulence proposed by Zakharov (1991) in terms of the separation of short and long scales is applied.

I. Rossby-Drift Wave Equations

In rotating fluids and magnetized plasmas the vortex state is an important structure that naturally arises. An example of such a long-lived structure is shown in Fig. 1, which is a weather map taken from the April 1, 1985 Los Angeles Times. The figure shows a cyclonic vortex with radius of 1000 km that produced severe cold weather for many days. In plasmas such vortex states are thought to be responsible for part of the anomalous transport measured in confinement systems. The fundamental equations for the dynamics of the slow neutral fluid flows on rotating planets and the $\mathbf{E} \times \mathbf{B}$ drift flows in inhomogeneous magnetized plasmas are isomorphic problems governed by what is now known as the Charney-Hasegawa-Mima equation (CHM). This important result was established by Hasegawa, McLennan and Kodama (1979) and Petviashvili (1980).

The reason for the coincidence of the two different systems and the conditions for the breakdown of the CHM are seen by considering the conservation laws

$$\frac{\partial n}{\partial t} + \mathbf{v} \cdot \nabla n + n \nabla \cdot \mathbf{v} = 0 \tag{1}$$

and

$$\frac{d\mathbf{v}}{dt} = -\nabla U + \mathbf{v} \times \mathbf{\Omega} \tag{2}$$

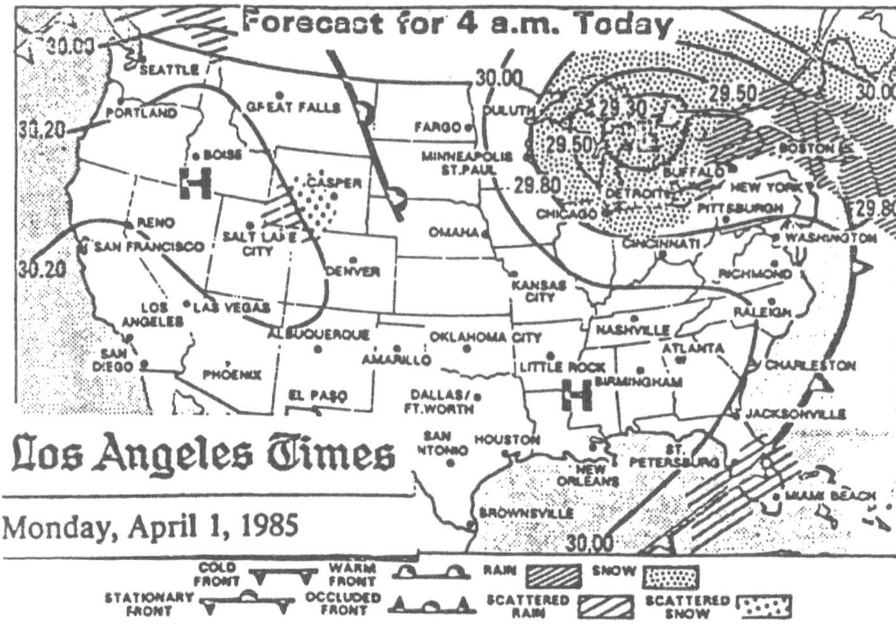

1. Large cyclonic vortex structure bringing heavy snow in late spring over the New York-Great Lakes area. The radius is approximately 1000 km and $h = \delta p/p \simeq 1/30$.

where for shallow water flows the conserved field $n \to H(x,y,t)$ — the depth of fluid layer and $U = gH(x,y,t)$ from the hydrostatic pressure $p = \rho g H$, while for plasmas, n is the ion density and $U = e\Phi/m_i$ where Φ is the electrostatic potential. On the rotating planet Ω is the Coriolis force parameter $\Omega = f = 2\Omega_p \sin\theta$, where θ is the latitude and $\Omega_p = 2\pi/T$ with T the period (day) for rotation, while for plasmas $\Omega = eB/m_i$ is the cyclotron frequency. Taking the rotational part of Eq. (2) and defining the vorticity as $\omega = \nabla \times \mathbf{v}$ we obtain

$$\left(\frac{\partial}{\partial t} + \mathbf{v}\cdot\nabla\right)(\Omega + \omega) = -(\Omega+\omega)(\nabla\cdot\mathbf{v}) + (\Omega+\omega)\cdot\nabla\mathbf{v}$$

and using Eq. (2) yields

$$= (\Omega+\omega)\frac{1}{n}\frac{dn}{dt} + (\Omega+\omega)\cdot\nabla\mathbf{v}\,. \tag{3}$$

The last term in Eq. (3) is the "vortex stretching term" that gives rise to the kinking and the reconnection of vortex filaments in the non-rotating or unmagnetized hydrodynamics. In the presence of large $|\Omega/\omega|$, however, the vortex filaments are forced to remain nearly straight (Taylor-Proudman Theorem) from the x,y-components of Eq. (3), and thus it is often a good approximation to drop this last term in Eq. (3), which describes the parallel compression of the fluid. Assuming $(\Omega + \omega) \cdot \nabla v_z \simeq \Omega\, \partial_z v_z$ is small, the parallel component of Eq. (3) gives Ertel's theorem

$$\frac{d}{dt}\left(\frac{\Omega + \omega_z}{n}\right) = 0 \tag{4}$$

where $\omega_z = \partial_x v_y - \partial_y v_x$. The conservation law (4) gives the CHM equation when ω_z is evaluated in the geostrophic or $\mathbf{E} \times \mathbf{B}$ drift approximation and only linear gradients are taken into account.

A. Ordering for slow flow motions

For flows that evolve slowly compared with Ω we introduce the small ordering parameter

$$\varepsilon = \frac{1}{\Omega}\frac{\partial}{\partial t} \sim \frac{\mathbf{v} \cdot \nabla}{\Omega} \ll 1 \,. \tag{5}$$

In the first order Eq. (2) yields

$$\mathbf{v} = \frac{\hat{\mathbf{z}} \times \nabla U(x, y, t)}{\Omega} \tag{6}$$

and the inertial acceleration $-d\mathbf{v}/dt$ correction gives

$$\mathbf{v} = \frac{\hat{\mathbf{z}} \times \nabla U}{\Omega} - \frac{1}{\Omega^2}\frac{d}{dt}\nabla U \,. \tag{7}$$

Using Eq. (6) to calculate ω_z and the convective derivative, Eq. (4), leads to the Charney equation

$$\frac{dq}{dt} = \left(\partial_t + \frac{gH_0}{\Omega}\{h, \}\right)\left[\frac{\left(\Omega + \frac{gH_0}{\Omega}\nabla^2 h\right)}{H_0(1+h)}\right] = 0 \tag{8}$$

where $h = (H - H_0)/H_0$ is the relative depth of the shallow fluid. In the case of the plasma we assume that the electrons quickly thermalize in the potential Φ giving $n = N(x)\exp(e\Phi/T_e)$ and the calculation of $\omega_z = (\Omega\, \rho_s^2\, e/T_e)\nabla^2\Phi$ so that the corresponding nonlinear p.d.e. is

$$\frac{dq}{dt} = \left(\partial_t + \frac{1}{B}\{\Phi, \}\right)\left[\frac{\Omega\left[1 + (\rho_s^2/T_e)\nabla^2(e\Phi)\right]}{N(x)\exp(e\Phi/T_e)}\right] = 0 \,. \tag{9}$$

Thus with the identification of $\rho_s^2 = m_i T_e/e^2 B^2$ with $L_d^2 = gH_0/\Omega^2$ and $e\Phi/T_e$ with h and the expansion of the Boltzman distribution as $\exp(e\Phi/T_e) \simeq 1 + e\Phi/T_e$ the two equations (8) and (9) are the same and can be written in units of $\rho_s \leftrightarrow L_d$ and $1/\Omega$ as

$$(\partial_t + \{\varphi, \quad \})(\nabla^2\varphi - \varphi + v_d x) = 0 \tag{10}$$

where $q \cong \nabla^2\varphi - \varphi + v_d x$, and $v_d = c_s^2/\Omega L_n \leftrightarrow gH_0/\Omega R_p$ is the linear long wave phase velocity. Here x is in the direction of the gradient (northward) and y in the direction of the linear wave propagation (westward).

Equation (10) is the CHM equation which has the following properties:

(1) conserves mass, energy, enstrophy and integrals of arbitrary functions of q

(2) translationally invariant in x, y

(3) possesses the exact Larichev-Reznick (1976) dipole vortex solutions.

The dipole vortex solution is a two-parameter family (r_0 = radius, u = speed) of solitary vortex solutions. The speed c of the wave components of Eq. (10) are in the range $0 < c < v_d$ and the speed u of the vortex solutions is either $u > v_d$ or $uv_d < 0$. Soliton-like or weakly inelastic collisions with zero impact parameter are given by Makino *et al.* (1982) and McWilliams and Zabusky (1982), and the strongly inelastic collisions that occur when the impact parameter $b \simeq r_0$ are shown in Horton (1989).

B. Inhomogeneous systems

Now when the size of the vortex r_0 is taken comparable with the variation of the inhomogeneous background of the medium, the structure of the Rossby-Drift Wave equation changes. Petviashvili (1977) and Tasso (1967) have shown that in a plasma the inhomogeneity introduces the KdV nonlinearity $\varphi\,\partial_y\varphi$ into Eq. (10). This change alone, however, spoils the conservation from (9) of the equation. The correct treatment of the nonlinear-inhomogeneous systems expands Eq. (9) to obtain the generalized q-conserving equation (Su *et al.* (1991))

$$\left(\frac{1}{T(x)} - \nabla^2\right)\frac{\partial\varphi}{\partial t} + (v_{d0} + v_d' x - K_T\varphi)\frac{\partial\varphi}{\partial y} - [\varphi, \nabla^2\varphi] = 0 \tag{11}$$

where $v_{d0} + v_d' x$ is the inhomogeneous drift-Rossby speed and $K_T = (d/dx)(1/T) = -T'/T^2$ is the inhomogeneity of the dispersion scale $\rho_s^2 = T/\Omega^2$ or $L_d^2 = gH_0/\Omega^2$. Here $T(x)$ is the dimensionless temperature profile normalized to unity at the center of the vortex. Equation (11) has the following important properties:

(1) mass, energy and moments of q are conserved, but not the usual enstrophy

(2) the dipole vortex is split into monopole vortices with only one sign (cyclone or anticyclone depending on the sign of v_d') being a long-lived vortex

(3) the equation is not translationally invariant in the x (north-south) direction — there is now a preferred direction for wave propagation which is toward the equatorial zone or to the hot plasma region.

We have shown in Su *et al.* (1991) that Eq. (11) has monopole vortex solutions given by

$$\nabla^2\varphi = k^2(u,x)\varphi + \frac{v'_{d0}}{2u^2}\varphi^2 \tag{12}$$

where $k^2 = (1/T(x) - v_d(x)/u) \sim \varepsilon$. The expansion of $k^2(u,x)$ about the location of the vortex leads to

$$k^2 = k_0^2 + \alpha x$$

where

$$\alpha = (K_T - v'_{d0}/u) \sim \varepsilon^2 \tag{13}$$

and the monopole vortex is given by

$$\varphi(x,y,t) \cong -\frac{4.8u^2\, k_0^2}{v'_{d0}} \cosh^{-4/3}\left(\frac{3}{4}\, k_0 \sqrt{x^2 + (y-ut)^2}\right) . \tag{14}$$

The speed u of the vortex depends on the amplitude φ_m with

$$u = \frac{1}{2}\left[v_{d0} + (v_{d0}^2 + 0.83\, v'_{d0}\,\varphi_m)^{1/2}\right] \approx v_{d0}(1 + 0.21 v'_{d0}\,\varphi_m) \tag{15}$$

for small, positive $v'_{d0}\,\varphi_m$. Thus the shear in the drift-Rossby speed $v_d(x)$ acts to change the solitary solution from a dipole to a monopole. One recalls that the monopole vortex is the natural solution in sheared flows (Horton *et al.*, 1987).

II. Propagation and Collisions of the Monopole Vortices

A. Vortex-vortex interactions

We have established that the monopole vortices in Eq. (14) can behave under collisions either as

(1) soliton-like collisions with the stronger vortex overtaking and passing through the weaker vortex

or

(2) point vortex-like interactions where two strong monopole vortices, which by Eq. (14) must be of the same sign, rotate about one another.

In Fig. 2 we show an example of the soliton-like pass-through collision. In Fig. 3 we show an example of the second case where two nearly equal strength monopole vortices interact like point-vortices rotating around one another.

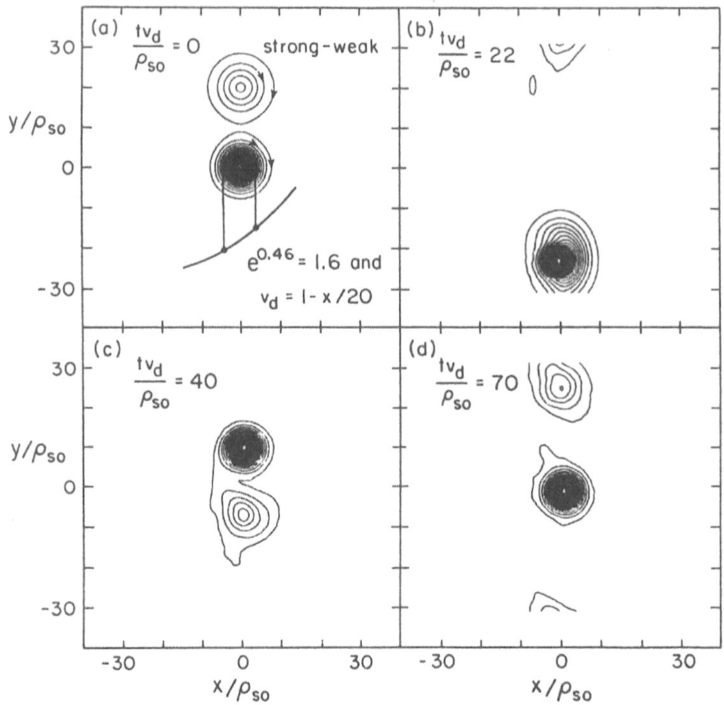

2. Nearly elastic overtaking collision of a strong monopole vortex with a weaker monopole. The profile of $1/T(x) = 1/L_d^2 = \exp(.046x)$ gives a variation of 1.6 over the core of the vortex. The gradient in the Rossby speed is $v_D(x) = 1 - x/20$. The solution conserves $q = \nabla^2\varphi - \varphi/T(x) + \int^x v_D\, dx'$. The speeds are $u_1 = 1.1$ and $u_2 = 1.7$ giving the expected collision time $\Delta\ell/\Delta u = 20/0.6 = 33$ compared with observed overlapping at $t = 22$ in frame (b). After separation (c)–(d) the weaker vortex still has 5 closed contours.

B. Wave radiation

As the amplitude φ_m of the vortices becomes small, the speed of propagation in Eq. (15) approaches the linear wave speed and the coupling to the wave field radiates energy from the vortex. Su *et al.* (1991) calculate this radiative decay of the vortex. The local energy conservation equation is

$$\frac{\partial \mathcal{E}}{\partial t} + \nabla \cdot \mathbf{S} = 0 \tag{16}$$

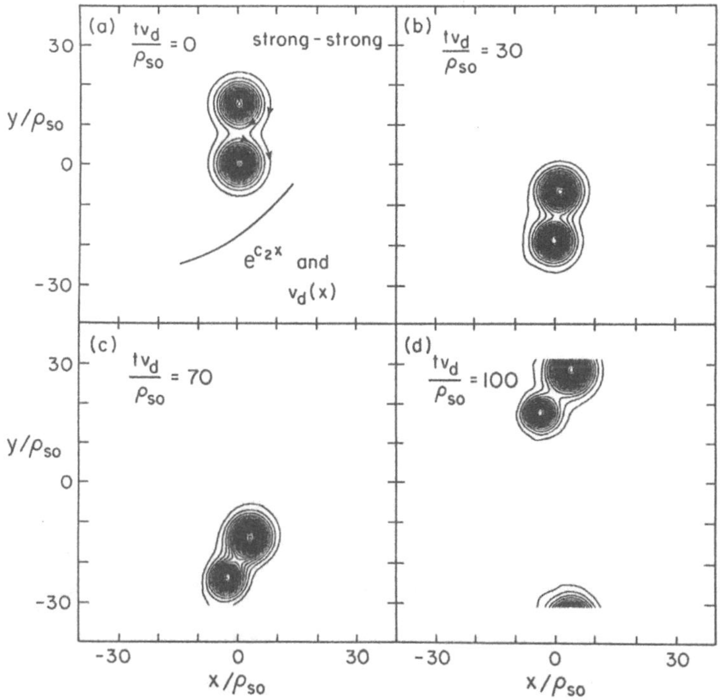

3. Point vortex type of interaction of two strong monopole vortices with speeds $u_1 = 1.30$ and $u_2 = 1.35$. Although merging might be expected here, instead the vortices rotate around one another after pulling together from the initial separation of $20\,L_d$.

where

$$\mathcal{E}(x,y,t) = \frac{1}{2}\left[\frac{\varphi^2}{T(x)} + (\nabla\varphi)^2\right] \tag{17}$$

and the "Poynting" flux is

$$\mathbf{S} = \left(\frac{1}{2}v_d(x)\,\varphi^2 - \frac{1}{3}K_T\,\varphi^3\right)\hat{\mathbf{y}} - \varphi\nabla\frac{\partial\varphi}{\partial t} - \nabla^2\varphi\hat{\mathbf{z}}\times\nabla(\varphi^2/2)\,. \tag{18}$$

The results of a lengthy calculation are that the decay of the vortex energy

$$E_v = \int \mathcal{E}\,d^2x = \frac{8.2\pi\,u^4\,k_0^2}{(v'_{d0})^2}\left(\frac{4}{3} + k_0^2\right) \tag{19}$$

is given by

$$\frac{dE_v}{d\tau} \cong -\frac{u|\alpha|\varphi_m^2}{16\pi\,k_0^3}\exp\left(-\frac{4(1-v_d/u)^{3/2}}{3|\alpha|}\right)\,. \tag{20}$$

The exponential decay factor is controlled by the strength of the inhomogeneity through α (Eq. (13)) and the closeness of the speed of propagation u to the drift speed v_d at the center of the vortex through $k_0^3 = (1 - v_{d0}/u)^{3/2}$. Simulations for $k_0^3 \gg \alpha$ and $k_0^3 \ll \alpha$ are shown in Su *et al.* (1991).

The decay of the vortex amplitude in Eq. (20) via the coupling to the radiation field will cause the speed u to decrease through Eq. (15), and as the speed $u(t)$ decreases the vortex decay rate increases exponentially through Eq. (20). Thus, the vortex will decay slowly initially and then suffer an abrupt death.

In the case of magnetic shear, which is another form of inhomogeneity that gives rise to a coupling to vertical v_z oscillations, Meiss and Horton (1983) show that the decay rate of the dipole vortex soliton is given by

$$\frac{dE_{dp}}{dt} = -2S\varphi_m^2 \left(1 - \frac{v_d}{u}\right)^{-1/2} \exp\left(-\frac{\pi L_s}{2L_n}\left(1 - \frac{v_d}{u}\right)\right) \tag{21}$$

where L_s/L_n is magnetic shear (inhomogeneity) length over the density gradient scale length L_n. We have also performed simulations for the shear induced decay of the vortex structures.

These vortex structures appear to be a natural or "self-organized" way in which the plasma can feed upon the free energy available in the density gradient and limit the radiation damping inherent in small amplitude waves. Recent simulations indicate that the vortex localization process in systems with rather different linear growth rates (due to damping caused by magnetic or velocity shear) can end up in similar final turbulent states when enough energy is fed into the system. This is because the localization to vortex structures essentially eliminates the shear damping mechanisms. This nonlinear dynamics and the shear damping introduces a form of hysteresis into the system, due to the slow decay rate of the vortices once they are formed.

III. Driving of the Large Scale Vortex Structures by the Small Scale Rossby-Drift Wave Turbulence

Finally, following the suggestion of Zakharov (1991), we consider the interaction of the small scale, weakly correlated Rossby-drift wave fluctuations φ (small scale) $\rightarrow \psi = \sum_{\mathbf{k}} \psi_{\mathbf{k}}(\varepsilon x, \varepsilon t) e^{i\mathbf{k}\cdot\mathbf{x} - i\omega_{\mathbf{k}} t}$ with

$$\langle \psi_{\mathbf{k}}(\varepsilon \mathbf{x}, t) \psi_{\mathbf{k}'}(\varepsilon \mathbf{x}, t) \rangle = \frac{\delta_{\mathbf{k},\mathbf{k}'} N_{\mathbf{k}}(\mathbf{X}, T)}{1 + k_\perp^2 \rho_s^2} \, , \tag{22}$$

where the wave density $N_{\mathbf{k}}(\mathbf{X}, T)$ satisfies

$$\frac{\partial N_{\mathbf{k}}}{\partial T} + \frac{\partial \tilde{\omega}_{\mathbf{k}}}{\partial \mathbf{k}} \cdot \frac{\partial N_{\mathbf{k}}}{\partial \mathbf{X}} - \frac{\partial \tilde{\omega}_{\mathbf{k}}}{\partial \mathbf{X}} \cdot \frac{\partial N_{\mathbf{k}}}{\partial \mathbf{k}} = 2\gamma_k N_{\mathbf{k}} + T_{\mathbf{k}}^{n\ell}(N_{\mathbf{k}}, N_{\mathbf{k}}) \tag{23}$$

as in Horton (1986), and the local drift wave frequency depends on the large scale $\varphi_L, \tilde{n}_L$ variations through

$$\tilde{\omega}_{\mathbf{k}} = \mathbf{k} \cdot \hat{\mathbf{z}} \times \nabla \varphi_L + \frac{\mathbf{k} \cdot \hat{\mathbf{z}} \times \nabla (n_0 + \tilde{n}_L)}{1 + k_\perp^2 \rho_s^2} ,$$

with large scale motions governed by φ (large scale) $\rightarrow \varphi(X, Y, T)$ that satisfies the CHM equation containing the average of $\langle \{\psi, \nabla^2 \psi\} \rangle$ over the small scale turbulence. The driven large scale CHM equation is given by

$$(1 - \nabla^2)\partial_T \varphi + v_d \partial_Y \varphi - \{\varphi, \nabla^2 \varphi\} = (\partial_X^2 - \partial_Y^2) A + \partial_{XY}^2 B \tag{24}$$

where

$$\begin{bmatrix} A(X,Y,T) \\ B(X,Y,) \end{bmatrix} = \sum_{\mathbf{k}} \begin{bmatrix} (k_x k_y) \\ (k_x^2 - k_y^2) \end{bmatrix} \frac{N_{\mathbf{k}}(X,Y,T)}{1 + k_\perp^2 \rho_s^2} . \tag{25}$$

This system of equations leads to the modulational growth of large scale structures from the inhomogeneity of the distribution of the small scale fluctuations. In a tokamak the small scale turbulence is known to have a strongly increasing strength toward the low density side and, at a given radius, an increase in strength toward the outside of the torus compared with the inside. In planetary atmospheric turbulence it may be expected that the intensity of the small scale turbulence is stronger in the equatorial zones than in the high latitude regions. It is clear from the structure of Eqs. (23)–(25) that when the basic assumptions of the scale separations are satisfied, that anisotropy and inhomogeneity in the small scale turbulence is a driving force on the large scale structures. We are in the process of investigating the driven CHM equation (24) and the propagation of the small scale turbulence by the nonlinear wave kinetic equation (23) for various systems. This separation of space-time scales appears to be an effective method for extending the study of Rossby-Drift Wave turbulence to more realistic inhomogeneous turbulent states compared with the previously studied homogeneous turbulent states, as for example, in Horton (1986).

Acknowledgment

The work was supported by the U.S. Department of Energy contract DE-FG05-80ET-53088.

References

A. Hasegawa, C.G. Maclennan, and Y. Kodama, Phys. Fluids **22**, 2122 (1979).

W. Horton, Phys. Fluids **29**, 1991 (1986).

W. Horton, T. Tajima, and T. Kamimura, Phys. Fluids **30**, 3485 (1987).

W. Horton, Phys. Fluids B **1**, 424 (1989).

V.D. Larichev and G.M. Reznik, Oceanology **16**, 547 (1976).

M. Makino, T. Kamimura, and T. Taniuti, J. Phys. Soc. Jpn. **50**, 980 (1981).

J.C. McWilliams and N.J. Zabusky, Geophys. Astrophys. Dyn. **19**, 207 (1982).

J.D. Meiss and W. Horton, Phys. Fluids **26**, 990 (1983).

V.I. Petviashvili, Fiz. Plazmy **3**, 270 (1977) [Sov. J. Plasma Phys. **3**, 150 (1977)].

V.I. Petviashvili, Pis'ma Zh. Eksp. Teor. Fiz. **32**, 632 (1980) [JETP Lett. **32**, 619 (1980)].

X.N. Su, W. Horton, and P.J. Morrison, Phys. Fluids B **3**, 921 (1991).

H. Tasso, Phys. Lett. A **24**, 618 (1967).

V.E. Zakharov, Review talk, Sherwood Theory Conference, Seattle, Washington, April, 1991.

Solitons on a Vortex Filament with Axial Flow

Y.H. Ichikawa, K. Konno, and H. Ohno

National Institute for Fusion Science, Chikusa-ku, Nagoya 464-01, Japan
Department of Physics, College of Science and Technology,
Nihon University, Chiyoda-ku, Tokyo 101, Japan

Studies of motion of vortex filament provide the crucial problems of mathematical physics. Here, the extended local induction equation, which accounts for the effect of axial flow, has been investigated on the basis of the inverse scattering method. The present analysis provides a two soliton solution with the intrinsic symmetry, while the solution constructed by the method of soliton surface does not show this property explicitely.

1 Introduction

During the past decades, a remarkable advancement in mathematical physics has been achieved in the field of nonlinear evolution equations, exploring various kinds of soliton systems in the framework of the inverse scattering transformation[1,5] Now, we are not only enjoying the solvability of the completely integrable nonlinear equations, but also have useful information on the inter-relationship among these soliton systems. Ishimori[6] and Wadati et al[7] have established the equivalence among various schemes of the inverse scattering transformations by a gauge transformation with appropriate transformation of independent variables. Sym and his collaborators have been developing unification of various soliton systems on the geometrical concept of soliton surfaces[8,9].

Turning to application of these mathematical methods to physical phenomena, studies of dynamical motion of vortex filament attract very active interest inspired by both experimental[10,12] and theoretical investigations[13,14]. With the help of auxiliary quantities of curvature and torsion, Hasimoto[13] reduced the local induction equation of the vortex filament [15] to the cubic nonlinear Schrodinger equation, and gave an explicit form of one soliton propagating along the vortex filament. Levi et al[16] have presented two solitons on the vortex filament by applying the concept of soliton surfaces.

Reducing the local induction equation directly to one of the completely integrable soliton equations, we have carried out a straightforward calculation of the inverse scattering transformation and analyzed the evolution of two vortex solitons[17]. We obtained a symmetric expressions with respect to the labeling of two solitons, while Levi-Sym-Wojciechowski's solution is not symmetric with respect to the parameters of the two solitons.

Recently, Fukumoto and Miyazaki[18] have given an equation of motion of three dimensional distortion of a vortex filament $\vec{X}(s,t)$ with axial flow

$$\frac{\partial}{\partial \tau}\vec{X} = \vec{X}_s \times \vec{X}_{ss} + w\vec{X}_{sss} + \frac{3}{2}\vec{X}_s \times (\vec{X}_s \times \vec{X}_{ss}) \tag{1}$$

where the suffix s stands for partial derivative with respect to the arclength s, and τ is a renormalized time. Adopting the Levi-Sym-Wojciechowski soliton solution, they have examined effect of axial flow on two soliton processes on the vortex filament. Since Eq. (1) has been shown to be equivalent to the hybrid Wadati-Konno- Ichikawa equation by Ishimori[6], we have study soliton solution on the vortex filament basing on the explicit calculation of the inverse scattering transformation for the hybrid Wadati-Konno- Ichikawa equation,

$$\frac{\partial}{\partial t}q_x + sgn(\frac{dx}{ds})\frac{\partial^2}{\partial x^2}(-i\frac{q_x}{\Phi} + w\frac{q_{xx}}{\Phi^3}) = 0 \tag{2}$$

where

$$\Phi = (1 + |\frac{\partial q}{\partial x}|^2)^{\frac{1}{2}} \tag{3}$$

Here, we have introduced a complex field variable $q(x,t)$ to express the y- and z- component of position vector $\vec{X}(s,t)$ as

$$\vec{X}(s,t) = (x, -\Im q(x,t), \Re q(x,t)) \tag{4}$$

where $\Re$ ($\Im$) designates the real (imaginary) part.

2 Equation of Motion of Vortex Filament with Axial Flow

Contrary to Ishimori's approach to show the equivalence between Eq. (1) and Eq. (2), where the gauge transformation has been applied to the A- K-N-S inverse scattering scheme, here let us reduce Eq. (2) directly to Eq. (1). First, we transform the independent variables (x,t) to (s,τ) by

$$s = x + E_+(x,t) \tag{5}$$

$$\tau = t \tag{6}$$

with the definition of

$$E_+(x,t) = \int_x^\infty (1 - sgn(\frac{dx}{ds})\Phi)dx \tag{7}$$

Introducing the abbreviation of $Q = \partial q/\partial x$, we have

$$\frac{\partial}{\partial x} = sgn(\frac{dx}{ds})\Phi\frac{\partial}{\partial s}$$

$$\frac{\partial}{\partial t} = \frac{\partial}{\partial \tau} + sgn(\frac{dx}{ds})\frac{i}{2}\Phi^{-1}\{Q^+Q_s - Q_s^+Q\}\frac{\partial}{\partial s}$$

$$+\frac{w}{2}\Phi^{-2}\{3(Q_s)^3 - |Q|^2_{ss} + \frac{3}{8}\Phi^{-2}(|Q_8|^2_s)^2\}\frac{\partial}{\partial s} \tag{8}$$

After lengthy, but straight forward calculation, we can express Eq. (2) in terms of the new independent variables as

$$\begin{aligned}&\frac{\partial}{\partial t}Q+i\frac{1}{2}sgn(\frac{dx}{ds})\Phi^{-3}[(2Q_{ss}-|Q|^2Q_{ss}-Q^2Q^*_{ss})\Phi^2\\&-2Q|Q|^2-2Q^*(Q_s)^2+Q^2(Q^*_s)^2-|Q|^2Q^*(Q)^2]\\&+w[\frac{3}{2}(|Q_s|^2Q_s-Q_s|Q|^2_{ss}-Q_{ss}|Q|^2_s\\&+\frac{5}{4}(|Q|^2_s)^2Q_s\Phi^{-2})\Phi^{-2}+Q_{sss}]=0\end{aligned}\tag{9}$$

As the second step, we transform the dependent variable $Q(s,t)$ into a three dimensional vector $\vec{A}=(a,b,c)$ as

$$a\equiv sgn(\frac{dx}{ds})\Phi^{-1}=\frac{\partial x}{\partial s}\tag{10a}$$

$$b\equiv i\ \ sgn(\frac{dx}{ds})\Phi^{-1}Q=i\frac{\partial}{\partial s}q\tag{10b}$$

$$c\equiv -i\ \ sgn(\frac{dx}{ds})\Phi^{-1}Q^*=-i\frac{\partial}{\partial s}q^*\tag{10c}$$

Then, we can show that Eq. (9) is directly transformed into

$$\frac{\partial a}{\partial t}=\frac{i}{2}(bc_{ss}-cb_{ss})-w[a_{sss}+\frac{3}{2}\frac{\partial}{\partial s}((a_s^2+b_sc_s)a)]\tag{11a}$$

$$\frac{\partial b}{\partial t}=-i(ab_{ss}-ba_{ss})-w[b_{sss}+\frac{3}{2}\frac{\partial}{\partial s}((a_s^2+b_sc_s)b)]\tag{11b}$$

$$\frac{\partial c}{\partial t}=+i(ca_{ss}-ac_{ss})-w[c_{sss}+\frac{3}{2}\frac{\partial}{\partial s}((a_s^2+b_sc_s)c)]\tag{11c}$$

Hence, with the definition of the tangent vector $\vec{T}$ as

$$\vec{T}\equiv\frac{\partial\vec{x}}{\partial s}=(\frac{\partial x}{\partial s},-\Im\frac{\partial q}{\partial s},\Re\frac{\partial q}{\partial s})=(a,\frac{1}{2}(b+c),-i\frac{1}{2}(b-c))\tag{12}$$

we establish that Eqs. (11a-11c) are reduced to the following expression of the equation of motion of the vortex filament with axial flow,

$$\frac{\partial}{\partial t}\vec{T}=\vec{T}\times\vec{T}_{ss}+w[\vec{T}_{sss}+\frac{3}{2}((\vec{T}_s)^2\vec{T})_s]\tag{13}$$

3 Soliton Solution by the Inverse Scattering Transformation

Referring to the work of Ishimori[6], we introduce the inverse scattering transformation to Eq. (2) by decomposing it into the Lax-pair equation,

$$\frac{\partial}{\partial x}U = \lambda\begin{pmatrix} -i & q_x \\ -q_x^* & i \end{pmatrix}U \tag{14}$$

and

$$\frac{\partial}{\partial t}U = \begin{pmatrix} A & B \\ C & -A \end{pmatrix}U \tag{15}$$

with the definition of

$$A = sgn(\frac{dx}{ds})\{-4\frac{w}{\Phi}i\lambda^3 - w\frac{q_x q_{xx}^* - q_x^* q_{xx}}{\Phi^3}\lambda^2 - 2\frac{i}{\Phi}\lambda^2\} \tag{16a}$$

$$B = sgn(\frac{dx}{ds})\{4w\frac{q_x}{\Phi}\lambda^3 + 2wi\frac{q_{xx}}{\Phi^3}\lambda^2 + 2\frac{q_x}{\Phi}\lambda^2 - w(\frac{q_{xx}}{\Phi^3})_x\lambda + i(\frac{q_x}{\Phi})_x\lambda\} \tag{16b}$$

$$C = sgn(\frac{dx}{ds})\{-4w\frac{q_x^*}{\Phi}\lambda^3 + 2wi\frac{q_{xx}^*}{\Phi^3}\lambda^2 - 2\frac{q_x^*}{\Phi}\lambda^2 + w(\frac{q_{xx}^*}{\Phi})_x\lambda - i(\frac{q_x^*}{\Phi})_x\lambda\} \tag{16c}$$

Here, we notice that inverse construction of the potential q_x from the eigenvalue problem Eq. (14) has been carried out in our previous study of soliton propagation along the vortex filament without axial flow[17]. Now the contribution of axial flow gives rise to modification of the temporal evolution of the scattering data through Eq. (15) with Eqs. (16a-16c).

Temporal evolution of the eigen function in the scattering region at $|x| \to \infty$ is determined by

$$\frac{\partial}{\partial t}U = A_\infty\begin{pmatrix} 1 & 0 \\ 0 & -1 \end{pmatrix}U \tag{17}$$

where

$$A_\infty = \lim_{|x|\to\infty} A = -i(4w\lambda^3 + 2\lambda^2) \tag{18a}$$

$$B_\infty = \lim_{|x|\to\infty} B = 0 \tag{18b}$$

$$C_\infty = \lim_{|x|\to\infty} C = 0 \tag{18c}$$

Therefore, the normalization constant of bound state $C_k(t)$ evolves in time according to

$$C_k(t) = C_k(0)\exp\{4i(2w\lambda_k^3 + \lambda_k^2)t\} \tag{19}$$

Referring to our previous results[17], we have one soliton with $\lambda = \xi + i\eta, \eta > 0$,

$$q = -i\frac{C^*(t)}{\lambda^{*2}}\frac{\exp(-2i\lambda^* s)}{1 - |C(t)|^2(\lambda_1 - \lambda_1^*)^{-2}\exp[2i(\lambda - \lambda^*)s]} \tag{20}$$

$$E_+ = \frac{-i|C(t)|^2|\lambda|^{-2}(\lambda - \lambda^*)^{-1}\exp[2i(\lambda - \lambda^*)s]}{1 - (\lambda - \lambda^*)^{-1}|C(t)|^2\exp[2i(\lambda - \lambda^*)s]} \tag{21}$$

where the laboratory coordinate x is defined by Eq. (5). Defining the initial position of soliton s_0 as

$$s_0 = (4\eta)^{-1}\log(|C(0)|^2/4\eta^2) \tag{22}$$

we can express the one soliton solution as

$$q = q_0 sech[2\eta(s - vt - s_0)]\exp\{-i[2\xi s + 4(\xi^2 - \eta^2 + 2w(\xi^2 - 3\eta^2)\xi)t]\} \tag{23}$$

$$E_+ = q_0\{\tanh[2\eta(s - vt - s_0)] - 1\} \tag{24}$$

where the amplitude of the vortex soliton q_0 is given by

$$q_0 = \frac{\eta}{\xi^2 + \eta^2} \tag{25}$$

and the velocity of the vortex soliton along the arc v is defined as

$$v = -4(\xi + w(3\xi^2 - \eta^2)) \tag{26}$$

Recalling the transformation of Eq. (5), we can express the one soliton solution in the laboratory coordinate as

$$\begin{aligned} q(x,t) = q_0 sech[2\eta(x - vt + E_+(x,t) - s_0)]\times \\ \exp\{-i[2\xi x + 4(\xi^2 - \eta^2 + 2w(\xi^2 - \eta^2)\xi)t + 2\xi E_+(x,t)]\} \end{aligned} \tag{27}$$

$$E_+(x,t) = q_0\{\tanh[2\eta(x - vt + E_+(x,t) - s_0)] - 1\} \tag{28}$$

The most important aspect of the vortex soliton is that although it propagates with the constant velocity v along the arc, the velocity with respect the laboratory coordinate is by no means constant. Also, we should notice that Eqs. (27) and (28) are nothing but the implicit soliton solution discussed by Hereman et al[19]. The relative size of ξ and η classifies the vortex soliton into the lump soliton ($|\xi| > \eta$), the cusp soliton ($|\xi| = \eta$), and the loop soliton ($|\xi| < \eta$), respectively.

4 Two Soliton Solution

Two soliton solution is given by

$$q = -i\frac{S}{R} \tag{29}$$

$$E_+ = -i\frac{T}{R} \tag{30}$$

where $R(s,t), S(s,t)$ and $T(s,t)$ are given by

$$R = 1 - \frac{|C_1|^2}{A_{11}^2}e^{2iA_{11}s} - \frac{|C_2|^2}{A_{22}^2}e^{2iA_{22}s}$$

$$-\frac{C_1C_2^*}{A_{12}^2}e^{2iA_{12}s} - \frac{C_1^*C_2}{A_{21}^2}e^{2iA_{21}s} - \frac{|C_1|^2|C^2|^2|\lambda_1-\lambda_2|^4}{(A_{11}A_{12}A_{21}A_{22})^2}e^{2i(A_{11}+A_{22})s} \tag{31a}$$

$$S = \frac{C_1^*}{\lambda_1^{*2}}e^{-2i\lambda_1^*s} - \frac{C_2^*}{\lambda_2^{*2}}e^{-2i\lambda_2^*s}$$

$$-\frac{|C_1|^2C_2^*(\lambda_1^*-\lambda_2^*)^2}{(\lambda_1^*\lambda_2^*A_{11}A_{12})^2}e^{2i(A_{11}-\lambda_2^*)s} - \frac{C_1^*|C_2|^2(\lambda_1^*-\lambda_2^*)^2}{(\lambda_1^*\lambda_2^*A_{21}A_{22})^2}e^{2i(A_{22}-\lambda_1^*)s} \tag{31b}$$

$$T = \frac{|C_1|^2}{|\lambda_1|^2A_{11}}e^{2iA_{11}s} + \frac{|C_2|^2}{|\lambda_2|^2A_{22}}e^{2iA_{22}s} + \frac{C_1C_2^*}{\lambda_1\lambda_2^*A_{12}}e^{2iA_{12}s}$$

$$+\frac{C_1^*C_2}{\lambda_2\lambda_1^*A_{21}}e^{2iA_{21}s} - \frac{|C_1|^2|C_2|^2(|\lambda_1|^2A_{22}+|\lambda_2|^2A_{11})|\lambda_1-\lambda_2|^4}{|\lambda_1|^2|\lambda_2|^2(A_{11}A_{12}A_{21}A_{22})^2}e^{2i(A_{11}+A_{22})s} \tag{31c}$$

respectively. $C_k(t)$ and A_{ij} are defined as

$$C_k(t) = C_k(0)\exp\{4i(\lambda_k^2+2w\lambda_k^3)t\}, k=1,2 \tag{32}$$

$$A_{ij} = (\lambda_i - \lambda_j^*) \tag{33}$$

Here, we notice that Eqs. (29) and (30) with the abbreviations of Eqs. (31)-(33) are symmetric with respect to the labeling of two solitons. In this connection we remark that the two soliton solution constructed by Levi et al[16] and adopted by Fukumoto and Miyazaki[18] appears to be not symmetric for the interchange of the two solitons.

It is interesting to observe that the axial flow simply modifies temporal variation of the soliton on the vortex filament so that the general feature of two soliton collision processes remains the same as our previous result[17]. Here, we illustrate two loop-soliton collision in Fig. 1.

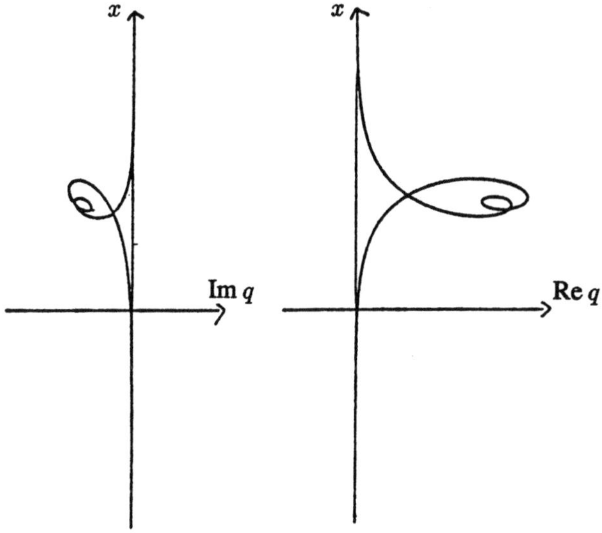

Fig. 1 – Collison of two loop solitons with axial flow (w=1). A large loop soliton ($\lambda_1 = 0.1 + i$, $C_1(0) = 3 + 4i$) and a small loop soliton ($\lambda_2 = -0.2 + 4i$, $C_2(0) = 2 + i$) at the instant of $t = 0.01$.

5 Conclusion

Transforming the extended local induction equation to the hybrid Wadati-Konno-Ichikawa equation, we have carried out explicit analysis of the associated inverse scattering transformation. We have obtained explicit expression for the two soliton solution which exhibits the intrinsic symmetry. So far we could not confirm analytically that the two solitons solution constructed by the method of soliton surfaces agrees with our solution.

References

[1] C.S. Gardner, J.M. Greene, M.D. Kruskal and R. Miura; Phys. Rev. Lett. 19, (1967) 1095 and; Comm. Pure Appl. Math. 27, (1974) 97

[2] V.E. Zakharov and A.B. Shabat; Soviet Phys. JEPT 34, (1972) 62

[3] M.J. Ablowitz, D.J. Kaup, A.C. Newell and H. Segur; Stud. Appl. Math. 53, (1974) 249

[4] D.J. Kaup and A.C. Newell; J. Math. Phys. 19, (1978) 798

[5] M. Wadati, K. Konno and Y.H. Ichikawa; J. Phys. Soc. Japan 47, (1979) 1698

[6] Y. Ishimori, J. Phys. Soc. Japan 51, (9182) 3036

[7] M. Wadati and K. Sogo; J. Phys. Soc. Japan 52, (1983) 394

[8] A. Sym; Lett. Nuovo Cimento 33, (1982) 394, ibid 36, (1983) 307, ibid 39, (1984) 193, ibid 40, (1984) 225, ibid 41, (1984) 33, ibid 41, (1984) 353

[9] A. Sym; Soliton Surfaces and their Applications, in Geometric Aspects of the Einstein Equations and Integrable Systems (ed. R. Martini), Lecture Notes in Physics, Springer, Vol. 239, (1985) 154

[10] E.J. Hopfinger, F.K. Browand and Y. Gagne; J. Fluid Mech. 125, (1982) 505

[11] T. Maxworthy, M. Mory and E.J. Hopfinger; Proc. AGARD Conf. on Aerodynamics of Vortical Type Flows in Three Dimensions, AGARD CPP- 342, (1985) paper 29

[12] T. Maxworthy, E.J. Hopfinger and L.G. Redekopp; J. Fluid Mech. 151, (1985) 141

[13] H. Hasimoto; J. Fluid Mech. 51, (1972) 477

[14] A. Aref and E.P. Flinchem; J. Fluid Mech. 148, (1983) 411

[15] R.H. Hama; Phys. Fluids 5, (1962) 1156

[16] D. Levi, A. Sym and S. Wojciechowsky; Phys. Lett. 94A, (1983) 408

[17] K. Konno, M. Mitcuhashi and Y.H. Ichikawa; Chaos, Solitons and Fractals 1, (1991) to be published, and see also Y.H. Ichikawa, K. Konno and M. Mitsuhashi; to appear in Research Trends in Physics, Chaotic Dynamics and Transport in Fluids and Plasmas, (Eds. V. Stephen et al.), American Institute of Physics Proceedings of Feb. 18-20, 1991 Meeting in La Jolla, CA.

[18] Y. Fukumoto and T. Miyazaki; J. Fluid Mech. 222, (1991) 369

[19] W. Hereman, P.P. Banerjee and M.R. Chatterjee, J. Phys. A, Math. Gen. 22, (1989) 241

Lagrangian Statistics of Turbulence

Y. Kaneda

Department of Applied Physics, Nagoya University,
Chikusa-ku, Nagoya 464-01, Japan

Abstract. A study is made of the Lagrangian velocity autocorrelation and of the Lagrangian statistics of the stretching and the curvature evolution of material line elements in turbulence. A discussion is presented on applications of the method of Lagrangian renormalized/perturbative expansions to the study of these statistics.

1. Introduction

The statistics of the dispersion, stretching and distortion of material elements in turbulence are determined by the Lagrangian statistics of the fluid velocity and velocity gradients. There are two kinds of problems in the study of these phenomena; one is kinematical, and the other is dynamical. The former is to estimate the Lagrangian statistics of the displacement (dispersion), evolution of the length and curvature, etc., of the material elements under given Lagrangian statistics of the velocity and velocity-gradients, and the latter is to estimate the Lagrangian statistics of the velocity and/or velocity gradients under given dynamics governing the evolution of the velocity field. Regarding the turbulent dispersion, there is a well-known relationship between the turbulent diffusivity and the Lagrangian velocity autocorrelation, so that we need solve only the dynamical problem for the estimation of the turbulent diffusivity. On the other hand, regarding the statistics of the evolution of the length and curvature, we have no such established kinematical relationship between the statistics and the statistics of the velocity and/or velocity gradients, so that both of the two kinds of problems remain to be solved.

In this paper, we discuss about two applications of the method of Lagrangian renormalized/perturbative expansions to the study of these problems. Both applications have a common feature in that a mapping S: $\mathbf{X} \to \widehat{\mathbf{X}}$ from the original field $\mathbf{X}$ to another field $\widehat{\mathbf{X}}$ plays a key role, where we write the mapping as

$$\widehat{\mathbf{X}} = S\mathbf{X}, \tag{1.1}$$

and S obeys a certain evolution equation which may be written in a symbolic notation as

$$\frac{\partial}{\partial t} S = f(S, \mathbf{X}). \tag{1.2}$$

The evolution of $\widehat{\mathbf{X}}$ can be known from (1.2) and the evolution equation of $\mathbf{X}$.

2. Lagrangian velocity correlation and Lagrangian renormalized approximation

Previously, I presented a two-point two-time closure approximation called Lagrangian renormalized approximation (LRA) for turbulence.[1] The LRA is constructed on the basis of the following mapping from the Eulerian velocity field $\mathbf{u}$ to a generalized field $\hat{\mathbf{u}}$:

$$\hat{\mathbf{u}}(\mathbf{x},t|s) = \int d^D\mathbf{x}'\psi(\mathbf{x}',s|\mathbf{x},t)\mathbf{u}(\mathbf{x}',s), \tag{2.1}$$

where $\psi(\mathbf{x},t|\mathbf{x}',t') \equiv \delta^D(\mathbf{x} - \mathbf{z}(\mathbf{x}',t'|t))$ obeys

$$\frac{\partial}{\partial t}\psi(\mathbf{x},t;\mathbf{x}',t') = -[\mathbf{u}(\mathbf{x},t)\cdot\nabla_{\mathbf{x}}]\psi(\mathbf{x},t;\mathbf{x}',t'). \tag{2.2}$$

Here D is the dimension of space, and $\mathbf{z}(\mathbf{x}',t'|t)$ is the position at time of the fluid element that was at $\mathbf{x}'$ at time t'. The mapping given by (2.1) and (2.2) is an example of mapping expressed by (1.1) and (1.2). A simple two-mode model to illustrate the advantage of introducing such a mapping as (2.1) is given in Ref.2.

Let us suppose that the evolution of the Eulerian field $\mathbf{u}$ is governed by the Navier-Stokes (NS) equations. Then, from the NS equations and (2.2) that determines the evolution of the mapping ψ in (2.1), we can know the Lagrangian evolution with respect to time s of the generalized field $\hat{\mathbf{u}}(\mathbf{x},t|s)$, and can construct renormalized expansions for the evolutions of a two-time two-point Lagrangian velocity correlation and a Lagrangian response function. A simple truncation of the expansions yields a closed set of equations called the LRA equations. The LRA gives approximations not only for single-time quantities such as the energy spectrum, but also for a two-time Lagrangian correlation function from which the Lagrangian velocity autocorrelation R_L can be estimated.

Regarding single-time quantities, comparisons have been made between the LRA values and existing data by experiments and direct numerical simulations (DNS) for two- and three- dimensional incompressible isotropic turbulences.[3] Recently, the LRA values of the two-time correlation R_L have been compared with those by DNS of two-dimensional Navier-Stokes turbulence with and without a mean flow of a simple shearing motion, as well as with existing data of DNS and laboratory experiments for three-dimensional isotropic turbulence.[4,5] These comparisons have shown good agreement of the LRA with DNS and experiments.

In order to increase the applicability of two-point closures to real anisotropic and inhomogeneous turbulences, it is desirable to develop an efficient method to solve the closure equations. Efforts in this direction are now being made. An application of a Monte-Carlo method has been tested.[6] A preliminary study of a non-uniform grid system in the wavevector space suggests that an approximate solution of closure equations may be obtained by quite a coarse grid system.

3. Stretching and curvature evolution of material line elements

An infinitesimal line element **e** moving with fluid velocity $\mathbf{u}(\mathbf{r},t)$ evolves according to

$$\frac{d}{dt}\mathbf{e} = \mathbf{W}\mathbf{e}, \qquad W_{ij}(t) = u_{i,j}(\mathbf{r},t). \tag{3.1}$$

It is straightforward to solve **e** in (3.1) in powers of **W**, and to obtain expansions of e^2 and $\zeta \equiv W_{ij}e_ie_j/e^2$ in powers of **W**. From these expansions we can obtain an expansion of the exponent γ_p defined by Drumond and Münch[7]:

$$\gamma_p \equiv \frac{1}{p < e^p >}\frac{d}{dt} < e^p > .$$

If the direction of **e** was initially isotropically distributed and the turbulence is isotropic, then discarding terms of $O(\mathbf{W}^3)$ in the expansion yields

$$\gamma_p \approx \alpha_p \int_0^t dt' < \mathrm{Tr}\ \mathbf{B}(t)\mathbf{B}(t') >, \tag{3.2}$$

where

$$\alpha_p = \frac{2}{D+2} + \frac{2p}{D(D+2)},$$

and **B** is the symmetric part of **W**. Kraichnan[8] showed that e is lognormally distributed when **W** is a white noise process in time. His result implies that (3.2) is exact in the case.

Now, let us consider a unitary transformation from **e** to $\hat{\mathbf{e}}$:

$$\hat{\mathbf{e}} = \mathbf{U}\mathbf{e}, \qquad \frac{\partial}{\partial t}\mathbf{U} = -\mathbf{U}\mathbf{A}, \tag{3.3}$$

where $\mathbf{U}(0) = \mathbf{I}$, **A** is an anti-symmetric matrix, and **I** is the identity matrix. Equation (3.3) is another example of a mapping expressed by (1.1) and (1.2). Equation (3.1) may be written as

$$\frac{\partial}{\partial t}\hat{\mathbf{e}} = \mathbf{V}\hat{\mathbf{e}}, \qquad \mathbf{V} = \mathbf{U}[\mathbf{W} - \mathbf{A}]\mathbf{U}^T, \tag{3.4a, b}$$

where $\mathbf{U}^T$ is the transpose of **U**. By the same way as in deriving (3.2) from (3.1), we obtain from (3.4),

$$\gamma_p \approx \alpha_p \int_0^t dt' < \mathrm{Tr}\ \mathbf{V}(t)\mathbf{V}(t') > . \tag{3.5}$$

If we put **A** to be the anti-symmetric part of **W**, then (3.4b) reduces to

$$\mathbf{V} = \mathbf{U}\mathbf{B}\mathbf{U}^T, \tag{3.6}$$

and (3.5) becomes just what Drumond and Münch[7] have derived.

They approximated (3.5) with (3.6) by (3.2), and found that the theoretical estimates of the ratios γ_1/γ_0, γ_2/γ_0, etc., are in good agreement with kinematical simulations using a simple model of turbulence. They used (3.2) in their theoretical estimations, and no comparison between the theory and simulations was made regarding the values of the γ_p's themselves. However, from the view point of renormalized expansions, we think that there may be a significant difference between (3.2) and (3.5) for the values of the γ_p's, even though they yield the same values for the ratios.

It is shown that combining (3.5) or (3.2) with another theoretical estimate of γ_p due to Girimaji and Pope[9] yields estimates of the average $<\zeta>$ and the second order cumulant ϕ of ζ. It is also shown by an analysis along the same lines as Drumond and Münch[10] that in the case of two-dimensions, unlike the case of three-dimensions studied by them, these statistics $<\zeta>$ and ϕ have a simple relation with the moments of the curvature of the line elements.

The theoretical estimates thus obtained have been compared with the results by DNS of two-dimensional incompressible turbulence. The estimates based on (3.5) with (3.6) were found to be much better than those based on (3.2).

The Lagrangian correlation $Q(s,t) \equiv < \mathrm{Tr}\mathbf{V}(s)\mathbf{V}(t) >$ in (3.5) cannot be estimated by purely kinematical consideration. For the estimation we need take into account the dynamics. From (2.1),(2.2) and the NS equations, we can obtain the expansion of $Q(s,t)$ for small $\tau = s - t$, and the expansion coefficients can be evaluated without difficulty under certain conditions. A rough estimate of the integral in (3.5) with (3.6) can be then obtained from these coefficients by assuming the τ-dependence of Q to be Gaussian. This estimate was found to be in fairly good agreement with the value by DNS in two-dimensional turbulence.

The detailed analysis of the above and comparison between the theory and DNS in two-dimensions will be presented in a paper being prepared in collaboration with T.Ishihara.

1. Y.Kaneda, J.Fluid Mech., 1981,**107**,131.
2. Y.Kaneda, in *Recent Studies on Turbulent Phenomena,* ed. T.Tatsumi, H.Maruo and H.Takami (Association for Science Documents Information, Tokyo, 1985)79.
3. T.Gotoh, Y.Kaneda and N.Bekki, J.Phys.Soc.Jpn. **57**,866(1988).
4. Y.Kaneda and T.Gotoh, to be published in Phys. Fluids A.
5. T.Gotoh and Y.Kaneda, to be published in Phys. Fluids A.
6. Y.Kaneda, to be published in *Chaotic Dynamics and Transport in Fluids and Plasmas,* ed. V.Stefan, W.Horton, Y.Ichikawa, I.Prigogine and G.Zaslavsky, (La Jolla International School of Physics Series: Research Trends in Physics, AIP(1991)).
7. I.T.Drummond and W. Münch, J.Fluid Mech., 1990,**215**,45.
8. R.H.Kraichnan, J.Fluid Mech., 1974,**64**,737.
9. S.S.Girimaji and S.B.Pope, J.Fluid Mech., 1990,**220**,427.
10. I.T.Drummond and W.Münch, J.Fluid Mech., 1991,**225**,529.

Pattern Formation Via Resonant Interactions in Fluid Flows

S. Leibovich and A. Mahalov

School of Mechanical and Aerospace Engineering, Cornell University, Ithaca, NY, USA

Pattern formation is a familiar element in the laminar-transition mechanism in fluid flows. Patterns also are clearly present in the disordered states existing in fully developed turbulent flows, where they are known as "coherent structures" Pattern formation is best understood in "closed flows" the prototype being thermal convection in fluid layers heated from below, because such problems are set in closed spatial domains in physical space and are therefore simpler mathematically than "open flows" which are typically set in domains idealized to be spatially infinite. Nevertheless, the experimental evidence of patterning is incontrovertible-even striking-in both closed and open flows, with the prototypical examples of the latter type being boundary layer flows or mixing layer flows. In the case of boundary layers and channel flows, the existence and importance of Tollmien-Schlichting waves has long been known. Helical waves are their analogues in rotating pipe flows. The dynamics leading to pattern formation dominate the characteristics of preturbulent flows, including the physically and technologically important transfer properties of the flow (such as heat, momentum, and mass transfer). In this presentation we describe some computational and analytical investigations of transition sequences and coherent structures in two classes of fluid flows, (i) rotating pipe flows, (ii) transitional boundary layers and channel flows.

(i) Rotating pipe flows

In the rotating pipe flow problem, we have studied 2D helical subspaces of the Navier-Stokes equations (Mahalov et al. (1990)) using analytical methods and weakly nonlinear analysis. A parallel search for interactions between 2D helical waves has revealed an infinite sequence of resonant triads corresponding to different combinations of the azimuthal wavenumbers. Allowing for spatial modulation, resonant interactions are governed by Three-Wave equations describing interactions of wave packets belonging to different invariant helical subspaces. Leibovich and Mahalov (1991) found that in the fast rotation regime of the rotating Hagen-Poiseuille flow, the equations describing waves in resonance approach an integrable limit. Zakharov and Manakov (1973, 1975) and Kaup(1976) have obtained solutions of the equations by the method of inverse scattering. These integrable equations admit solutions that are localized in space. They still have complicated temporal signatures and closely resemble localized states observed in the experiment of Hegseth et al. (1989) on a variant of the Taylor-Couette flow. We explore the three-dimensional flows resulting from interactions between 2D waves residing in helical subspaces.

(ii) Boundary layers and channel flows

Resonant interactions are possible between two symmetrically oriented three-dimensional Tollmien-Schlichting (TS) waves and a two-dimensional TS wave in plane channel flow. The symmetry of the problem requires quadratic coupling coefficients to vanish identically. We break this symmetry by sliding one boundary of the channel with respect to the other with a constant velocity. If the amount of sliding is small then the quadratic and cubic terms enter amplitude equations at the same order. We discuss possible equilibrium states, their stabilities and possible interactions in the corresponding amplitude equations. Recently, Corke and Mangano (1989) and Kachanov (1989) have published new experimental data on the laminar-turbulent transition in boundary layers. It appears that resonant interactions of Tollmien-Schlichting waves play a major role in this transition. We derived the corresponding amplitude equations allowing for spatial modulation and cubic interaction terms. We have proposed that doubling, tripling etc. of spikes observed in experiments may be attributes to the splitting of a wave into solitons. This phenomena is known to exist for completely integrable equations such as the nonlinear Schrodinger equation, resonant three-wave equations, and others.

References

1. T.C. Corke, R.A. Mangano (1989), Resonant Growth of Three-Dimensional Modes in Transitioning Blasius Boundary Layers, *J. of Fluid Mech.*,**209**, p. 93-115.

2. J. Hegseth, C.D. Andereck, F. Hayot, Y. Pomeau (1990), Spiral Turbulence and Phase Dynamics, *Phys. Rev. Lett.*, **62**, No. 3, p. 257-260.

3. Yu. Kachanov (1990), Secondary and Cascade Resonant Instabilities of Boundary Layers, preprint.

4. D.J. Kaup (1976), Stud. Appl. Math. **55**, 9.

5. S. Leibovich, A. Mahalov (1991), Spatio-temporal Dynamics in Hagen-Poiseuille Problem, In preparation.

6. A. Mahalov, E.S. Titi, S. Leibovich (1990), Invariant Helical Subspaces for the Navier-Stokes Equations, *Arch. Rational Mech. Anal.*, **112**, No. 3, p. 193-222.

7. V.E. Zakharov and S.V. Manakov (1973), Zh. Eksp. Teor. Fiz. Pis'ma Red. **18**, 413 [Sov. Phys. – JETP Lett. **18**, 243 (1973)].

8. V.E. Zakharov and S.V. Manakov (1975), Zh. Eksp. Teor. Fiz. **69** 1654 [Sov. Phys. – JETP **42**, 842 (1976)].

Length Scale of Vortices and Mode Competition in Quasi 2D Shear Flows

D. Yu. Manin

Institute of Atmospheric Physics, Moscow, Russia

Strictly 2D flows constitute a broad and useful class of Navier—Stokes solutions, but in practice they are quickly destroyed by 3D instability. A class of *quasi two-dimensional (Q2D)* flows can be defined[1] in physical terms as flows whose approximate two-dimensionality is due to the presence of flat boundaries (bottom and/or top) so that vertical velocity component is negligible. However the vertical dependence of horizontal velocity is essential in that it provides most of the dissipation. This results in the Rayleigh friction term, $-\lambda\Delta\Psi$ in the governing Q2D equation for stream function Ψ

$$\frac{\partial}{\partial t}\Delta\Psi + [\Delta\Psi, \Psi] = \nu\Delta^2\Psi - \lambda\Delta\Psi + F, \tag{1}$$

where F is the forcing, and the specific expression for λ depends on the physical situation[1]. In particular, large-scale atmospheric flows can be regarded Q2D with λ being the inverse Ekman time scale.

The Rayleigh, or *external* friction drastically changes both weakly and strongly nonlinear behavior of shear flows. For the latter, it cuts off the inverse energy cascade at large scales, thus providing the characteristic length scale of instability-generated vortical structures[2].

This contribution presents results of numerical simulation of a Q2D shear flow at moderate Reynolds numbers, when inverse energy cascade leads to competition of vortical modes with different length scales. Results are compared with experimental data.

REFERENCES

1. Dolzhanskii, F.V. *et al.* Sov. Phys. Uspekhi, 1990, **7** (30), 495.
2. Manin, D.Yu. Preprint No.6, Inst. Atmos. Phys., Moscow, 1989.

Dynamics of Vortex-Current Filaments in MHD Plasma

V.I. Petviashvili

Kurchatov Institute of Atomic Energy, 123182 Moscow, Russia

The MHD analog of vortex filaments in ordinary hydrodynamics is shown to be vortex-current filaments with circulation of both velocity and magnetic field around them, like in the Alfven waves, polarized circularly. So they can be treated as singular Alfven waves. The evolutionary equations of such filaments are derived from ideal MHD equations. The filaments interact if their cross-helicities have opposite sign.

A vorticity field in incompressible fluid during relaxation has the tendency to break down into set of vortex filaments. Equations governing dynamics of vortex points (in 2D case) and vortex filaments (in 3D case) are well known in ordinary hydrodynamics [1]. The Magnetohydrodynamic (MHD) analog of these equations was derived in [2,3]. Important new feature appear in this case: the velocity circulation around MHD filaments is necessarily accompanied by the circulation of the magnetic field. This results in interesting new symmetry properties and some peculiarities with interaction, in particular the absence of self-action.

The formation of filaments obviously is connected with the filamentationary instability of homogeneous current in the plasma [4]. In some papers the filamentary structures observed in solar and interstellar plasma are treated as magnetic field lines. In view of the theory proposed here the structures seem more natural to be treated as vortex-current filaments.

We start from equations of incompressible one-fluid MHD:

$$d_t\mathbf{v} = -\nabla P/\rho + (\nabla \times \mathbf{c_A}) \times \mathbf{c_A}; \quad \mathbf{c_A} = \mathbf{B}(4\pi\rho)^{-\frac{1}{2}}; \tag{1}$$

$$\partial_t\mathbf{c_A} = \nabla \times (\mathbf{v} \times \mathbf{c_A}); \quad \nabla \cdot \mathbf{v} = \nabla \cdot \mathbf{c_A} = 0; \tag{2}$$

where $\mathbf{B}$ is the magnetic field; $\mathbf{v}$ - velocity, ρ - density, P - pressure of the plasma and $\mathbf{c_A}$ is the Alfven velocity.

Instead of $\mathbf{v}$ and $\mathbf{c_A}$, Elsasser [5] introduced new variables:

$$\mathbf{u} = \mathbf{v} + \mathbf{c_A}; \quad \mathbf{w} = \mathbf{v} - \mathbf{c_A}; \tag{3}$$

Then (1,2) takes the symmetric form:

$$\partial_t\mathbf{u} + (\mathbf{w} \cdot \nabla)\mathbf{u} = -\nabla Q; \quad \nabla \cdot \mathbf{u} = \nabla \cdot \mathbf{w} = 0; \tag{4}$$

$$\partial_t\mathbf{w} + (\mathbf{u} \cdot \nabla)\mathbf{w} = -\nabla Q; \quad \nabla^2 Q = -\nabla \cdot (\mathbf{u} \cdot \nabla)\mathbf{w} \tag{5}$$

Equations (4,5) look like Euler equations and become these when $u = w$ i.e. when the magnetic field is absent.

Characteristic equations of the set (4,5) are:

$$\partial_t \mathbf{r} = \mathbf{w}(t, r); \partial_t \mathbf{r} = \mathbf{u}(t, r) \tag{6}$$

Here $\mathbf{r}(t)$ are trajectories of small perturbations. We look for the approximate solutions of the set (4,5) in the singular Biot-Savart contour integral form:

$$\mathbf{u} = \sum_{m=1}^{M} G_m \oint \mathbf{R}_m \times \mathbf{r}'_m ds; \quad \mathbf{R}_m = (\mathbf{r} - \mathbf{r}_m)|\mathbf{r} - \mathbf{r}_m|^{-3} \tag{7}$$

$$\mathbf{w} = \sum_{n=1}^{N} G_n \oint \mathbf{R}_n \times \mathbf{r}'_n ds; \tag{8}$$

Here $\mathbf{r}_m(t, s)$ is the coordinate of the m-th filament on which the field is singular $\mathbf{r}'_m = \partial_s \mathbf{r}_m$ is tangent vector to the filament, depending on time t and curve parameter s. We see that curl of (7,8) are localized on the filaments as 2D Dirac functions. Substituting (7,8) into (6) we obtain integro-differential evolutionary equation of 3D vortex-current filaments:

$$\partial_t \mathbf{r}_m = \sum_{n=1}^{N} G_n \oint \mathbf{R}_{m,n} \times \mathbf{r}'_n ds; \quad m = 1, .., M \tag{9}$$

$$\partial_t \mathbf{r}_n = \sum_{m=1}^{M} G_m \oint \mathbf{R}_{n,m} \times \mathbf{r}'_m ds; \quad n = 1, .., N \tag{10}$$

$$\mathbf{R}_{m,n} = (\mathbf{r}_m - \mathbf{r}_n)|\mathbf{r}_m - \mathbf{r}_n|^{-3}$$

We have two types of filaments labeled respectively by m and n. They differ by the sign of cross helicity $H = \mathbf{v} \cdot \mathbf{c_A}$ - the scalar product of hydrodynamical and Alfven velocities circulating around the filament. As we see from (9,10) the interaction occurs only between filaments with opposite sign of H. The constants G_m and G_n are proportional to electric current in the respective filaments. We see that the interaction of electric currents in plasma is completely different form that in vacuum. The striking feature of vortex-current filaments is the equality of values of circulating hydrodynamical and Alfven velocities. The system (9,10) is hamiltonian and conserves energy and mechanical dipole moment:

$$E = \sum \sum G_m G_n \oint \oint |\mathbf{r}_m - \mathbf{r}_n|^{-1} d\mathbf{r}_m \cdot d\mathbf{r}_n; \quad d\mathbf{r}_m = \mathbf{r}'_m ds$$

$$T = T_u + T_w; T_u = \sum G_m \oint \mathbf{r}_m \times d\mathbf{r}_m; \quad T_w = \sum G_n \oint \mathbf{r}_n \times d\mathbf{r}_n$$

The magnetic dipole moment equals to $T_u - T_w$ and is not conserved. This leads to large fluctuations of magnetic field in the system of vortex current filaments [2].

If we have strong constant magnetic field $\mathbf{B}_0$ in a plasma then oscillations of components of $\mathbf{u}, \mathbf{w}$ along it may be neglected and (4,5) are reduced to:

$$(\partial_t - c_0\partial_z)\mathbf{u}_\perp + (\mathbf{w}_\perp \cdot \nabla)\mathbf{u}_\perp = -\nabla_\perp P; c_0 = (4\pi\rho)^{-\frac{1}{2}}B_0; \qquad (11)$$

$$(\partial_t + c_0\partial_z)\mathbf{w}_\perp + (\mathbf{u}_\perp \cdot \nabla)\mathbf{w}_\perp = -\nabla_\perp P; w_\perp, u_\perp << c_0 \qquad (12)$$

Here we suppose that $\mathbf{B}_0$ is directed along z and that the dependence on z is weak: $c_0\partial_z \sim \mathbf{w}_\perp \cdot \nabla \sim \mathbf{u}_\perp \cdot \nabla$. 2D vectors $\mathbf{u}_\perp$ and $\mathbf{w}_\perp$ are components of $\mathbf{u}, \mathbf{w}$ on x,y plane. We suppose now that filament-like singularities exist in plasma; so we take

$$\mathbf{u}_\perp = c_0 \sum_{m=1}^{M} G_m \mathbf{R}_m + \mathbf{c}_\perp; \quad \mathbf{R}_m = (\mathbf{r} - \mathbf{r}_m) \times \mathbf{e}_z |\mathbf{r} - \mathbf{r}_m|^{-2} \qquad (13)$$

$$\mathbf{w}_\perp = c_0 \sum_{n=1}^{N} G_n \mathbf{R}_n - \mathbf{c}_\perp; \quad \mathbf{c}_\perp = (4\pi\rho)^{-\frac{1}{2}}\mathbf{B}_\perp; \qquad (14)$$

Here $\mathbf{r}_m, \mathbf{r}_n$ are vector coordinates of singularities on x,y plane $\mathbf{r}_m = \mathbf{r}_m(t, z)$, corresponding to upstream and downstream perturbations relative to $\mathbf{B}_0, \mathbf{B}_\perp$ is a small poloidal external magnetic field, $\mathbf{e}_z$ is the unit vector along $\mathbf{B}_0$. We have M filaments of downstream type and N filaments of upstream type. Substituting (13,14) in equation of characteristics of (11,12) we obtain the equations of motions for coordinates of singularities (cf (9,10)):

$$(c_0^{-1}\partial_t - \partial_z)\mathbf{r}_m = \sum_{n=1}^{N} G_n \mathbf{R}_{m,n} - \mathbf{b}(\mathbf{r}_m); m = 1, .., M \qquad (15)$$

$$(c_0^{-1}\partial_t + \partial_z)\mathbf{r}_n = \sum_{m=1}^{M} G_m \mathbf{R}_{n,m} + \mathbf{b}(\mathbf{r}_n); n = 1, .., N \qquad (16)$$

$$\mathbf{R}_{m,n} = (\mathbf{r}_m - \mathbf{r}_n) \times \mathbf{e}_z |\mathbf{r}_m - \mathbf{r}_n|^{-2}; \mathbf{b} = \mathbf{B}_\perp / B_0 = \mathbf{c}_\perp / c_0$$

The type of vortex is determined by the sign of cross helicity H. The signs of constants G_m cannot be determined in MHD except in special cases. For instance if the filament is created by a source of fast electrons then $G_m, G_n < 0$. The set (15,16) can be easily reduced to convenient complex form.

Equations (9,10) and (15,16) are valid in the limit of vanishing cross- sections of filaments. When the radius of the filament approaches the ion cyclotron radius the dispersive corrections appear, considered in [6].

The set (15,16) can be used as a first approximation for electron beams propagating in Tokamaks with toroidal field $\mathbf{B}_0$. In this case z is the periodic toroidal coordinate, $\mathbf{r}$ is the vector coordinate on the cross-section of the torus. The beams can travel many times around torus and this must be taken into account in the sums (15,16). Let us consider two sources of thin electron beams at the edge of toroidal plasma. The beams interact if they have opposite sign of cross-helicity. As a result of interaction one of the beams after several revolutions around torus can penetrate into the central region of the plasma, while the second stays at the periphery. This can be used as a simply way of generating stationary toroidal current in a Tokamak.

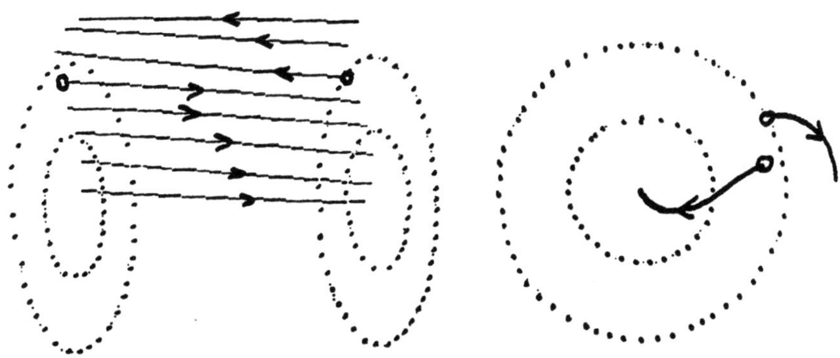

Fig. 1 – Trajectories of two electron beams with 5 revolutions of first and 3 revolutions of second around 250cm long and 10cm radius periodical cylinder. Sources are marked as 0. (left: side view, right: projection on the cross-section). $G_1 = .03, G_2 = .03, \mathbf{b}(r) = .003(1 - .005r^2)\mathbf{r} \times \mathbf{e}_z$

References

[1] Batchelor G.K., "Introduction to Fluid Mechanics" Cambridge U.P. New York 1967.

[2] Petviashvili N.V., Proceedings of the joint Varenna-Abastumani- ESA-Nagoya- Potsdam "Workshop on Plasma Astrophysics" in Telavi, USSR, 1990, p. 33.

[3] Petviashvili V.I., Sov. J. Plasma Physics, 1991, v. 17, #2.

[4] Priest E.R., "Solar Magnetohydrodynamics", Dordrecht, D. Reidel Publ. Co.

[5] Elsasser W.M., Phys. Rev., 1950, v. 79, p. 183.

[6] Petviashvili V.I., Pokhotelov O.A., "Solitary Waves in Plasma and Atmosphere" Gordon & Breach Sc. Publ. (1991).

Success of Arnol'd's Method in a Hierarchy of Ocean Models

P. Ripa

C.I.C.E.S.E., 22800 Ensenada, B.C., México

The method of Arnol'd[1,2] is used to derive stability conditions of Hamiltonian systems with singular Poisson brackets; notably, those of fluid mechanics:[3,4,5,6,7] Let the field(s) $\varphi(\mathbf{x},t)$ represent the state variables, and the *Hamiltonian* $\mathcal{H}[\varphi]$ and *Poisson bracket* $\{.,.\}$ be such that $\partial_t\varphi = \{\varphi,\mathcal{H}\}$. The bracket must have the usual properties;[5] in addition, it may be *singular*, *i.e.* there may exist non-trivial functional(s) of state $\mathcal{C}[\varphi]$ -called *Casimir(s)*- such that $\{\mathcal{C},\mathcal{F}\} \equiv 0$, for **any** admissible[8] functional of state $\mathcal{F}[\varphi]$.

Now, let $\varphi = \Phi(\mathbf{x})$ be a steady solution of the equations of motion: If the Poisson bracket is singular -as it is, in the cases of interest here- $\{\Phi,\mathcal{H}\} = 0$ does not necessarily imply that $\mathcal{H}$ has an extremum at $\varphi = \Phi$; however, there may exist a Casimir $\mathcal{C}$ such that $\delta(\mathcal{H}+\mathcal{C}) \equiv 0$, where $\Delta\mathcal{S} := \mathcal{S}[\Phi+\delta\varphi]-\mathcal{S}[\Phi] = \delta\mathcal{S} + \frac{1}{2}\delta^2\mathcal{S} + \ldots$ and $\delta^n\mathcal{S} \sim O(\delta\varphi)^n$ for any $\mathcal{S}[\varphi]$. Arnol'd's **first theorem** for **Formal Stability** represents finding sufficient conditions on the basic state Φ such that

$$\delta(\mathcal{H}+\mathcal{C}) \equiv 0 \ \& \ \delta^2(\mathcal{H}+\mathcal{C}) > 0; \qquad (F1)$$

whereas the **second theorem** is used to derive conditions such that

$$\delta(\mathcal{H}+\mathcal{C}) \equiv 0 \ \& \ \delta^2(\mathcal{H}+\mathcal{C}) < 0. \qquad (F2)$$

Nonlinear Stability implies the existence of a *norm* of the disturbance $\|\delta\varphi\|$ which may bounded by a multiple of its initial value. For instance, this can be easily proved when there are two constants a & A such that

$$A\|\delta\varphi\|^2 \geq \Delta(\mathcal{H}+\mathcal{C}) \geq a\|\delta\varphi\|^2 > 0, \qquad (N1)$$

(first theorem), or

$$A\|\delta\varphi\|^2 \leq \Delta(\mathcal{H}+\mathcal{C}) \leq a\|\delta\varphi\|^2 < 0, \qquad (N2)$$

(second one), and thus $\|\delta\varphi\|_{t>0} \leq \sqrt{(A/a)}\|\delta\varphi\|_{t=0}$. There are, then, four possibilities: First & second theorem (indicated by 1 & 2), and formal & nonlinear Arnol'd stability (indicated by F & N); N has stronger consequences than F, *e.g.*, *a priori* finite-amplitude bounds on the saturation of the disturbance.[9,10,11] These stability conditions apply, in principle, to quite general basic states $\Phi(\mathbf{x})$; however, if the problem is invariant under a certain transformation, then Andrews' theorem[12,13,14] requires that Φ have that symmetry (or that the allowed perturbations be modulo that transformation of the basic state, which is easier said than done).

Two families of "classical" ocean models are considered here: **primitive equations** (*PEM*)[15] and **quasi-geostrophic** (*QGM*)[16] ones; all these models are Hamiltonian with a singular (non-canonical) Poisson bracket. Within each family, models are classified by their vertical structure, which consists of a stack of homogeneous layers; top and bottom boundaries may be either rigid or free. A model with n layers and one or two free boundaries is found to have some properties in between those with rigid boundaries and n & n+1 layers; for this reason that model is denoted as an "$n\frac{1}{2}$-layer" one. Each *QGM* is an approximation to the slow manifold of the corresponding *PEM*.

Applications of Arnol'd's method have been more numerous -and successful- to the *QGM*[17,18,19] than to the *PEM*.[20] This difference is exemplified next with one model from each family.

The simplest of the *PEM* (the $1\frac{1}{2}$-layer case) is but the *Shallow Water Equations*, for which $\varphi = \varphi(x,y,t) := (h,u,v)$; h is the (variable) layer depth and (u,v) are the velocity components along the horizontal directions (x,y). In the absence of topographic effects, the evolution equations are $\dot{h} = -h(u_x+v_y)$, $\dot{u} = fv - gh_x$, and $\dot{v} = -fu - gh_y$, where $(\dot{\ }) := (\ )_t + u(\ )_x + v(\ )_t$, f $(:= f_0+\beta y)$ is the Coriolis parameter, and g is the (appropriate) gravity. From these equations it is easy to prove that the Kelvin circulation γ^i in each disconnected part of the horizontal boundary are constant, and that $\dot{q} = 0$, where q $(:= (f+v_x-u_y)/h)$ is the *potential vorticity*.

For a steady basic state, $\Phi = (H,U,V)$, it must be $HU = -\Psi_y$, $HV = \Psi_x$ & $Q = Q(\Psi)$. Of the four possibilities mentioned above, only formal stability in the sense of Arnol'd's first theorem can be proved; the conditions are[21,22]

$$dQ/d\Psi > 0 \quad \& \quad U^2+V^2 < gH; \qquad (F1)'$$

the first (second) of these conditions guarantees that the wave Casimir (energy) is positive definite. As the number of layers is increased, the first condition remains the same, but the corresponding to the second condition becomes harder and harder to satisfy. In the limit of continuous stratification, Arnol'd's method fails to provide a formal stability condition,[15,23] even though there are normal modes stability criteria.[24,25]

The simplest case in the *QGM* family corresponds to the 1-layer model (rigid top & bottom boundaries), *i.e.*, Euler equations in the plane; this is the model for which most results have been derived.[1,2,5] However, in order to make the comparison with the shallow water equations more transparent, I'll use the next case ($1\frac{1}{2}$-layer), usually known as the "equivalent barotropic". In this model it is $\varphi = \varphi(x,y,t) := (q,\gamma^1)$, and the evolution equations are simply $\gamma^1_t = 0$ & $\dot{q} = 0$. In order to solve the last one, the height and velocity fields are calculated, from (q,γ^1), as $(h, u, v) = (H_0+f_0g^{-1}\psi, -\psi_y, \psi_x)$, and $\nabla^2\psi - R^{-2}\psi + \beta y = q$, where H_0 is a constant and R $(:= |f_0|^{-1}\sqrt{gH_0})$ is the *Rossby deformation radius*.[16,26]

For a steady solution, it must be $Q = Q(\Psi)$; in the region of phase space where this *QGM* is a good approximation to the *PEM* it is $(\Psi,Q)_{QGM} \approx H_0(\Psi,Q)_{SWM}$. All four possibilities of Arnol'd stability can be proved in this case, namely, formal first theorem (wave pseudoenergy positive definite)[1,2]

$$dQ/d\Psi > 0, \qquad (F1)''$$

and second theorem (wave pseudoenergy negative definite)[16]

$$-\lambda^2 - R^{-2} < dQ/d\Psi < 0 \qquad (F2)''$$

(where λ^2 is the smallest eigenvalue of Helmholtz equation with vanishing boundary condition), as well as nonlinear first theorem

$$A \geq dQ/d\Psi \geq a > 0, \qquad (N1)''$$

and second one

$$-\lambda^2 - R^{-2} < A \leq dQ/d\Psi \leq a < 0. \qquad (N2)''$$

Furthermore, unlike with the case of condition (*F*1)′ for the *PEM*, all these four possibilities are easily generalizable to cases with more layers, including the continuously stratified one.

The reason for this difference is that, in the *QGM* the Casimirs equal the integral of an ordinary function of q, and (h,u,v) are a **linear** (although non-local) functional of q. In the *PEM*, on the other hand, the Casimirs' density involve both q and h, and q is **nonlinear** functional of (h,u,v).

[1] Arnol'd, V.I., 1965. *Condition for nonlinear stationary plane curvilinear flows of an ideal fluid*, **Dokl. Akad. Nauk. USSR, 162:** 975-978; (English transl: **Soviet Math., 6:** 773-777, 1965).

[2] Arnol'd, V.I., 1966. *On an apriori estimate in the theory of hydrodynamical stability*, **Izv. Vyssh. Uchebn. Zaved. Matematika, 54:** 3-5; (English transl. **Amer. Math. Soc. Transl., Series 2 79:** 267-269, 1969).

[3] Holm, D.D., J.E. Marsden, T. Ratiu, and A. Weinsten, 1985. *Nonlinear stability of fluid and plasma equilibria*, **Phys. Reports, 123:** 1-116.

[4] Abarbanel, H.D.I., D.D. Holm, J.E. Marsden and T. Ratiu, 1986. *Nonlinear stability of stratified fluid equilibria*, **Phil. Trans. Roy. Soc. London, A 318:** 349-409.

[5] McIntyre, M.E., and T.G. Shepherd, 1987. *An exact local conservation theorem for finite-amplitude disturbances to non-parallel shear flows, with remarks on Hamiltonian structure and on Arnol'd's stability theorems*, **J. Fluid Mech., 181:** 527-565.

[6] Shepherd, T.G., 1990. *Symmetries, conservation laws, and hamiltonian structure in geophysical fluid dynamics*, **Adv. Geophys., 32:** 287-338.

[7] Shepherd, T.G., 1990. *Arnol'd stability applied to fluid flow: successes and failures*, in **Nonlinear phenomena in atmospheric and oceanic sciences**, G.F. Carnevale & R.T. Pierrehumbert, eds.

[8] Admissibility in the sense of properties of the functional derivatives at the boundaries.

[9] Shepherd, T.G., 1988. *Rigorous bounds on the nonlinear saturation of instabilities to parallel shear flows*, **J. Fluid Mech., 196:** 291-322.

[10] Shepherd, T.G., 1988. *On the nonlinear saturation of baroclinic instability. Part I: The two layer model*, **J. Atmos. Sci., 45:** 2014-2025.

[11] Shepherd, T.G., 1989. *Nonlinear saturation of baroclinic instability. Part II: continuously stratified fluid*, **J. Atmos. Sci., 46:** 888-907.

[12] Andrews, D.G., 1984. *On the existence of nonzonal flows satisfying sufficient conditions for stability*, **Geophys. Astrophys. Fluid Dyn., 28:** 243-256.

[13] Carnevale, G.F., and T.G. Shepherd, 1990. *On the interpretation of Andrews' theorem*, **Geophys. Astrophys. Fluid Dyn., 51:** 1-17.

[14] Ripa, P., 1991. *A tale of three theorems*, **Rev. Méx. Fís.**, submitted.

[15] Ripa, P., 1991. *General stability conditions for a multi-layer model*, **J. Fluid Mech., 222:** 119-137.

[16] Ripa, P., 1991. *Wave energy-momentum and pseudo energy-momentum conservation for the layered quasi-geostrophic instability problem*, **J. Fluid Mech.**, in press.

[17] Blumen, W., 1968. *On the Stability of Quasi-Geostrophic Flow*, **J. Atmos. Sci., 25:** 929-931.

[18] Salusti, E., and F. Zirilli, 1986. *A Large class of Non-zonal oceanic flows satisfying the Arnold-Blumen sufficient condition for Stability*, **Geophys. Astrophys. Fluid Dyn., 35:** 157-171.

[19] Swaters, G.E., 1986. *A nonlinear stability theorem for baroclinic quasigeostrophic flow*, **Phys. Fluids, 29:** 5-6.

[20] Blumen, W., 1973. *Stability of a two-layer fluid model to nongeostrophic disturbances*, **Tellus, 25:** 12-19.

[21] Holm, D.D., J.E. Marsden, T. Ratiu, and A. Weinstein, 1983. *Nonlinear stability conditions and a priori estimates for barotropic hydrodynamics*, **Phys. Lett., 98 A:** 15-21.

[22] Ripa, P., 1987. *On the stability of elliptical vortex solution of the shallow-water equations*, **J. Fluid. Mech., 183:** 343-363.

[23] Ripa, P., 1990. *Positive, negative and zero wave energy and the flow stability problem, in the Eulerian and Lagrangian-Eulerian descriptions*, **Pure Appl. Geophys., 133:** 713-732.

[24] Miles, J.W., 1961. *On the stability of heterogeneous shear flows*, **J. Fluid Mech., 10:** 496.

[25] Howard, L.N., 1961. *Note on paper of John W. Miles*, **J. Fluid Mech., 10:** 509.

[26] Benzi, R., S. Pierini, A. Vulpiani, and E. Salusti, 1982. *On nonlinear hydrodynamic stability of planetary vortices*, **Geophys. Astrophys. Fluid Dyn., 20:** 293-306.

Wavelets and Two Dimensional Turbulence

J. Weiss

Aware, Inc., One Memorial Drive, Cambridge, MA 02142, USA

The time evolution of an inviscid, incompressible fluid is governed by Euler's equations for the velocity field, $\hat{v}$:

$$\hat{v}_t + \hat{v} \cdot \nabla \hat{v} = \nabla p \tag{1}$$

where

$$\begin{aligned} &\hat{v} && = (u,v)(\hat{x},t) \\ &\nabla \cdot \hat{v} && = 0 \\ &\hat{x} && = (x,y) \in D \in R^2 \\ &\hat{v} \cdot \hat{m} && = 0 \ \text{ for } \ \hat{x} \in \partial D \\ &\hat{m} && \perp \partial D \\ &t = 0; && \hat{v} = \hat{v}_0(\hat{x}). \end{aligned}$$

It is known [8] that there exist smooth, global solutions to (1) where $\hat{v}_0$ and D are smooth. Furthermore, these solutions conserve the total energy:

$$E = \int_D \frac{\hat{v} \cdot \hat{v}}{2} = \int_D \frac{\hat{v}_0 \cdot \hat{v}_0}{2} \tag{2}$$

and the total enstrophy (squared vorticity):

$$\Omega = \int_D \frac{\text{curl}\hat{v} \cdot \text{curl}\hat{v}}{2} = \int_D \frac{\text{curl}\hat{v}_0 \cdot \text{curl}\hat{v}_0}{2}. \tag{3}$$

It has been conjectured [9] that the existence of these dual invariants implies a transfer of energy toward the large scales of motion across a $k^{-\frac{5}{3}}$ spectral range; and corresponding transfer of enstrophy toward the small scales across a k^{-3} energy spectrum. This conjecture is supported by numerous closure [7, 15], and simulation [12, 5] results. The transfer of energy to the large-scales is thought to explain the tendency of numerical simulations to evolve from random velocity fields toward states consisting of a few large scale regions of like-signed vorticity [5]. On the other hand, the transfer of vorticity (enstrophy) to the small scales of motion is less well understood and the subject of some controversy [17, 10]. Since numerical simulations are limited by finite degree truncations and the use of eddy viscosities that distort the inviscid behavior at small scales, the process of enstrophy transfer has remained somewhat obscure.

With regard to the above it is useful to note that following a fluid element, the vorticity is conserved. That is, if:

$$c = v_x - u_y \tag{4}$$

then

$$c_t + \hat{v} \cdot \nabla c = \frac{dc}{dt} = 0. \tag{5}$$

Thus:

$$c(\hat{x}(x_0, t), t) = c_0(\hat{x}_0). \tag{6}$$

The vorticity is advected by the velocity field as a passive scalar. However, unlike a passive scalar, the vorticity is dynamically related to (determines) the velocity field, and changes in the distribution of the vorticity imply changes in the advecting velocity field. This intrinsic feedback process can be better understood by introducing the stress tensor, A:

$$A = \nabla \hat{v} \ ; \ A_{ij} = \frac{\partial v_j}{\partial x_i}. \tag{7}$$

The gradients of vorticity, ∇c, evolve according to the equation

$$\frac{d}{dt} \nabla c = -A \nabla c. \tag{8}$$

Since $A^2 = \frac{trA^2}{2} I$ we expect, under the assumption that A is slowly varying with respect to ∇c, exponential growth of the gradients of vortitcity when $trA^2 > 0$, and oscillatory behavior when $trA^2 < 0$. See references [19, 20, 1, 13].

Previously, we have examined the evolution of the gradients of vorticity as these relate to the geometry of two-dimensional Euler flow and the phenomena of two-dimensional turbulence [19].

We have further examined [20] two-dimensional Euler flows by developing a numerical method based on compactly supported scaling functions. The translates of the scaling function provide the basis for a Galerkin approximation to the solutions of Eulers equations.

The combination of wavelet-Galerkin space discretization and implicit time differencing allows the development of essentially inviscid approximations for two-dimensional Euler flow. We find that Backward-Euler time differencing will damp the high-frequency components of the solution. This makes Backward-Euler time differencing quite useful for inviscid calculations

We have integrate the Euler flow using Backward-Euler time differencing for 7500 time steps. The flow evolves from the neighborhood of an unstable steady state to the neighborhood of a stable steady state. The stable steady state is categorized by a classical cats-eye structure, with vorticity assembled into positive and negative regions.

We repeat the above calculation, but use a conservative time differencing and a *hyperviscosity* to tame the small scales near the cutoff (grid) scales. In experiment two the effective Reynolds number is 12800 with a D6 scaling function Galerkin element.

A basic difference between the inviscid calculation and the calculations using a hyperviscosity clearly emerges over the long term integration. The inviscid method preserves the structure of the hyperbolic stagnation points of the flow. These appear to be essential to the long time organization (limit) of the Euler flow. In effect, the hyperbolic points feed energy to the large scale vortex structures by creating local vortex *wakes*. Vortex pair merging [2, 3] in the hyperbolic wakes add by accretion to the large scale vortex structures.

The hyperviscosity annuls the hyperbolic stagnation points. As a consequence the large scale vortex structures are smaller and dynamically isolated. This result suggests that the hyperbolic regions are essential to the structural morphogenesis of inviscid flows.

Again, the inviscid method appears to provide a better closure to the unresolved fine scales. We conjecture that the unresolved structures represent a sharpening of the the structures already observed and would introduce no qualitatively new phenomena. For instance, the *terracing* of vorticity observed at large times would be steeper if better resolved. A reasonable conjecture is that for infinite time the terracing vorticity approaches a piecewise constant limit state that could be asymptotically modeled by vortex methods [16].

Compactly supported wavelets

Ingrid Daubechies defined the class of compactly supported wavelets [4, 18]. Briefly, let φ be a solution of the *scaling* relation

$$\varphi(x) = \sum_{k=0}^{N} a_k \varphi(2x - k).$$

The a_k are a collection of coefficients that categorize the specific wavelet basis. The expression φ is called the scaling function.

The normalization $\int \varphi dx = 1$ of the scaling function obtains the condition

$$\sum a_k = 2.$$

The translates of φ are required to be orthonormal

$$\int \varphi(x - k)\varphi(x - m) = \delta_{k,m}.$$

¿From the scaling relation this implies the condition

$$\sum_{k=0}^{N} a_k a_{k-2m} = \delta_{0m}.$$

For coefficients verifying the above two conditions, the functions consisting of translates and dilations of the scaling function, $\varphi(2^j x - k)$, form a complete, orthogonal basis for square integrable functions on the real line, $L^2(R)$.

If only a finite number of the a_k are nonzero then φ will have compact support.

Smooth scaling functions arise as a consequence of the degree of approximation of the translates. The conditions that the polynomials $1, x, \cdots, x^{p-1}$ be expressed as linear combinations of the translates of $\varphi(x-k)$ is implied by the condition

$$\sum(-1)^k k^m a_k = 0$$

for $m = 0, 1, \cdots, p-1$.

We apply compactly supported wavelets to the numerical resolution of two-dimensional turbulence. These have several advantages for the numerical approximation of solutions of differential equations, including exact representation of polynomials of certain degrees and compact-support. For nonlinear systems (Euler equations) with solutions that may develop sharp gradients the primary advantage seems to be that wavelets can accurately approximate the smooth component to a solution while correctly resolving the components associated with strong gradients. The reason for this is that wavelets are less smooth than their order of approximation and therefore are less *stiff* than other higher order methods, i.e. Fourier or spline bases. For instance, the six term Daubechies scaling function (D6) can exactly represent polynomials through the second degree. However, the actual scaling function has only a continuous 1.06 derivative.

The two dimensional Euler equations

The two dimensional Euler equations in the stream function-vorticity formulation are

$$c_t + \psi_y c_x - \psi_x c_y = 0$$

$$\Delta\psi = -c.$$

The velocity field is $u = \psi_y$ and $v = -\psi_x$.

In applying the Wavelet-Galerkin method to the two dimensional Euler equations in the stream function-vorticity formulation we expand and approximate the stream function and vorticity in terms of the scaling function φ,

$$\psi(x,y) = \sum_{j=1}^{N}\sum_{k=1}^{N} \Psi_{j,k}\varphi(x-j)\varphi(y-k),$$

$$c(x,y) = \sum_{j=1}^{N}\sum_{k=1}^{N} C_{j,k}\varphi(x-j)\varphi(y-k).$$

Since we assume periodic boundary conditions there is a periodic wrap around in (x,y) and we let the period scale with the number of terms in the expansion. This allows neglect of the dilation factor in the scaling function. Substituting into the equation and projecting the result onto the subspace spanned by $\{\varphi(x-j)\varphi(y-k) : \ j = 1, \cdots, N; \ k = 1, \cdots, N\}$ requires evaluating terms of the form

$$\int \varphi_{xx}(x)\varphi_x(x-j)\varphi(x-k)dx.$$

We use an exact method of evaluation developed in reference [11].

This uniquely determines the $\Psi_{j,k}$ and $C_{j,k}$ as solutions of the Wavelet-Galerkin ordinary differential equations. Define the *connection coefficients*

$$\Omega_j^{02} = \int \varphi(x)\varphi_{xx}(x-j)dx$$

$$\Omega_{jl}^{100} = \int \varphi_x(x)\varphi(x-j)\varphi(x-l)dx$$

$$\Omega_{km}^{001} = \int \varphi(x)\varphi(x-k)\varphi_x(x-m)dx.$$

With the summation convention on the indices (p, q, j, l, k, m) the Wavelet-Galerkin equations are

$$C_t(p,q) + C(j+p, k+q)\Psi(l+p, m+q)\left(\Omega_{j,l}^{100}\Omega_{k,m}^{001} - \Omega_{j,l}^{001}\Omega_{k,m}^{100}\right) =$$

$$\sigma\left(\Omega_j^{02}C(j+p,q) + \Omega_k^{02}C(p,k+q)\right)$$

and

$$C(p,q) = -\Omega_j^{02}\Psi(j+p,q) - \Omega_k^{02}\Psi(p,k+q).$$

The approximation of the Jacobian nonlinearity depends on the vorticity and stream function fields and is a conservative approximation to the Jacobian of these fields. It can be easily shown, using the orthogonality of the translates of the scaling function and the properties of the Jacobian operator that,

$$\sum_p\sum_q C(p,q)\Theta(p,q) = 0$$

$$\sum_p\sum_q \Psi(p,q)\Theta(p,q) = 0.$$

These relations directly imply that the Wavelet-Galerkin approximation of the Jacobian will conserve both vorticity and energy.

The first experiment

We integrate the Euler equations with Backward-Euler time differencing

$$(c_{n+1} - c_n)/dt + J(c_{n+1}, \psi_{n+1}) = 0$$

$$\Delta\psi_n = -c_n$$

and a D6 Wavelet-Galerkin spatial discretization. The Backward-Euler differencing does not conserve the Energy or the Enstrophy. However, it does strongly damp the high frequency components of the solution. This makes Backward-Euler differencing quite useful for inviscid calculations since without the specific inclusion of viscosity it tames the small-scale terms near the resolution cut-off.

At each time step we solved the implicit difference scheme by iteration

$$v_{j+1} = c_n - dt \cdot J(v_j, \Delta^{-1}v_j)$$

until

$$\sup |v_{j+1} - v_j| < \epsilon.$$

In this experiment $\epsilon = 1.e - 3$, $dt = .025$ and the maximum initial Vorticity has a magnitude of 1.4017.

We integrated the equations for 7500 time steps of magnitude .025. In this period of time the solution undergoes a sharp transition from an unstable steady-state toward, what appears to be, a stable,steady-state solution. During this period of time the algorithm approximately conserves energy while transferring about one-half of the enstrophy out of the solution. The related gradients of vorticity are largest just before the event transferring the vorticity occurs.

The second experiment.

We integrate the Euler equations with a conservative time differencing of the advection and a Backward-Euler time differencing of the hyperviscosity

$$(c_{n+1} - c_n)/dt + J((c_{n+1} + c_n)/2, (\psi_{n+1} + \psi_n)/2) = \sigma \Delta^2 c_{n+1}$$

$$\Delta \psi_n = -c_n.$$

and a D6 wavelet-Galerkin spatial discretization.

The conservation of enstrophy and energy is implied by the conservative approximation of the Jacobian in the wavelet-Galerkin discretization.

Again, we use iteration to solve the implicit scheme at each time step

$$v_{j+1} = (I - \sigma dt \cdot \Delta^2)^{-1}(c_n - dt \cdot J((v_j + c_n)/2, -\Delta^{-1}(v_j + c_n)/2)$$

until

$$\sup |v_{j+1} - v_j| < \epsilon.$$

In this experiment $\epsilon = 1.e - 3$, $dt = .025$ and the initial vorticity has a magnitude of 1.4017. The initial vorticity for this experiment corresponds to the vorticity at time step 2000 in the first experiment. After time step 2000 the Euler flow evolves rapidly.

We integrated the solution from time step 2000 to time step 7000.

Numerical results.

Figure 1 shows the evolution of energy, the evolution of enstrophy, and the evolution of the gradients of vorticity for both experiments.

Figures 2 shows the evolution of the vorticity field for the first experiment. We show a field produced by a three point smoothing in the x and then the y direction. Even though the numerical solution has a considerable oscillatory component, the tensor product three point smoothing produces a consistent and smooth result.

Figures 3 shows the actual (calculated) vorticity fields of the second experiment at time steps 2000, 3000, 4000, 5000, 6000 and 7000.

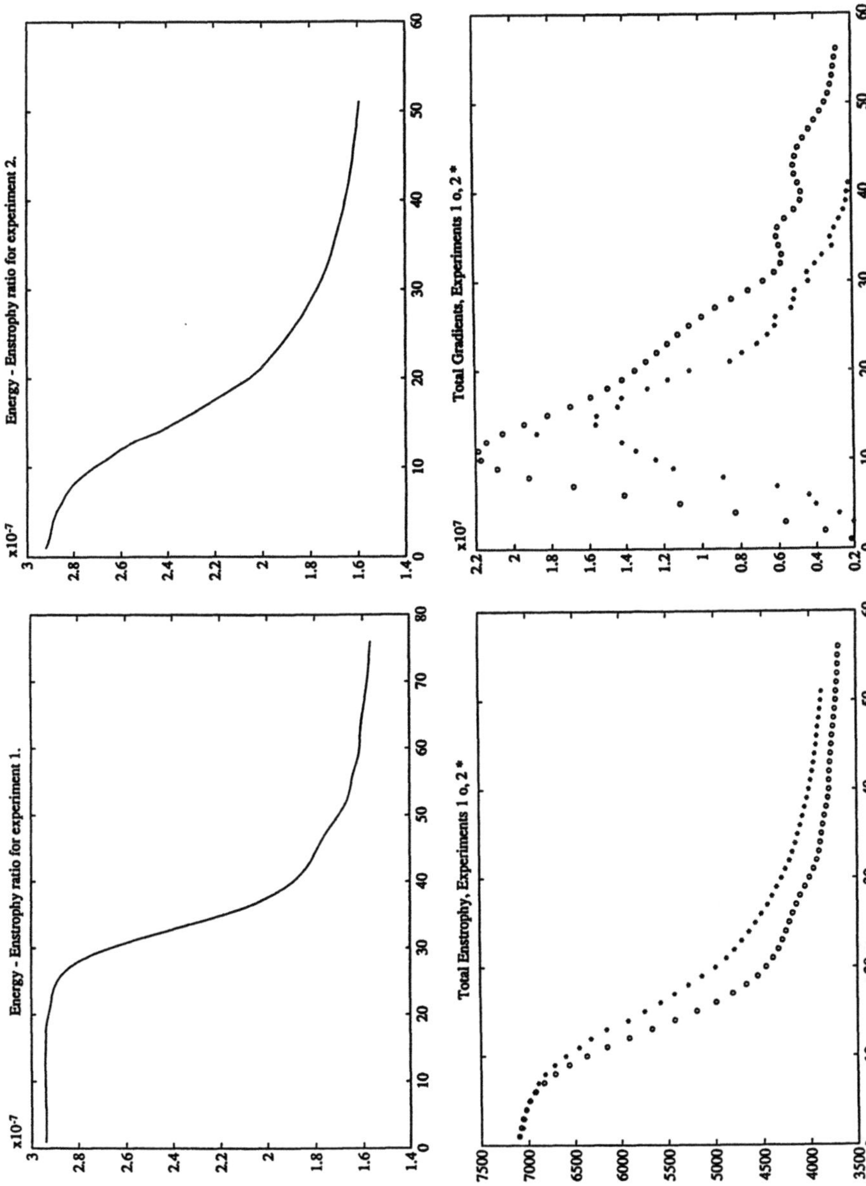
Energy - Enstrophy ratio for experiment 1.
x10-7
Energy - Enstrophy ratio for experiment 2.
x10-7
Total Enstrophy, Experiments 1 o, 2 *
Total Gradients, Experiments 1 o, 2 *
x10⁷

Figure 1

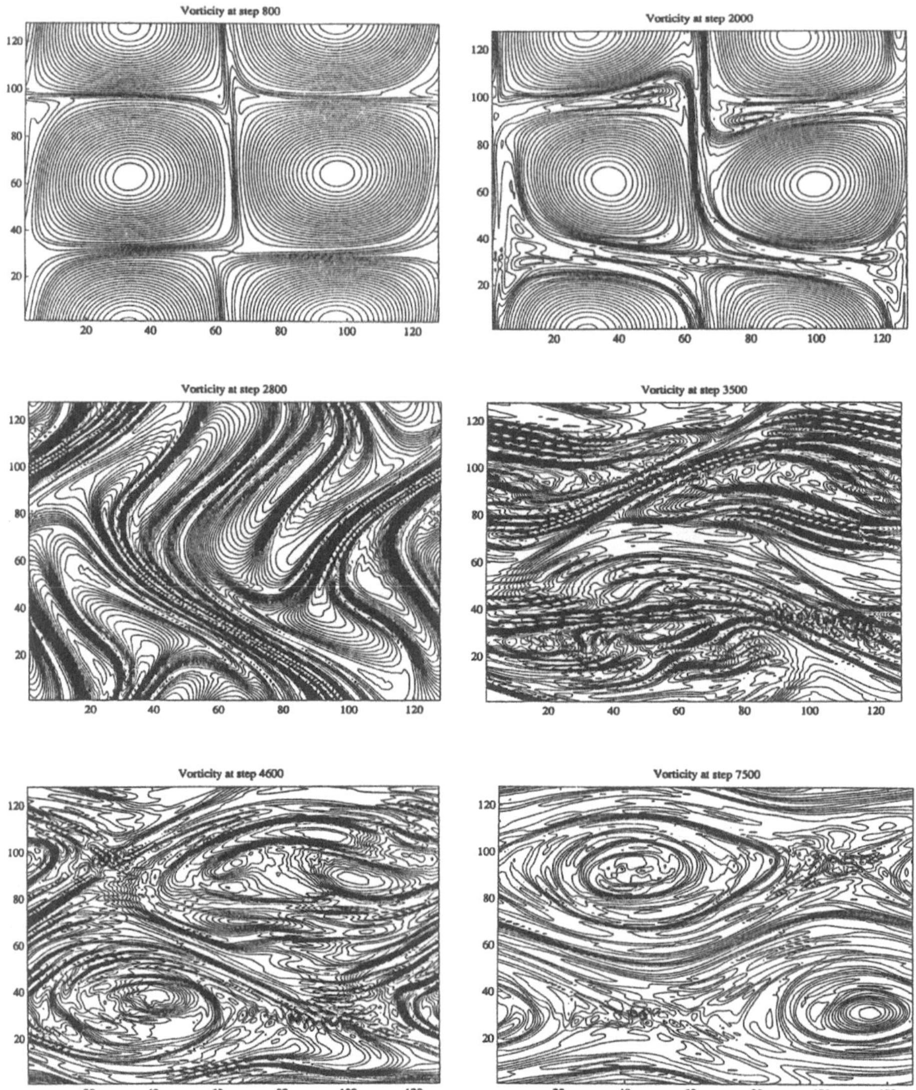
Vorticity at step 800
Vorticity at step 2000
Vorticity at step 2800
Vorticity at step 3500
Vorticity at step 4600
Vorticity at step 7500
20
40
60
80
100
120

Figure 2

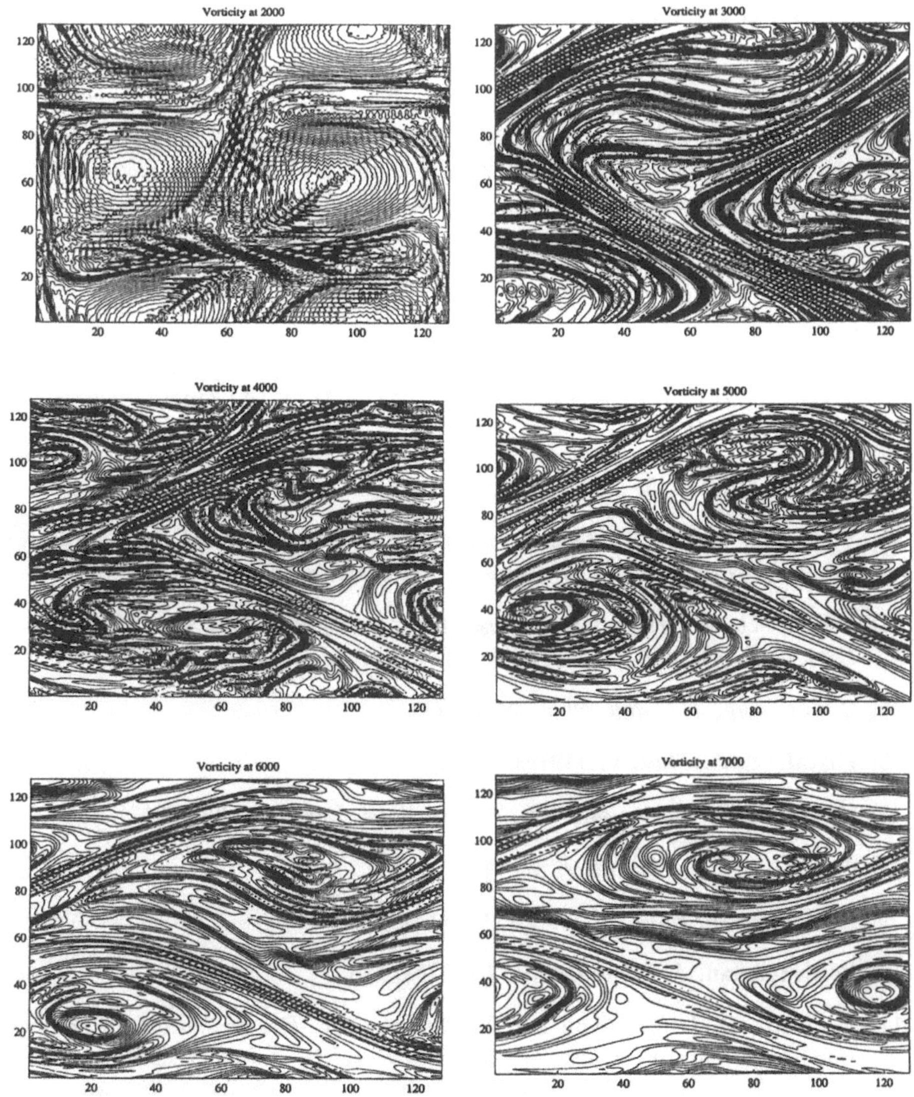

Vorticity at 2000
Vorticity at 3000
Vorticity at 4000
Vorticity at 5000
Vorticity at 6000
Vorticity at 7000

Figure 3

Acknowledgement

This research was supported in part by the Advanced Research Projects Agency of the Department of Defense and was monitored by the Air Force Office of Scientific Research under Contract No. F49620-89-C-0125. The United States Government is authorized to reproduce and distribute reprints for governmental purposes notwithstanding any copyright notation herein.

References

[1] M. Brachet, M. Meneguzzi, H. Politano and P. Sulem, *The dynamics of freely decaying two-dimensional turbulence*, *J Fluid Mech* **194** (1988), 333–349.

[2] Y. Couder, *Two-dimensional grid turbulence in a liquid film*, *J. Physique Lett.* **45** (1984), 353–366.

[3] Y. Couder and C. Basdevant, *Experimental and numerical study of vortex couples in two-dimensional flows*, *J. Fluid Mech.* **173** (1986), 225–251.

[4] I. Daubechies, *Orthonormal bases of compactly supported wavelets*, *Comm. Pure Appl. Math.* **41** (1988), 906-966.

[5] B. Fornberg, *J. Comp. Phys.* **25** (1977)

[6] Fox, *J. Math. Phys.* **17** (1976), 1148.

[7] U. Frisch, M. Lesieur and P.L. Sulem, *Phys. Rev. Lett.* **37** (1967), 895.

[8] T. Kato, *Arch. Ratl. Mech.* **25** (1967), 188.

[9] R. Kraichnan, *Phys. Fluids* **10** (1967), 1417.

[10] R. Kraichnan, *J. Fluid Mech.* **67** (1975), 155.

[11] A. Latto, H. L. Resnikoff and E. Tenenbaum, *The evaluation of connection coefficients of compactly supportes wavelets*, Aware Technical Report AD910708 (July 1991) and the proceedings of the French - USA workshop on *Wavelets and Turbulence*, Princeton University, June 1991, Ed. Y. Maday, Springer-Verlag (1992).

[12] D. Lilly, *J. Fluid Mech.* **45** (1971), 395.

[13] K. Ohkitani, *Wavenumber space dynamics of enstrophy cascade in a forced two-dimensional turbulence*, *Phys. Fluids* **A3** (1991), 1598–1611.

[14] S. Orszag and M. Israeli, *Ann. Rev. Fluid Mech.* **6** (1974), 281.

[15] A. Pouquet, M. Lesieur, J.C. Andre and C. Basdevant, *J. Fluid Mech.* **72** (1975), 305.

[16] D.I. Pullin, P.A. Jacobs, R.H.J. Grimshaw and P.G. Saffman, *Instability and filamentation of finite-amplitude waves on vortex layers of finite thickness*, *J. Fluid Mech.* **209** (1989), 359–384.

[17] P. Saffman, *Stud. Appl. Math.* **50** (1971), 377.

[18] G. Strang, *Wavelets and dilation equations: A brief introduction*, *SIAM Review* **31** (1989), 614-627.

[19] J. Weiss, *The dynamics of enstrophy transfer in two-dimensional hydrodynamics*, *Physica* **48D** (1991), 273-294.

[20] J. Weiss, *Wavelets and the study of two-dimensional turbulence*, Aware Technical report AD910628 and the proceedings of the French - USA workshop on *Wavelets and Turbulence*, Princeton University, June 1991, Ed. Y. Maday, Springer-Verlag (1992).

Part VI

Inverse Scattering Problems

Spectral Theory of Linear λ-Matrices and the Solution of Certain Nonlinear Algebraic and Functional Equations

B.A. Dubrovin[1], *A.S. Fokas*[2], *and P.M. Santini*[3]

[1]Department of Mechanics and Mathematics, Moscow State University, 119899 Moscow, Russia

[2] Department of Mathematics and Computer Science and the Institute for Nonlinear Studies, Clarkson University, Potsdam, NY 13699-5815, USA

[3]Dipartimento di Fisica, Università "La Sapienza", I-00185 Roma, Italy

1. Introduction

We present a short review of some of the results contained in [1]. In §2 we consider a direct and an inverse problem associated with the linear λ - matrix $L_A := \lambda A + U$, where A is a given nxn diagonal matrix and U is an nxn matrix with entries u_{ij}. We show that given a certain set of n^2 generic inverse data, the n^2 u_{ij}'s can be reconstructed in terms of Riemann theta functions of genus $(n-1)(n-2)/2$. These inverse data are polynomial expressions of the u_{ij}'s; therefore, using this result, the solution of certain systems of nonlinear algebraic equations can be expressed in terms of Riemann - theta functions. In §3 we construct a d - dimensional $(d \leq n)$ commutative algebra of λ - matrices of the same form and we give explicit parametrization of such an algebra when $d = n$ and $d = n-1$, in terms of rational and elliptic functions respectively. These formulae can be used to solve certain functional equations.

2. Spectral Theory on linear λ matrices and the solution of certain nonlinear algebraic equations

The basic object of our investigation is the linear eigenvalue problem

$$L_A\varphi = (\lambda A + U)\varphi = \mu\varphi, \tag{1}$$

$$A = diag(a_1, a_2, .., a_n), \quad a_i \neq a_j, \quad i \neq j, \quad (U)_{ij} = u_{ij}, \quad i,j = 1,..,n,$$

where A is considered given, λ, $\mu \in C$, and φ is some eigenvector. Associated with equation (1) we consider the Riemann surface Γ defined by the algebraic equation

$$\Gamma: \quad \Delta(\mu, \lambda) = det(\mu I - L) = 0,$$

where I is the identity matrix. The surface Γ is an n - sheeted covering of the complex λ plane which possesses, as $\lambda \to \infty$, n distinguished points denoted by $\infty_1,..,\infty_n$: $\infty_j = (\infty, a_j\infty)$. For generic u_{ij}'s Γ possesses the following properties: i) Γ has $n(n-1)$ branch points. ii) The genus of Γ is given by $g = (n-1)(n-2)/2$.

In the direct problem we characterize an appropriate eigevector $\varphi = (1, \varphi_2, .., \varphi_n)^T$ (T denotes transposition). φ_j, $2 \le j \le n$, have the following properties: a) it is meromorphic on Γ; b) it has g finite zeros, g finite poles (which are common for all φ'_js), one zero at ∞_1 and one pole at ∞_j; c) its asymptotic behaviour for large λ is given by:

$$\varphi_j \sim \frac{u_{j1}}{a_1 - a_j}\frac{1}{\lambda}; \qquad \varphi_j \sim \frac{a_j - a_1}{u_{1j}}\lambda; \qquad \varphi_j \sim \frac{u_{jk}}{u_{1k}}\frac{a_k - a_1}{a_k - a_j}, \quad k \ne 1, j,$$

where $2 \le j \le n$.

The n^2 spectral data consist of: i) the coefficients σ_i, $i = 1,.,n(n+1)/2$ of the polynomial $\Delta(\mu,\lambda) = 0$; ii) the finite poles $Q_i = (\delta_i, \mu_i)$, $i = 1, 2, .., g$, (with $\Delta(\mu_i, \delta_i) = 0$) of the φ_j's; iii) the norming constants $\nu_i := u_{i1}$, $i = 2, .., n$.

In the inverse problem we reconstruct φ, and hence U, in terms of the n^2 spectral data using Riemann - theta functions Θ of genus g and differentials of third kind, through the following expressions:

$$\varphi_j = c_j \frac{\Theta(A(P) - A(D) - A(\infty_j) + A(\infty_1) - K)}{\Theta(A(P) - A(D) - K)} exp(\int_{P_0}^{P} \Omega_{\infty_1\infty_j}), \tag{2a}$$

$$c_j = \frac{\nu_j}{a_1 - a_j} \frac{\Theta(A(\infty_j) - A(D) - K)}{\Theta(2A(\infty_1) - A(\infty_j) - A(D) - K)} \tag{2b}$$

for the nonspecial divisor $D = Q_1 Q_2 .. Q_g$, where the Abel maps of $P \in \Gamma$ and of D are given by

$$A(P) = (A_1(P), .., A_g(P))^T, \quad A_k(P) := \int_{P_0}^{P} \omega_k, \qquad k = 1, .., g,$$

$$A(D) = (A_1(D), .., A_g(D))^T, \quad A_k(D) := \sum_{l=1}^{g} \int_{P_0}^{P_l} \omega_k, \qquad k = 1, .., g.$$

Furthermore K is the Riemann vector, ω_j and $\Omega_{\infty_1\infty_j}$ are holomorphic differentials of the first and third kind respectively [2].

The direct and inverse problem can also be solved using an eigenvector φ with a different normalization than the one used above. The direct and inverse problem with $\varphi = (\varphi_1, \varphi_2, .., \varphi_n)^T$, where $\sum_{j=1}^{n} \varphi_j = 1$ was solved earlier by one of the authors (B. Dubrovin) [3].

It turns out that the spectral data are certain nonlinear algebraic equations of the u_{ij}'s. Therefore the solution of the inverse problem provides the solution of certain systems of nonlinear, nonhomogeneous algebraic equations. The particular form of the algebraic equations depends on the normalization of the eigenvector φ. For example, if $n = 3$, $\varphi = (1, \varphi_2, .., \varphi_n)^T$, $u_{ii} = 0, i = 1, 2, 3$, then $g = 1$, Γ is specified by 3 parameters and the finite pole $Q_1 = (\delta_1, \mu_1)$ of φ is given by

$$\delta_1 = \frac{1}{a_3 - a_2}\left(\frac{u_{13}u_{32}}{u_{12}} - \frac{u_{12}u_{23}}{u_{13}}\right), \quad \mu_1 = \mu(\delta_1) = \frac{1}{a_2 - a_3}\left(a_3\frac{u_{12}u_{23}}{u_{13}} - a_2\frac{u_{13}u_{32}}{u_{12}}\right).$$

Therefore the system of the 6 nonlinear algebraic equations

$$\sigma_1 = u_{12}u_{21} + u_{23}u_{32} + u_{13}u_{31}, \quad \sigma_2 = u_{12}u_{23}u_{31} + u_{21}u_{32}u_{13},$$

$$\sigma_3 = a_1u_{23}u_{32} + a_2u_{13}u_{31} + a_3u_{12}u_{21},$$

$$\nu_2 = u_{21}, \quad \nu_3 = u_{31};$$

$$(a_3 - a_2)u_{12}u_{13}\delta_1 + {u_{12}}^2u_{23} - {u_{13}}^2u_{32} = 0,$$

can be solved in terms of Θ functions of genus 1 (i.e. in terms of elliptic integrals).

It is worthwhile to remark that these nonlinear algebraic equations are **constructed** expressing the data in terms of the u_{ij}'s, through the direct problem associated with (1), and are **solved** through the solution (2) of the corresponding inverse problem. Since they are associated with the linear eigenvalue problem (1) only, they are conceptually different from soliton equations, which arise from the compatibility of a pair of linear operators (the Lax pair) of the type

$$L\varphi = \mu\varphi, \qquad M\varphi = z\varphi.$$

Following a Lax pair philosophy, a different class of nonlinear algebraic equations for the u_{ij}'s has been recently associated with the eigenvalue problem (1). Indeed one of the authors (P.M.Santini) [4] has shown that the compatibility of (1) and of

$$M\varphi = z\varphi, \qquad M := \sum_{i,s}\sum_{j=0}^{s} c_{is}z^j\Gamma^{(i+j-s,s-j)}(A, U)$$

yields the nonlinear algebraic equations

$$[L, M] = \sum_{i,j} c_{ij}[A, \Gamma^{(i-j-1,j+1)}(A, U)] = 0,$$

where the $\Gamma^{(i,j)}$'s are symmetric nonabelian generalizations of scalar monomials defined by

$$(\lambda A + U)^k = \sum_{j=0}^{k} \lambda^j\Gamma^{(j,k-j)}(A, U).$$

It was also observed in [4] that these equations are the most elementary instances of nonlinear systems associated with a Lax pair and therefore solvable through the inverse problem associated with (1).

3. Commuting algebras of λ matrices and the solution of certain functional equations

If M is linear in λ, like the operator (1), then the construction of the two commuting matrices L and M is a particular case of constructing a d - dimensional ($d \leq n$) algebra $\mathcal{A}$ of commuting matrices, linear in λ. Let $L = \lambda A + U_A \in \mathcal{A}$ be an element of such an Abelian algebra, and let $\mathbf{A} \subset C^n$ be the $d-$ dimensional space of diagonal leading terms $A \in \mathbf{A}$ (the coefficients of λ). Then it is easy to show that every element $L \in \mathcal{A}$ has the form

$$L_A = \lambda A + [A, Q] + D_A, \quad D_A = diag(d_{1A}, d_{2A}, .., d_{nA}),$$

$$d_{iA} = \sum_{j=1}^{n} (R)_{ij}(a_j - a_i),$$

where Q and R are off-diagonal matrices whose entries q_{ij} and r_{ij} satisfy the "functional equations"

$$\frac{q_{ik}q_{kj}}{q_{ij}} - r_{ik} + r_{jk} = \sum_{\alpha=1}^{n-d} \gamma_k^{(\alpha)} p_{ij}^{(\alpha)}, \tag{3}$$

for some matrices $p_{ij}^{(\alpha)}$'s, where $\sum_{i=1}^{n} \gamma_i^{(\alpha)} a_i = 0, \quad \alpha = 1, .., n-d, \quad \forall A \in \mathbf{A}$.
If $d = n$, equation (3) becomes

$$\frac{q_{ik}q_{kj}}{q_{ij}} - r_{ik} + r_{jk} = 0; \tag{4}$$

if $d = n - 1$, equation (3) becomes

$$\frac{q_{ik}q_{kj}}{q_{ij}} - r_{ik} + r_{jk} = \gamma_k p_{ij}, \tag{5a}$$

which is equivalent to the following constraint between Q and R:

$$\gamma_l\left(\frac{q_{ik}q_{kj}}{q_{ij}} - r_{ik} + r_{jk}\right) = \gamma_k\left(\frac{q_{il}q_{lj}}{q_{ij}} - r_{il} + r_{jl}\right), \quad i \neq j \neq k \neq l. \tag{5b}$$

A slightly less general version of equation (5a) was constructed and solved, in a different context, in [5].

If $d = n$, it can be shown that the spectral curve associated with $L \in \mathcal{A}$ is a rational curve, thus the commutative algebra of λ matrices can be parametrized in terms of rational functions, and the inverse spectral tranform

$$q_{ij} = \frac{\nu_i}{\nu_j}\frac{\lambda_j}{z_i - z_j}, \qquad r_{ij} = \frac{\lambda_j}{z_i - z_j}, \qquad i \neq j$$

solves the constraint (4) in terms of the spectral data $\nu_i, \lambda_i, z_i, i = 1, .., n$.

If $d = n - 1$, the spectral curve associated with $L \in \mathcal{A}$ is an elliptic curve, the commutative algebra of λ matrices can be parametrized in terms of elliptic functions and the inverse spectral transform

$$q_{ij} = \lambda_j \frac{\nu_i}{\nu_j}\frac{\sigma(z_i - z_j + z_0)}{\sigma(z_0)\sigma(z_i - z_j)}, \qquad r_{ij} = \lambda_j \zeta(z_i - z_j), \quad i \neq j$$

solves the constraint (5) in terms of Weierstrass σ - and ζ - functions, where $\nu_i, \lambda_i, z_i,\ i = 1, .., n$ are suitable spectral data and $\lambda_i = \gamma_i$. In addition we have that $p_{ij} = \zeta(z_0) - \zeta(z_i - z_j + z_0)$.

References

1. B. A. Dubrovin, A. S. Fokas and P. M. Santini, "Inverse Spectral Transform and the Solution of Certain Nonlinear Algebraic Equations", Preprint Dipartimento di Fisica, Università di Roma I, 1991.
2. G. Springer, *Introduction to Riemann surfaces*, Addison Wesley, (1957).
3. B. A. Dubrovin, Func. Anal. Appl., **11**, 265 (1977).
4. P. M. Santini, Inverse Problems, **6**, 665 (1990).
5. M. Bruschi and F. Calogero, SIAM J. Math. Anal. **21** (1990).

Nonlinear Acoustic Tomography

P. Monk

Department of Mathematical Sciences, University of Delaware,
Newark, DE 19716, USA

1 Introduction.

The use of acoustic waves as probes to determine otherwise inaccessible properties of a medium is extremely widespread (see [1] for a complete review of the current theory). Applications include sonar, medical imaging and non–destructive testing. Despite the importance of the applications, there is as yet no acceptable method for solving the full non–linear problem at resonance frequencies (frequencies at which the size of the features under investigation are approximately the wavelength of the incident acoustic field). The medical imaging problem, which consists in trying to determine the sound speed, density and absorption properties of a bounded inhomogeneous medium from scattered acoustic waves [2] is the motivation for the investigation described in this paper and more fully in [4]. We shall present a solution technique for a standard model inverse acoustic scattering problem which consists of reconstructing the refractive index of an inhomogeneity from given far field data (far field data is essentially the measured scattered field at considerable distance from the inhomogeneity). This model inverse problem simplifies the inhomogeneity by neglecting density and absorption but includes two important features of the real problem: nonlinearity and ill–posedness. Furthermore the method we present can easily be extended to more general problems [3].

Methods for the full non–linear inverse problem seem to fit into two basic categories [4]. In the first category regularized non–linear optimization techniques are used to compute an inhomogeneity whose far field pattern best fits the given far field pattern (cf. [10, 11, 16, 12, 14, 17]). The second category of methods, advocated by the author and D. Colton, attempts to find an inhomogeneity such that special weighted averages of the far field pattern attain prescribed values [5, 6, 7, 9, 8, 3, 4]. The rational for such an approach is the following: The far field pattern has a Fourier series in which the coefficients of higher order modes decay very rapidly. Thus, due to noise in the measured data, each far field pattern contains only a few reliable modes. Our method allows simultaneously the use of a large number of far field patterns and control over the way in which the reliable data is extracted. The method advocated here should have significant advantages over the first type methods when many far field patterns are needed to successfully solve the inverse problem.

The method presented here is a modification of the method in [6] and is described more fully in [4]. As mentioned previously, this investigation is carried out at resonance frequencies. Thus neither low nor high frequency asymptotic methods are applicable.

2. The Inverse Scattering Problem. Let us now describe in more detail the inverse scattering problem and our method for it's resolution. First we describe a well–posed forward problem. This is the standard problem of predicting a far field pattern from a knowledge of the inhomogeneity and the incoming wave. We consider the problem of the propagation of a time harmonic acoustic wave through an infinite inhomogeneous cylinder such that the refractive index and incident field are independent of the axial coordinate. Let $n(x), x \in \mathbf{R}^2$ denote the refractive index of the inhomogeneity. We shall assume that the inhomogeneity is bounded so that there is a disc Ω_b of radius b such that the

$$\{x \in \mathbf{R}^2 | n(x) \neq 1\} \subset \Omega_b.$$

Let k denote the wave number of the incident field. Then the forward problem is to determine the velocity potential $u(x)$ such that

$$\Delta u + k^2 n\, u = 0 \quad \text{in} \quad \mathbf{R}^2 \tag{1}$$

(2) $$u(x) = e^{ikr\cos(\theta-\alpha)} + u^s(x) \quad \text{in} \quad \mathbf{R}^2$$

(3) $$\lim_{r\to\infty} \sqrt{r}\left(\frac{\partial u^s}{\partial r} - iku^s\right) = 0$$

where (r, θ) are the polar coordinates of x, u^s is the scattered field and $e^{ikr\cos(\theta-\alpha)}$ is the incoming plane wave. Equation (3) is the Sommerfeld radiation condition which is assumed to hold uniformly in θ. Problem (1)–(3) is linear and well–posed [5].

The asymptotics of the scattered field u^s are such that

(4) $$u^s(x) = \frac{e^{ikr}}{\sqrt{r}} F(\theta; k, \alpha) + 0\left(\frac{1}{r^{3/2}}\right)$$

where $F(\theta; k, \alpha)$ is the far field pattern of the scattered field. Given n, k and α the numerical approximation of F is a well understood problem (cf. [13]).

The inverse problem we shall consider is to reconstruct $n(x)$ from a knowledge of the far field pattern. We shall assume that $F(\theta; k, \alpha)$ is known for all observation angles $\theta \in [-\pi, \pi]$, all incident directions $\alpha \in [-\pi, \pi]$ and a discrete set of wave numbers $k_1, k_2, \cdots, k_{N_k}$. Of course in our numerical scheme only a discrete set of observation angles and incident directions are used.

Our inverse algorithm, which is described and justified more fully in [4], consists of two steps which we now describe in detail.

(1) **Computation of the Herglotz Kernels.** To describe this process, we first introduce an auxiliary problem. Let $F_\lambda(\theta; k, \alpha)$ be the far field pattern of the scattered field h^s satisfying the following exterior impedance problem in which λ is a free parameter at our disposal:

(5) $$\Delta h + k^2 h = 0 \quad \text{in} \quad \mathbf{R}^2 \backslash \overline{\Omega}_b$$

(6) $$h(x) = e^{ikr\cos(\theta-\alpha)} + h^s(x) \quad \text{in} \quad \mathbf{R}^2 \backslash \overline{\Omega}_b$$

(7) $$\frac{\partial h}{\partial r} + i\lambda h = 0 \quad \text{on} \quad \partial\Omega_b$$

(8) $$\lim_{r\to\infty} \sqrt{r}\left(\frac{\partial h^s}{\partial r} - ikh^s\right) = 0$$

F_λ can be computed explicitly using cylindrical wave functions. Recall that Ω_b is a priori known to contain the region in $\mathbf{R}^2$ in which $n(x) \neq 1$.

For each p with $-P \le p \le P$ and each k_l, $1 \le l \le N_k$ we compute the Herglotz kernels $\{g_P(\theta; k)\}_{p=-P}^{P}$ by solving the equation

(9) $$\int_{-\pi}^{\pi} [F(\theta; k_l, \alpha) - F_\lambda(\theta; k_l, \alpha)] g_p(\theta; k_l) d\theta = (-1)^p \sqrt{\frac{2}{\pi k_l}}\, e^{-i\left(p\frac{\pi}{2} + \frac{\pi}{4} - p\alpha\right)}$$

for all $\alpha \in [-\pi, \pi]$. In practice we solve a finite dimensional version of (9) obtained by approximating g_p via a finite Fourier series. Furthermore, we find that (9) is very ill–conditioned and so regularize by Tikhonov regularization. Note however that the problem is linear so that a wide variety of methods exist for regularizing the problem, and in practice the problem is cheap to solve. In the numerical algorithm, far field data is given at discrete values of θ and α. Adding more θ or α data points does not greatly increase computational costs. Even adding data for more wave numbers is not expensive at this stage since each problem in (9) is small and linear.

The parameter P controls the amount of information extracted from the far field pattern. In the axisymmetric case ($n(x) \equiv n(|x|)$) we can show that $g_p(\theta; k)$ is entirely determined by the $-p^{th}$ Fourier coefficient of F (as well as the pth Fourier coefficient of F_λ). Thus, in the axisymmetric case, P is chosen to reflect the number of reliable Fourier coefficients in F (see [3] for results varying P). In the general case g_p is no longer directly related to the Pth Fourier coefficients of F but the set $\{g_p\}_{p=-P}^{P}$ is still indirectly related to the number of reliable modes of F.

The free parameter λ is chosen to eliminate certain solvability problems associated with (9) (see [8]). If n is real (as in the case of this paper) we should choose $\lambda > 0$, whereas if the inhomogeneity is absorbing (hence n complex) we can choose $\lambda = 0$.

We remark that by the reciprocity principle (9) is equivalent to weighting the far field patterns so that the weighted far field pattern is given by the right hand side of (9) [4].

Once the Herglotz kernels have been computed, we can compute the corresponding Herglotz wave functions $\{v_p(r,\theta;k)\}_{p=-P}^{P}$ defined by

$$v_p(r,\theta;k) = \int_{-\pi}^{\pi} g_p(\phi;k)e^{-ikr\cos(\theta-\phi)}d\phi. \tag{10}$$

(2) **Optimization to compute $n(x)$.** The next stage of the algorithm uses the Herglotz wave functions in (10) to find an appropriate $n(x)$. Using the theory of Herglotz wave functions, we can show that if g_p and v_p are given by (9) and (10), then the pair (W_p, v_p) satisfies the following system of equations:

$$\Delta W_p + k^2 n W_p = 0 \quad \text{in} \quad \Omega_b \tag{11}$$

$$\left(\frac{\partial}{\partial r} + i\lambda\right)\left(W_p - H_p^{(1)}(kr)e^{ip\theta}\right) = 0 \quad \text{on} \quad \partial\Omega_b \tag{12}$$

$$W_p = v_p + W_p^s \quad \text{in} \quad \Omega_b \tag{13}$$

$$\Delta W_p^s + k^2 W_p^s = 0 \quad \text{in} \quad \mathbf{R}^2\backslash\overline{\Omega}_b \tag{14}$$

$$\lim_{r\to\infty} \sqrt{r}\left(\frac{\partial}{\partial r}W_p^s - ikW_p^s\right) = 0. \tag{15}$$

Strictly speaking (9)–(15) only hold exactly for certain special coefficients $n(x)$ (for example asymmetric functions $n(x) = n(|x|)$), but in general a solution W_p to (11)–(12) always exists and the decomposition (13)–(15) is always valid where v_p can be approximated in $C^1(\overline{\Omega}_b)$ by a Herglotz wave function and (9) holds to arbitrary accuracy. This is sufficient for our purposes.

The numerical method for approximating n is based on (11)–(15). Given a candidate function n, we can approximate W_p by a cubic isoparametric finite element function W_p^h that satisfies (11)–(12) in the usual finite element Galerkin sense. We also can compute an approximation to W_p^s from (12), (14) and (15) in $\mathbf{R}^2\backslash\overline{\Omega}_b$ using cylindrical wave functions. Then a Levenberg–Marquardt optimization scheme is used to adjust the candidate function n so that (13) is satisfied in a regularized least squares sense on $\partial\Omega_b$. Again Tikhonov regularization is necessary since the problem of computing n is ill–posed. For details of each step described above see [4].

The optimization scheme described above is quite time consuming. In particular the cost of each step of the optimization scheme increases rapidly with N_k. An advantage of the method in this paper compared to our old scheme in [6] is that N_k can be traded off against P. Indeed we shall show numerical results for a problem in which $N_k = 1$. Thus would not have been possible with our old method.

3. Numerical Results. Here we present two sets of numerical results. For other results, including the effect of noise in the data on reconstruction accuracy, see [4]. For each example, we compute synthetic far field data by solving (1)–(4) for a known inhomogeneity n using the method of [13], then we use this synthetic data in the inverse algorithm.

Our first example ("Twin peaks" from [4]) has an exact $n(x)$ given as follows: Let $r_1 = \sqrt{(x+1/2)^2+y^2}$ and $r_2 = \sqrt{(x-1/2)^2+y^2}$ then

$$n(x) = \begin{cases} (3+\cos(2\pi r_1))/2 & \text{if } r_1 < 1/2 \\ (3+\cos(2\pi r_2))/2 & \text{if } r_2 < 1/2 \\ 1 & \text{otherwise.} \end{cases} \tag{16}$$

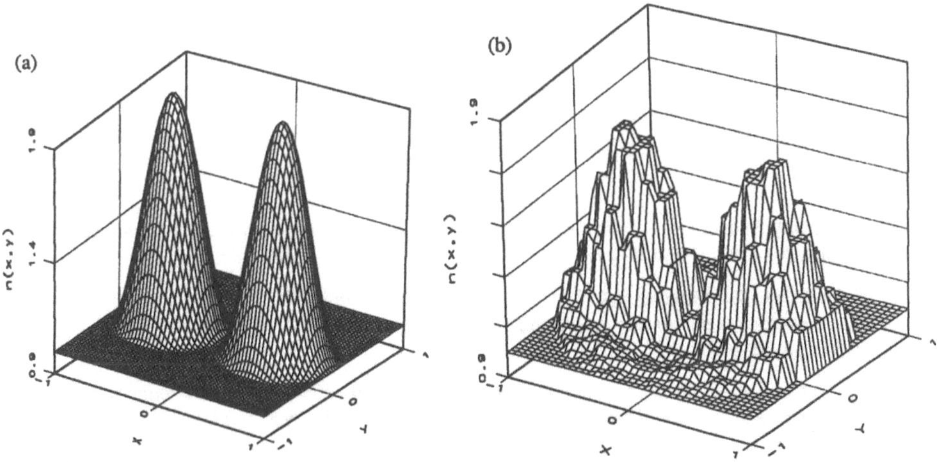

Figure 1: Graphs of the original n and reconstruction n_h^* for Example 1 in which n is given by (16). In (a) we show the exact n. In (b), we give the reconstruction using parameter values given in the text. Data for only one wave number is used, yet the reconstruction is of comparable quality to the reconstruction in [4] which uses data at four different wave numbers.

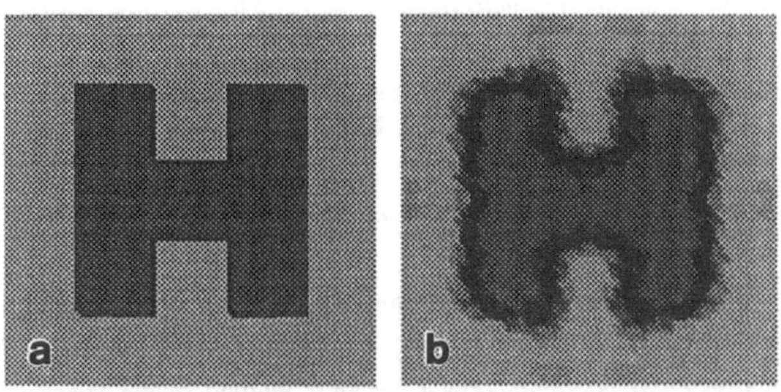

Figure 2: Graphs of the original n and reconstruction n_h^* for H. The exact solution is $n = 2$ in the H shaped region in (a) and $n = 1$ otherwise. The reconstruction is shown in (b). Here we use a density plot of 24 gray levels between 1 and 2 to display the value of the function n and it's reconstruction. The square region is the domian $[-1, 1] \times [-1, 1]$.

The reconstruction is shown in Figure 1 and was performed using data for a single wave number ($N_k = 1$), $k_1 = 7$, $P = 2$, and 51 incoming waves measured at 128 observation points. Despite using only one wave number, the discrete relative error in the reconstruction is only about twice the error reported in [4] using four wave numbers ($N_k = 4$). The use of a single wave number increases the sensitivity of the problem to the choice of regularization parameters and the optimization algorithm must take more steps to converge (nevertheless, the total time for the optimization algorithm is less then the time with $N_k = 4$). This example shows that our method can work with data at a single frequency.

Our second example is shown in Figure 2 using a density plot. The exact refractive index is shown in Figure 2a and is $n = 2$ inside H and $n = 1$ elsewhere. The reconstruction using $N_k = 4$ ($k_1 = 1, k_2 = 3, k_3 = 5, k_4 = 7$), $P = 2$ and 51 incoming waves is shown in Figure 2b. In performing the reconstruction we do not align the finite element mesh with the discontinuity in n.

Much remains to be done to obtain a robust and reliable inverse solver. The present algorithm is computer intensive and a better way of inferring n from (11)–(15) is needed. Methods for accurately locating discontinuities in the unknown refractive index would be useful in the development of the finite element grid (see [15] for work in this direction). Finally an understanding of the nature of continuous dependence in the regularized problem is needed.

References

[1] R. Bates, V. Smith, and R. Murch, *Manageable multidimensional inverse scattering theory*, Physics Reports, 201 (1991), pp. 185–277.

[2] T. Cavicchi, S. Johnson, and W. O'Brien Jr., *Application of the sinc basis moment method to the reconstruction of infinite circular cylinders*, IEEE Trans. on Ultrason. Ferroelect. Freq. Contr., 35 (1988), pp. 22–33.

[3] D. Colton and P. Monk, *A comparison of two methods for solving the inverse scattering problem for acoustic waves in an inhomogeneous medium*. To appear in *J. Comp. Appl. Math.*

[4] ——, *The numerical solution of an inverse scattering problem for acoustic waves*. Submitted for publication.

[5] ——, *The inverse scattering problem for time harmonic acoustic waves in an inhomogeneous medium*, Q. J. Mech. appl. Math., 41 (1988), pp. 97–125.

[6] ——, *The inverse scattering problem for time harmonic waves in an inhomogeneous medium: Numerical experiments*, IMA J. Appl. Math., 42 (1989), pp. 77–95.

[7] ——, *A new method for solving the inverse scattering problem for acoustic waves in an inhomogeneous medium*, Inverse Problems, 5 (1989), pp. 1013–1026.

[8] ——, *The inverse scattering problem for time harmonic acoustic waves in an inhomogeneous medium*, in Inverse Problems in Partial Differential Equations, D. Colton, R. Ewing, and W. Rundell, eds., Philadelphia, 1990, SIAM Publications, pp. 73–84.

[9] ——, *A new method for solving the inverse scattering problem for acoustic waves in an inhomogeneous medium II*, Inverse Problems, 6 (1990), pp. 935–947.

[10] S. Gutman and M. Klibanov, *Newton-Kantorovich method for three dimensional inverse scattering problem at fixed frequencies*. To appear.

[11] S. Johnson and M. Tracey, *Inverse scattering solutions by a sinc basis, multiple source, moment method – Part I: Theory*, Ultrason. Imaging, 5 (1983), pp. 361–375.

[12] S. Johnson, Y. Zhou, M. Tracey, M. Berggren, and F. Stenger, *Inverse scattering solutions by a sinc basis, multiple source, moment method – Part III: Fast algorithms*, Ultrason. Imaging, 6 (1984), pp. 103–116.

[13] A. Kirsch and P. Monk, *An analysis of the coupling of finite element and Nyström methods in acoustic scattering*. In preparation.

[14] R. Kleinman and P. van den Berg, *A hybrid method for two dimensional problems in tomography*. To appear in *J. Comp. Appl. Math.*

[15] L. Päivärinta and E. Somersalo, *Inversion of discontinuities of the Schrödinger equation in three dimensions*, SIAM J. Math. Anal., 22 (1991), pp. 480–499.

[16] M. Tracey and S. Johnson, *Inverse scattering solutions by a sinc basis, multiple source, moment method – Part II: Numerical evaluations*, Ultrason. Imaging, 5 (1983), pp. 376–393.

[17] V. Weston, *Multifrequency inverse problem for the reduced wave equation with sparse data*, J. Math. Physics, 25 (1984), pp. 1382–1390.

Spectral Transform for Nonlinear Evolution Equations with N Space Dimensions

P.C. Sabatier

Laboratoire de Physique Mathématique, F-34095 Montpellier Cedex 05, France

1. Introduction

We study a special case of a method we visited recently[1] and which generalizes well-known methods for constructing solutions of non linear evolution equations. The input is an integral equation which is restricted here to the form

$$\psi_j(k,x,t) = I_j + \int \frac{d\sigma(\lambda,\Lambda)}{-k+\Lambda} T(\lambda,\Lambda,x,t)\ \psi_j(\lambda,x,t) \tag{1.1}$$

where $k,\lambda,\Lambda, \in \mathbb{C}$, ψ_j, $I_j \in \mathbb{C}^N$, $x \in \mathbb{C}^N$, $t \in \mathbb{R}$, T is a linear mapping of $\mathbb{C}^N$ into itself, $d\sigma$ is a measure in $\mathbb{C}^2$. For $j = 1,2...,N$, I_j are known column vectors, defined as $(I_j)_k = \delta_{jk}$

The essential assumption of the method is that, for the chosen "reflection coefficient" T, and for $j = 1,2..N$, the ψ_j's are uniquely determined by (1.1). They depend on x, t because T does. A linear differential operator L acting on ψ_j yields on the right-hand side of (1) a new free term J and, inside the integral, $TL\varphi$ plus additional terms that we called "*parasite terms*"[1]. A differential operator L which cancels parasite terms for a given dependence of T on x, t, is "*compatible*" with the "*dispersion*" of T. In this case, if it follows from the assumptions that

$$L\psi = J + \oint TL\psi \tag{1.2}$$

where we dropped the index j and we used $\oint$ as a reduced writing of $\int \frac{d\sigma}{-k+\Lambda}$, has at most one solution, and if we can construct a solution in a parallel way, say $M(\psi)$, where M usually is nonlinear, we can get a nonlinear relation $L\psi = M\psi$. Now, for an arbitrary fixed matrix B, and two components x,y for x, it is easy to derive from (1) the following relations

$$B\frac{\partial\psi}{\partial x} = \oint TB\frac{\partial\psi}{\partial x} + \oint \left\{[B,T]\frac{\partial\psi}{\partial x} + B\frac{\partial T}{\partial x}\psi\right\}$$

$$B\frac{\partial^2\psi}{\partial x\partial y} = \oint TB\frac{\partial^2\psi}{\partial x\partial y} + \oint\left\{[B,T]\frac{\partial^2\psi}{\partial x\partial y} + B\frac{\partial T}{\partial x}\frac{\partial\psi}{\partial y} + B\frac{\partial T}{\partial y}\frac{\partial\psi}{\partial x} + B\frac{\partial^2 T}{\partial x\partial y}\psi\right\} \tag{1.3}$$

$$kB\frac{\partial\psi}{\partial x} = B\frac{\partial\psi^{(1)}}{\partial x} + \oint\left\{\lambda TB\frac{\partial\psi}{\partial x} + \oint\lambda\ [B,T]\frac{\partial\psi}{\partial x} + (\Lambda-\lambda)\ BT\frac{\partial\psi}{\partial x} + \Lambda B\frac{\partial T}{\partial x}\psi\right\}$$

and, if $BI = 0$:

$$B\psi = \oint TB\psi + \oint [B,T]\psi \tag{1.4}$$

$$kB\psi = B\psi^{(1)} + \oint \lambda TB\psi + \oint \{(\Lambda-\lambda)\ TB\psi + \Lambda[B,T]\psi\} \tag{1.5}$$

where $\psi^{(1)}$ appears in the asymptotic expansion of ψ versus k :

$$\psi_j = I_j + k^{-1}\ \psi_j^{(1)} + k^{-2}\ \psi_j^{(2)} + \ldots \tag{1.6}$$

The equation (1.2) with J replaced by I, and the equations (1.3) to (1.6) hold if ψ is replaced by the matrix $\Psi = (\psi_1, \psi_2, \ldots \psi_N)$. In addition, it is obvious that multiplying both sides of (1.3), (1.4), (1.5) by a fixed matrix is freely allowed *on the right*.

2. Differential forms and evolution equations

Let C and $\Gamma \in (\mathcal{M}_N)^N$, $\mathcal{M}_N$ being the set of N×N matrices, $\Psi \in \mathcal{M}_N$, Δ and grad be the laplacian and the gradient operator with respect to the space variables $X_1, X_2, \ldots X_N$. We introduce the basic 2nd order differential form :

$$K(\Psi) = -\frac{1}{2}\Delta\,\Psi + C.\text{grad}\,\Psi + k\,\Gamma.\text{grad}\,\Psi \tag{2.1}$$

Cancelling the grad T parasite terms of (2.1) is possible if the space dispersion relations hold :

$$\text{grad}\ T = [C,T] + \Lambda\Gamma T - \lambda T\Gamma \tag{2.2}$$

If we impose that the second derivatives of T be continuous, all the components of C and T must commute with each other, and since they are N×N matrices, they must be a polynomial of $d^o(N-1)$ of a matrix S. For the sake of simplicity, we assume here that S is diagonal, so that

$$\Gamma = \sum_{\ell=1}^{N} g_\ell\ \eta_\ell \tag{2.3}$$

where η_ℓ is the matrix whose ℓ,ℓ element is 1, the others are zero, and $g_\ell \in \mathbb{C}^N$. A similar expansion holds for C. We can now define the differential form $\mathcal{F}(\Psi) = -\theta\,\frac{\partial\Psi}{\partial t} + K(\Psi)$ which is compatible with the space dispersion relation (2.2) and the time dispersion relation

$$\theta\,\frac{\partial T}{\partial t} = \frac{1}{2}\,[Q.QT - TR.R] \tag{2.4}$$

where $Q = C+\Lambda\Gamma$, $R = C+\lambda\Gamma$. The free term of $\mathcal{F}(\Psi)$ is $\Gamma.\text{grad}\ \Psi^{(1)}$. If (1.1) is written for the matrix $\Psi = (\psi_1, \psi_2, \ldots, \psi_N)$, multiplying it on the right by the $f(\Psi^{(1)})$, and using the uniqueness assumption we get the first nonlinear relation

$$\mathcal{F}(\Psi) = F(\Psi) + k\ f(\Psi) = \Psi\ f(\Psi^{(1)}) \tag{2.5}$$

We study now other compatible forms. Some of them may yield linear differential equations for Ψ (called "ghost equations") - a case which is better to prevent. Hence we are led to introducing the most general linear forms, which are constructed by means of the endomorphisms of $\mathbb{C}^{N^2}$ into $\mathbb{C}^{N^2}$, generally written for N×N matrices as

$$\Psi \longrightarrow \Phi = \sum_{i,j} B_{ij} \, \Psi \, \tau_{ij} \tag{2.6}$$

where τ_{ij}'s element ℓm is $\delta_i^\ell \, \delta_j^m$, and the B_{ij}'s are N×N matrices - $B_{ij} = 0$ iff the endomorphism is small. Thus the most general first order linear differential form acting on Ψ and without free term, is

$$G(\Psi) = \sum_{i,j,\ell} B_{ij}^\ell \, \frac{\partial\Psi}{\partial X_\ell} \, \tau_{ij} + \sum_{i,j} M_{ij} \, \Psi \, \tau_{ij} \tag{2.7}$$

where B_{ij}^ℓ, $M_{ij} \in \mathcal{M}_N$. $G(\Psi)$ vanishes if $\sum_{i,j} M_{ij}\tau_{ij}$ does, together with the parasite terms, for any T, any Ψ. It is easy to show that the condition for which this happens only if all B_{ij}^ℓ and M_{ij} vanish is :

$$\det[g_1, \, g_2, \ldots, \, g_N] \neq 0 \tag{2.8}$$

This condition is also sufficient to discard second order ghost equations. If we assume it, and if, for the sake of simplicity, we also assume C=0, we can show that the most general first order form compatible with dispersion relation is an arbitrary linear combination of

$$\mathcal{G}_\ell(\Psi) \equiv G_\ell(\Psi) + k \, g_\ell(\Psi) = - \operatorname{grad} \Psi \, . \, q_\ell(1-\eta_\ell) + k\eta_\ell \, \Psi \, (1-\eta_\ell) \tag{2.9}$$

where the vector q_ℓ's are defined to make a bi-orthogonal system with the g_ℓ's (i.e. $q_\ell . g_m = \delta_{\ell m}$). The free term of $\mathcal{G}_\ell(\Psi)$ is $g_\ell(\Psi^{(1)})$, so that, as above :

$$\mathcal{G}_\ell(\Psi) = \Psi \, g_\ell(\Psi^{(1)}) \tag{2.10}$$

It is possible to derive nonlinear evolution equations and nonlinear differential constraints for $\Psi^{(1)}$ by eliminating the terms showing k between (2.5) and (2.10), and between the N equations (2.10). It is also possible to do it by first setting in (2.5) and (2.10) $\Psi = \exp[k\Gamma.X] \, \varphi^T$ where X is the vector $(X_1, \, X_2, \ldots, \, X_N)$. Setting $\Phi^T = \Psi^{(1)}$, we obtain the spectral problems

$$T_o\varphi \equiv \left(\theta \, \frac{\partial}{\partial t} + \frac{1}{2} \, \Delta\right) \varphi + (\operatorname{grad} \Phi.\Gamma) \, \varphi = \frac{1}{2} \, k^2 \, \varphi \, \Gamma.\Gamma \tag{2.11}$$

$$T_\ell\varphi = (1-\eta_\ell) \, [q_\ell . \operatorname{grad} \Phi + \Phi \, \eta_\ell \, \Phi] \, (1-\eta_\ell) = 0 \tag{2.12}$$

whose commutators yield the constraints ($\ell = 1,2,\ldots,N$)

$$(1-\eta_\ell).[q_\ell.\text{grad}\ \Phi + \Phi\ \eta_\ell\Phi]\ (1-\eta_\ell) = 0 \tag{2.13}$$

and the evolution equations for the off-diagonal eelements ($\ell'\neq\ell$) of Φ :

$$q_\ell.\text{grad}\ (g_\ell.\text{grad}\ \Phi_{\ell'\ell}) - \frac{1}{2}\Delta\ \Phi_{\ell'\ell} - \theta\ \frac{\partial}{\partial t}\Phi_{\ell'\ell}$$

$$+\ g_\ell.\text{grad}\ [\Phi_{\ell'\ell}\ \Phi_{\ell\ell}] - \sum_m \text{grad}\ \Phi_{\ell' m}.g_m\ \Phi_{m\ell} = 0 \tag{2.14}$$

The constraint (2.14) relate the diagonal elements to the off diagonal ones, and, in the case $N \geqslant 3$, the off-diagonal ones to each other. It results that for any given off-diagonal element, only two space derivatives remain free. Hence these motions are on two-dimension manifolds, and coupled with each other inside the N-dimensional space.

Extensions to nondiagonal S, to vectorial cases coupled in a different way, to cases giving up the condition (2.8) will be published later, together with special solutions, dimensional reductions and extensions to higher order forms.

(1) Preprint Montpellier PM 91/11, May 1991.

Index of Contributors

ND | Springer Series in **Nonlinear Dynamics**

Editors

Francesco Calogero	Dipartimento di Fisica Università di Roma "La Sapienza" Piazzale Aldo Moro 2 I-00185 Rome, Italy
Benno Fuchssteiner	Fachbereich Mathematik Universität Paderborn Postfach 16 21 W-4790 Paderborn, Fed. Rep. of Germany
George Rowlands	Department of Physics University of Warwick Coventry CV4 7AL, United Kingdom
Miki Wadati	Institute of Physics College of General Education University of Tokyo Komaba 3-8-1, Meguro-ku Tokyo 153, Japan
Vladimir E. Zakharov	Landau Institute for Theoretical Physics Russian Academy of Sciences ul. Kosygina 2 117334 Moscow, Russia

Springer

MIX
Papier aus verantwortungsvollen Quellen
Paper from responsible sources
FSC® C105338

If you have any concerns about our products,
you can contact us on
ProductSafety@springernature.com

In case Publisher is established outside the EU,
the EU authorized representative is:
Springer Nature Customer Service Center GmbH
Europaplatz 3, 69115 Heidelberg, Germany

Printed by Libri Plureos GmbH
in Hamburg, Germany